ATOMIC AND MOLECULAR DATA AND THEIR APPLICATIONS

Previous Proceedings in the Series of Conferences on Atomic and Molecular Data and Their Applications

	Year	Held in	Publisher	ISBN
2nd	2000	Oxford, England	AIP Conf. Proceedings vol. 543	1-56396-971-8
1st	1997	Gaithersburg, Maryland	AIP Conf. Proceedings vol. 434	1-56396-751-0

Other Related Titles from AIP Conference Proceedings

635 Atomic Processes in Plasmas: 13th APS Topical Conference on Atomic Processes in Plasmas
Edited by David R. Schultz, Fred W. Meyer, and Fay Ownby, October 2002, 0-7354-0090-3

551 Atomic Physics 17: XVII International Conference on Atomic Physics; ICAP 2000
Edited by Ennio Arimondo, Paolo De Natale, and Massimo Inguscio, February 2001, 1-56396-982-3

547 Atomic Processes in Plasmas: Twelfth Topical Conference
Edited by Roberto C. Mancini and Ronald A. Phaneuf, December 2000, 1-56396-976-9

525 Multiphoton Processes: ICOMP VIII: 8th International Conference
Edited by Louis F. DiMauro, Richard R. Freeman, and Kenneth Kulander, June 2000, 1-56396-946-7

500 The Physics of Electronic and Atomic Collisions: XXI International Conference
Edited by Yukikazu Itikawa, Kazuhiko Okuno, Hiroshi Tanaka, Akira Yagishita, and Michio Matsuzawa, February 2000, 1-56396-777-4

To learn more about these titles, or the AIP Conference Proceedings Series, please visit the webpage http://proceedings.aip.org/proceedings

ATOMIC AND MOLECULAR DATA AND THEIR APPLICATIONS

3rd International Conference on Atomic and Molecular Data and Their Applications
ICAMDATA

Gatlinburg, Tennessee   24–27 April 2002

EDITORS
David R. Schultz
Predrag S. Krstić
Fay Ownby
Oak Ridge National Laboratory
Oak Ridge, Tennessee

SPONSORING ORGANIZATIONS
U.S. DOE Office of Fusion Energy Sciences
ORNL Controlled Fusion Atomic Data Center
International Union of Pure and Applied Physics
ORNL Physics Division
National Institute of Standards and Technology
UT/Battelle
ORNL Physical Sciences Directorate

Melville, New York, 2002
AIP CONFERENCE PROCEEDINGS ■ VOLUME 636

Editors:

David R. Schultz
Predrag S. Krstić
Fay Ownby

Oak Ridge National Laboratory
P.O. Box 2008
Oak Ridge, TN 37831-6372
USA

E-mail: schultzd@ornl.gov
    krstic@mail.phy.ornl.gov
    ownbyfm@ornl.gov

The articles on pp. 5-13 and 134-143 were authored by U.S. Government employees and are not covered by the below mentioned copyright.

Authorization to photocopy items for internal or personal use, beyond the free copying permitted under the 1978 U.S. Copyright Law (see statement below), is granted by the American Institute of Physics for users registered with the Copyright Clearance Center (CCC) Transactional Reporting Service, provided that the base fee of $19.00 per copy is paid directly to CCC, 222 Rosewood Drive, Danvers, MA 01923. For those organizations that have been granted a photocopy license by CCC, a separate system of payment has been arranged. The fee code for users of the Transactional Reporting Service is: 0-7354-0091-1/02/$19.00.

© 2002 American Institute of Physics

Individual readers of this volume and nonprofit libraries, acting for them, are permitted to make fair use of the material in it, such as copying an article for use in teaching or research. Permission is granted to quote from this volume in scientific work with the customary acknowledgment of the source. To reprint a figure, table, or other excerpt requires the consent of one of the original authors and notification to AIP. Republication or systematic or multiple reproduction of any material in this volume is permitted only under license from AIP. Address inquiries to Office of Rights and Permissions, Suite 1NO1, 2 Huntington Quadrangle, Melville, N.Y. 11747-4502; phone: 516-576-2268; fax: 516-576-2450; e-mail: rights@aip.org.

L.C. Catalog Card No. 2002112523
ISBN 0-7354-0091-1
ISSN 0094-243X
Printed in the United States of America

Contents

Preface ix
Program and Local Committees x
Sponsors xi
Meetings in This Series xii

I. APPLICATIONS OF ATOMIC AND MOLECULAR DATA

A. Radiation Physics

The Use of Atomic Data in Applications Involving Ionizing Radiation 5
    P. M. *Bergstrom, Jr.*
**Application of Radiation Track in Radiation Biophysics
and Dosimetry** 14
    H. *Nikjoo* and S. Uehara
**Laboratory Data Needs and Applications for Assessing Radiation
Effects in Biological Materials** 23
    L. H. *Toburen* and J. L. Shinpaugh

B. Lighting

Physical Aspects of Mercury-Free High Pressure Discharge Lamps 35
    M. *Born*
**The Physics of Fluorescent Lamps: Do We Understand the
Atomic Processes?** 48
    G. G. *Lister*

C. Etching, Plasma Displays, and Plasma Processing

**Modeling of Moderate Pressure Microwave Plasmas Used
for Diamond Deposition: Collisional Data Required for
Process Simulation** 61
    K. *Hassouni* and A. Gicquel
Ultraviolet Production Efficiency of AC-PDPs and Ways to Increase It 75
    K. *Suzuki*, N. Uemura, S. Ho, and M. Shiiki
Atomic and Molecular Data Needs in Thermal Plasmas 85
    P. *Fauchais*, V. Rat, J. Aubreton, and M. F. Elchinger
**New Gas Chemistry for High-Performance SiO_2 Patterning in
Sub-0.1 μm ULSIs** 95
    S. *Samukawa*

*Italicized names indicate the authors who presented the papers.

D. Atmospheres

Atmospheric Pollutant Removal by Non-Thermal Plasmas: Basic Data Needs for Understanding and Optimization of the Process 111
    S. Pasquiers, M. Cormier, and O. Motret

E. Astrophysics

The Effect of High-Lying Configurations and Ionization and Recombination Processes on Analyses of Solar and Stellar Coronal Spectra 125
    R. Doron, E. Behar, G. A. Doschek, and U. Feldman

Atomic and Molecular Data Needs for Astrophysics 134
    R. L. Kurucz

Charge Transfer Data Needs for Cometary X-Ray Emission Modeling 144
    P. C. Stancil, J. G. Wang, M. J. Raković, D. R. Schultz, and R. Ali

Interstellar Molecules: The New Frontiers for Molecular Data 154
    L. M. Ziurys and A. J. Apponi

F. Fusion Energy

Atomic and Molecular Processes for Heat and Particle Control in Tokamaks 161
    H. Kubo

Atomic Physics Processes Important to the Understanding of the Scrape-Off Layer of Tokamaks 171
    W. P. West, B. Goldsmith, T. E. Evans, and R. E. Olson

II. ATOMIC AND MOLECULAR PHYSICS AND DATA

A. Atomic and Molecular Physics

Atomic Processes in Plasmas—An Overview 185
    H. R. Griem

Benchmark Calculations for Electron Collisions with Complex Atoms 192
    K. Bartschat

Experimental Data for Electron-Ion Collisions 202
    A. Müller

Electron-Molecule Collisions in the Static-Exchange Correlation-Polarization Approximation 213
    R. R. Lucchese and F. A. Gianturco

Theoretical Atomic Data: Universality and Precision 221
    Z. R. Rudzikas

*Italicized names indicate the authors who presented the papers.

Electron Collision Data for Polyatomic Molecules in Plasma Processing and Environmental Processes 233
    H. Tanaka, M. Kitajima, and H. Cho
Developing Cross Section Sets for Fluorocarbon Etchants 241
    C. Winstead and V. McKoy

B. Atomic and Molecular Databases

Spectr-W^3 Online Database on Atomic Properties of Atoms and Ions 253
    A. Y. Faenov, A. I. Magunov, T. A. Pikuz, I. Y. Skobelev, P. A. Loboda,
    N. N. Bakshayev, S. V. Gagarin, V. V. Komosko, K. S. Kuznetsov,
    S. A. Markelenkov, S. A. Petunin, and V. V. Popova
Electron-Molecule Cross Section Data for Hydrogen Plasma Applications 263
    R. Celiberto, A. Laricchiuta, R. K. Janev, and M. Capitelli
Atomic and Molecular Databases for Fusion Divertor Plasma 277
    P. S. Krstić and D. R. Schultz

Participants 287
Author Index 291

*Italicized names indicate the authors who presented the papers.

Preface

The International Conference on Atomic and Molecular Data and Their Applications (ICAMDATA) is an ongoing series of meetings that brings together scientists from around the world who produce, collect, evaluate, disseminate, and utilize atomic and molecular data. This forum is intended to facilitate presentation of the latest results in these areas as well as to serve as a conduit for communication among the various specialties concerned with all aspects of the process from creation to use of data. In fact, the conference grew out of a workshop held at the Harvard Smithsonian Center for Astrophysics in 1996 where it was recognized that a periodically convened, international gathering was needed to raise awareness of the particular needs for atomic and molecular data and to stimulate communication between producers and users. The first ICAMDATA was held at the National Institute of Standards and Technology, Gaithersburg, Maryland, USA, in 1997 and the second at Keble College, Oxford, England, in 2000.

The 3rd ICAMDATA, held April 24-27, 2002, in Gatlinburg, Tennessee, USA, consisted of thirty invited oral presentations, four panel sessions, and fifty-two poster presentations, and was attended by over one hundred participants from fourteen countries. The present proceedings contains written manuscripts for twenty six of the oral presentations, organized into groupings that reflect the themes and disciplines represented at the conference such as applications in lighting, astrophysics, and radiation physics as well as atomic and molecular physics and databases and data centers.

Special thanks is owed to the conference secretary Ms. Fay Ownby, who also served as chief editor of this volume, and to Ms. Carlene Stewart, the conference co-secretary. In addition, the conference chair and co-chairs wish to express their thanks to all of the participants, the local conference committee, the Oak Ridge Associated Universities for their fostering of university – national laboratory interchange through this meeting, and all the sponsors of the conference. We particularly acknowledge support of the publication of this volume by the US DOE Office of Fusion Energy Sciences. We look forward to the continued success of ICAMDATA at the 4th meeting in 2004 to be held in Japan.

David R. Schultz, Chair
Predrag S. Krstic, Co-Chair
H. Kennon Carter, Co-Chair
Oak Ridge, Tennessee
July 2002

International Program Committee

R. K. Janev, Chair (Macedonia)
G. Lister, Vice-Chair (USA)
K. A. Berrington, Secretary
  (United Kingdom)
D. R. Schultz, Treasurer (USA)
J. F. Babb (USA)
K. L. Bell (Northern Ireland)
L. Brown (USA)
M. Capitelli (Italy)
R.E.H. Clark (Austria)
R. Flannery (USA)

T. Kato (Japan)
M. J. Kushner (USA)
H. E. Mason (United Kingdom)
W. L. Morgan (USA)
K. Niemax (Germany)
L. C. Pitchford (France)
L. Presnyakov (Russia)
E. Roueff (France)
Z. Rudzikas (Lithuania)
W. Wiese (USA)

International Advisory Board

J. N. Bardsley (USA)
P. G. Burke (Northern Ireland)
R. W. Crompton (Australia)
A. Dalgarno (USA)
J.-L. Delcroix (France)
F. Gianturco (Italy)
Y. Hatano (Japan)
R. Hulse (USA)

Y. Itikawa (Japan)
J. M. Li (China)
I. Martinson (Sweden)
C. Mendoza (USA)
J. Rumble (USA)
E. Salzborn (Germany)
M. Seaton (United Kingdom)
I. I. Sobelman (Russia)

Local Committee

David Schultz, Chair
Predrag Krstic, Co-Chair
Ken Carter, Co-Chair
Fay Ownby, Secretary
Carlene Stewart, Co-Secretary
Mark Bannister
Charles Havener

Herb Krause
Joseph Macek
Fred Meyer
Serge Ovchinnikov
Carlos Reinhold
Lynda Saddiq
Randy Vane

Sponsors

US DOE Office of Fusion Energy Sciences

The ORNL Controlled Fusion Atomic Data Center

The International Union of Pure and Applied Physics (IUPAP)

The ORNL Physics Division

The National Institute of Standards and Technology

UT/Battelle, managers of ORNL

The ORNL Physical Sciences Directorate

Meetings in This Series

1. Gaithersburg, Maryland — September 29-October 2, 1997

2. Oxford, England — March 26-30, 2000

3. Gatlinburg, Tennessee — April 24-27, 2002

I. APPLICATIONS OF ATOMIC AND MOLECULAR DATA

A. Radiation Physics

The Use of Atomic Data in Applications Involving Ionizing Radiation

P. M. Bergstrom, Jr.

Photon and Charged Particle Data Center, National Institute of Standards and Technology, Gaithersburg, MD 20899-8460

Abstract. Ionizing radiation is utilized in many industrial, medical and research applications. The term ionizing radiation implies that the interaction of the radiation with the object of interest occurs at the atomic level, through the removal of electrons from atoms and molecules. In trying to understand, enhance and develop technologies that utilize ionizing radiation, atomic data and tools to utilize these data sets are essential. In this paper some current applications of ionizing radiation are discussed. The computational tools applied to these situations are outlined. Currently available data sets are reviewed. Data needs are discussed as are some of the efforts underway at the National Institute of Standards and Technology (NIST) to enhance both data and tools.

GENERAL CONSIDERATIONS

Applications of ionizing radiation are highly dependent on the availability of quality data. The data needed depends on the type of ionizing radiation, on the application and on the level of analysis. In this paper, we are concerned with β and γ radiation at energies where they primarily interact with the electronic structure of the atom or molecule rather than the nucleus. Because we are dealing with ionizing radiation, problems are not completely separable into electron-only or photon-only problems. An understanding of the trajectories of the electrons ionized by photons may be necessary to provide an adequate description of a problem involving a photon source. Similarly, through the bremsstrahlung process, an electron source generates secondary photons which may need to be considered for an accurate solution in a particular application. In this paper we will categorize problems according to source type, keeping in mind the possible need to couple different particle types. The level of accuracy may also depend on the energies at which one seeks to understand the radiation processes. For example, in biological systems, much of the damage done by radiation is at low energies and involves collisions between electrons and molecules. However, if one is just interested in the dose and not details of the biological effects, higher energy atomic data is likely sufficient. In this paper, we concentrate on the data utilized at higher energies as the lower energy effects are discussed elsewhere in this volume.

There are myriad possible uses for ionizing radiation. Most of the applications at high energy fall into several broad categories. Imaging applications rely on a balance of the ability of the radiation to penetrate the object being imaged, without being so penetrating that there are not enough interactions to provide the necessary contrast. Shielding applications are characterized by the need to confine the radiation to a region where its deleterious effects can be avoided. Dose deposition applications, where one is relying on the radiation to deliver energy to an object, are those that will be discussed here. Typical dose deposition applications are measurement, sterilization of items and processing that changes other properties of materials.

The understanding of radiation processes by theoretical or computational means has expanded greatly with the advancing power of computers. This increase in computational capacity has enabled the use of the tool of choice, the Monte Carlo simulation of particle transport, to evolve from providing qualitative data for aiding the decision making process into a more quantitative role. Monte Carlo transport simulates a physical system by following the particles through a geometrical description of the system, sampling the laws of physics expressed in terms of the particles' mean free paths and differential cross sections to determine where those particles interact, what secondaries they produce, and the properties of the particles emerging from the collisions. Typical simulations involve only millions or billions of the trillions of particles used in the corresponding application. The goal is to obtain a statistical sample that adequately represents the process. The vast increase in computer memory and disk space has allowed larger problems with more refined geometrical descriptions to be considered. The increase in speed enables more particle trajectories to be simulated per unit time, lowering the time necessary to reach a given statistical precision.

The application of Monte Carlo methods to these problems challenges commonly available sources of data. In addition to the total cross sections needed in more approximate calculations, Monte Carlo codes can make use of all detailed information about the interactions and the resulting secondary particles. The level of detail to which this data is used depends on the quantity of interest in the calculations. One issue that the developer and user of Monte Carlo codes must face is the need to choose between seemingly different databases. As we will see, these differences can be illusory as much of the same data is utilized by different database developers.

In the next section we we turn to an application that makes use of photon data, the maintenance of the national standard for dose and exposure, and consider the tools and data relevant to it. Then we consider an application that is heavily dependent on the availability of quality electron data. We discuss the application (industrial ebeam processing), the tools utilized to understand the application and the data utilized by those tools

PHOTON DATA

In the area of radiation, NIST maintains the standards for absorbed dose and exposure for the United States. The determination and dissemination of these standards is performed by the Radiation Interactions and Dosimetry Group of the Ionizing Radiation Division. The instruments used by the group for this purpose vary according to the energy of the radiation involved. For lower energy X-rays, a free-air chamber is used. At higher energies, for example those typical of Cs-137 and Co-60 sources, a series of graphite-walled ionization chambers are used. These chambers are modeled using the Bragg-Gray cavity theory[1] as modified by Spencer and Attix[2]. Two of the assumptions of this theory are that the cavity is small and that it is surrounded by enough material to exclude externally produced electrons. There are deviations for real chambers from cavity theory. Considering air kerma K, one may write

$$K = \frac{Q_{air}}{V\rho_{air}} \frac{(W/e)}{1-\bar{g}} \left[\frac{(\bar{S}/\rho)_{graphite}}{(\bar{S}/\rho)_{air}} \right] \left[\frac{(\bar{\mu}_{en}/\rho)_{air}}{(\bar{\mu}_{en}/\rho)_{graphite}} \right] \Pi k_i. \qquad (1)$$

The correction factors k_i are used to model departures from the idealizations of cavity theory. One specific correction is for the effects of the wall surrounding the cavity that are not accounted for in the theory. In the past, this correction was made experimentally by varying the thickness of the wall for a given cavity and extrapolating to zero wall thickness[3]. The consensus is that this method is flawed and that the wall correction can be obtained accurately using Monte Carlo calculations.

The Monte Carlo transport of photons usually is performed in what is referred to as a single scatter or analog method. In this method, the photon follows a straight-line trajectory between collisions. The frequency of collisions are determined by the mean free path of the photon in the transport medium. The mean free path needs to be known absolutely as it must be compared to the physical dimensions of the system. The mean free path may be determined independently, for example in attenuation experiments. However, it is usually constructed from the total cross sections for the major photon-atom interaction processes. In the energy range of interest here, these processes are photoeffect, elastic and inelastic photon scattering and pair and triplet production. A number of large databases exist that tabulate the total cross sections for these processes. Some of the more commonly used ones are those of Storm and Israel [4], XCOM [5] and EPDL [6]. Once it is determined that an interaction should be simulated, the ratio of the total cross section for a given process to the total cross section for all processes determines the relative frequency of a given process. One must then find the energy and momentum (direction) of all particles emerging from the interaction. These are determined by differential cross sections, conservation laws and rules of thumb.

In the energy range of interest for a graphite ionization chamber in a Co-60 beam, inelastic scattering is the main photon-atom intereaction mechanism. In the remainder of this section, we concentrate on databases appropriate to this process. Of the three databases discussed above, XCOM and EPDL contain the same inelastic photon scattering data. This inelastic scattering data was obtained from the incoherent scattering function tabulation of Hubbell et al [7]. In the incoherent scattering function approximation one starts from the nonrelativistic A^2 matrix element for the photon-electron interaction, summing it over all final states. Once this summation has been performed, the resulting cross section includes all inelastic channels, not just ionization. At this point, one has an expression where the free scattering angular distribution, which is the Thomson cross section, is multiplied by the incoherent scattering function. The incoherent scattering function is a function of the scattering charge and of the momentum transfer. In this approximation the energy of the scattered photon is taken to be that for scattering from a free and stationary electron, i.e. the value given by Compton's formula. The ad hoc replacement of the Thomson cross section by the Klein Nishina cross section is necessary to produce the correct high energy behavior. Hubbell and coworkers tabulated both the scattering factors and the corresponding total cross sections calculated using them. Storm and Israel present their own calculations of total cross sections within this formalism. While the Hubbell et al and the Storm and Israel tabulations each present independent calculations for the total cross sections, these calculations are based on the same underlying data. For elements with nuclear charge greater than that of carbon, these tabulations utilize the same incoherent scattering functions, those of Cromer and coworkers [8].

The atomic data directly appearing in Eq. 1 are the mass energy-absorption coefficients μ_{en} and the electronic stopping powers S. Stopping powers will be discussed in the next section. The mass energy-absorption coefficient is based on the mass energy-transfer coefficient through

$$\mu_{en} = (1-g)\mu_{tr} \qquad (2)$$

and is a function of the material and of the incident photon energy. For inelastic photon-atom scattering one may express the mass energy-transfer coefficient in terms of the cross section as

$$\mu_{tr}^{compt} = \left[1 - \frac{<\hbar\omega_f> + x_{fl}}{\hbar\omega_i}\right] \frac{\sigma_{compt}}{uA}. \qquad (3)$$

Clearly, the mass energy-absorption coefficient reflects the locally deposited energy rather than that carried away by the scattered photon, $\hbar\omega_f$, the fluorescence photons, x_{fl}, and bremsstrahlung, g. Storm and Israel differ from Hubbell et al in their presentation of inelastic photon-atom scattering data in that they do not present the scattering factors. They do, however, give a quantity that they call the absorption cross section. Their absorption cross section does not take into account fluorescence and bremsstrahlung.

As systems are modeled to finer and finer resolution in space or energy, improvements to the available databases are needed. In the case of inelastic photon-atom scattering, the assumption that the energy of the scattered photon obeys Compton's relation is no longer valid as the scattering electrons are not free and stationary. There is a spread in the possible values of energy for the scattered photon due to the momentum of the bound electrons. Modifications to the usual inelastic photon-atom scattering treatments have been pursued by several groups [9, 10, 11] to reflect this. To date, no single source of data is available that consistently treats inelastic photon-atom scattering from the level of the cross section doubly differential in scattered photon energy and angle through the total cross section.

Work is underway at NIST to develop a consistent database for Compton scattering of photons from electrons bound in atoms. The Compton scattering cross section is the ionization component of the inelastic photon-atom scattering cross section and is usually the quantity of interest in applications. This work is being pursued in the impulse approximation [12, 13] and starts from the cross section doubly differential in scattered photon energy and angle. The database will contain these doubly differential cross sections and other differential data such as the angular distribution of scattered

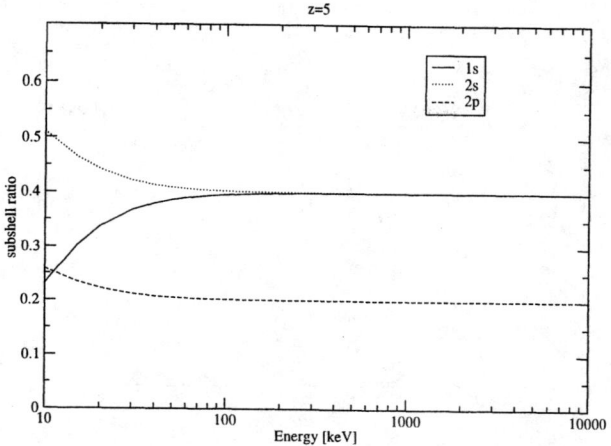

FIGURE 1. Impulse approximation calculation of the relative ionization of the subshells of Boron by Compton scattering.

photons, subshell ionization ratios and integrated data such as total cross sections and mass energy-transfer and mass energy-absorption coefficients.

In Figure 1, data obtained within the impulse approximation for Compton scattering of photons from Boron shows the relative contributions of the various subshells to the overall cross section. These calculations use the Compton profiles of Biggs, Mendelsohn and Mann [14]. At low energy, the least tightly bound subshells provide the dominant contribution. As the energy of the incident photons exceeds the binding energy of a given subshell, that subshell starts to contribute to the overall cross section for ionization. At higher energies, the relative contribution for a given subshell is given by the ratio of the number of electrons in that shell to the total number of electrons in the atom.

Figure 2 demonstrates the difference between total cross sections obtained in the impulse approximation (IA) and the relativistic impulse approximation (RIA) with those contained in XCOM for photons scattering from Boron. Clearly, most of the differences occur at lower energies. The IA and the RIA cross sections are smaller than the ISF cross section in this region, reflecting the inclusion of additional channels, primarily excitation, included in the ISF. Surprisingly, the IA results are somewhat closer to the ISF than to the RIA here.

ELECTRON DATA

Industrial processing of goods using radiation is increasingly common. In particular, Co-60 and electron beams are applied to a wide variety of products. The purpose of this processing varies from sterilization as in the case of medical devices and supplies

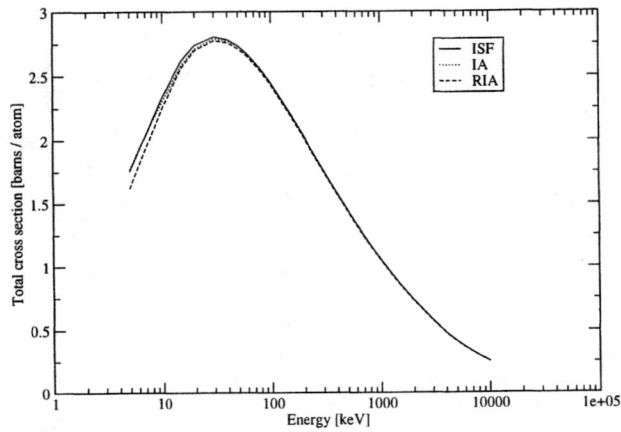

FIGURE 2. Comparison of total cross sections for Compton scattering of photons by Boron in the incoherent scattering factor (ISF), the impulse approximation (IA) and the relativistic impulse approximation (RIA).

to changing material properties as in the case of tires. Treatment of food items kills harmful bacteria and prolongs product shelf life. Naturally, the use of radiation is subject to public acceptance, particularly in the case of consumer goods. Part of this acceptance requires that treated items not be "radioactive", limiting the energies of the electron beams and of the photons that they might produce. Facilities containing large quantities of radionuclides, such as Co-60, are not generally welcomed by their neighbors.

In October of 2001, letters containing anthrax were sent through the United States Mail to members of the media and of the United States government. Several people who handled the mail died and an effort was undertaken by the Postal Service to determine how to sanitize potentially contaminated mail and what efforts should be taken to prevent similar events in the future. The developed state of the radiation processing industry enabled immediate action to be taken. Atomic data utilized in Monte Carlo transport codes was of critical importance in the analyses that led to critical decisions. At present, electron beam technology is being used to process mail sent to Zip codes in Washington where government offices are located.

The computational studies performed by NIST in support of the USPS attempts to sanitize potentially contaminated mail had two limits. They were that the dose delivered should provide reasonable assurance that anthrax contaminating the mail would be killed and that the dose not be so high that the contents of the mail would be damaged or set afire. If one considers first class letters, mainly consisting of paper and air, one can use Monte Carlo studies to determine, within the confines of the operational parameters of existing electron beam facilities, how to package those letters for optimal throughput. The emphasis of these studies was on determining the maximum packing density of the letters that permitted a reasonably uniform dose to be delivered. Packages are not as straightforward to analyze. Depending on their contents, it may be possible to treat them with e-beams. However, for thick or dense packages, it may be necessary to convert the e-beams to bremsstrahlung sources. All of these scenarios may be modeled using Monte Carlo codes.

The Monte Carlo transport method for electrons (or positrons) does not typically follow the trajectories of the particles on a collision-by-collision basis. It was recognized long ago [15] that this was impractical in most instances due to the many collisions that charged particles undergo while they slow down in matter. Instead, the effects of many collisions are condensed into one pseudostep. This condensed history method relies on accurate data to describe the effects of multiple collisions on the energy loss and the angular deflections of the electrons. In addition to these multiple-scattering data, it is necessary to include the effects of discrete interactions above a certain threshold energy. These interactions enable one to account for the nonlocal deposition of energy from the electrons as they slow down via emission of bremsstrahlung or through the knock-on process.

Electron transport codes are divided into two general classes depending on how they treat the discrete processes. In the class I approach, applied to inelastic electron scattering in such codes as ETRAN[16], ITS [17] and MCNP [18] the properties of the incident electron are left unchanged. This appears to violate the laws of conservation of energy and of momentum. However, the mean energy loss is accounted for through the stopping power with fluctuations about that mean taken into account through straggling distributions. The class II approach utilizes all available information to compute the properties of both the incident and secondary particles emerging from a collision. This approach has been utilized in several codes[10, 11, 19]. Pains must be taken to modify the stopping power to remove the energy transferred from the incident particle to the secondaries so that it is not counted twice. The full simulation of these above-threshold discrete events accounts for some of the particle straggling about the energy loss given by the stopping power. Usually, the remaining straggling contributions are ignored. In all cases, the bremsstrahlung process is treated in the class II approach. That is that both the incident electron and emitted photon's post-collision properties are affected by the collision. Substantial simplifications are often invoked in determining these postcollision properties. In some cases, the consequences of these simplifications are hardly noticeable. This is often the case in the angular distribution of the ejected electrons. Subsequent multiple scattering and relatively short ranges can wash out all but the most gross details of these angular distributions. In other cases, the simplifications can be problematic. For example, in the code EGS4[19], the angular distribution of bremsstrahlung photons is simply the electron rest mass energy divided by the incident electron energy. These photons are usually more penetrating and typically undergo far fewer collisions than electrons and retain a greater memory of the initial angular distribution.

Stopping powers, taken from the work of Seltzer and Berger, are available from the International Commission on Radiation Units and Measurement Publication 37[20]. Some stopping power data for water, taken from these tables, is given in Figure 3. The stopping power is large at low energies as a result of electrons expending their remaining energy in inelatic collisions. There is a broad intermediate energy range where the stopping power is relatively constant. At higher energies, radiative losses due

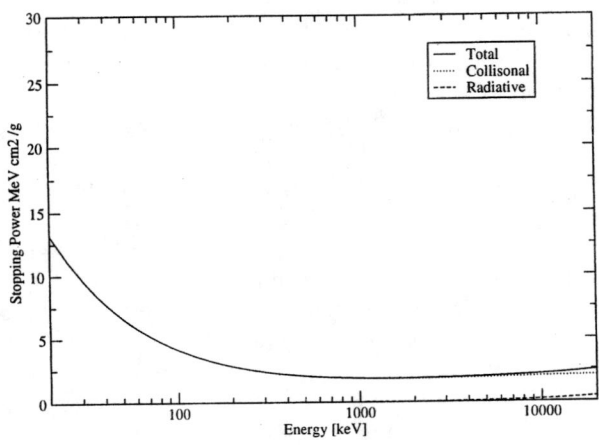

FIGURE 3. Stopping powers for electrons in water.

to the bremsstrahlung process take hold. The collisional part of the stopping power is determined using the theory of Bethe[21] for the passage of charged particles through matter. The radiative stopping powers utilize bremsstrahlung cross sections taken from the work of Tseng and Pratt[22] at low energies (< 2 MeV) and interpolated smoothly through a range of intermediate energies to 50 MeV where high energy formulae are valid.

The angular deflection of the electrons is determined from the multiple scattering theories of Moliere[23] or of Goudsmit and Saunderson [24]. Moliere's theory is valid for small angles and can be recovered in this limit from the more general theory of Goudsmit and Saunderson. The GS distribution can also utilize elastic scattering cross sections calculated in any potential. However, most implementations of the GS distribution have been computationally unwieldy and this led to the use of the more approximate Moliere distribution in a number of codes. Recently, Kawrakow and Bielajew[25] developed a simpler sampling scheme for the GS distribution. Unfortunately, the elastic scattering cross sections in use are still of the screened Rutherford type.

The Photon and Charged Particle Data Center at NIST also continues to work towards improving their data for electron and positron physics. Among the recent developments in electron-atom interactions are a new database for elastic electron-atom scattering by Berger and Seltzer [26]. This database makes fully relativistic partial-wave elastic scattering cross sections calculated in Dirac-Fock potentials available for energies between 1 keV and 10 MeV. At higher energies, a WKB approach is applied. Total cross sections and first and second transport cross sections are provided. In addition, correction factors for exchange and finite nuclear size are tabulated. These cross sections are on a grid suitable for direct input into the ETRAN code. They could also be incorporated into the Goudsmit-Saunderson theory to provide better multiple scattering distributions.

CONCLUDING REMARKS

In summary, the databases currently in use for applications involving ionizing radiation were developed several decades ago. The advances in computing in that time frame now enables routine computations, using Monte Carlo codes, to perform accurate simulations of these applications. These codes have put increasing demand for more accurate and more differential cross section data. The new database efforts discussed here for Compton scattering of photons by electrons bound in atoms and for the elastic scattering of electrons and positrons from atoms represent NIST's initial attempts to deal with these needs.

REFERENCES

1. Gray, L. H., *Proceedings of the Royal Society A*, **156**, 578–596 (1936).
2. Spencer, L. V., and Attix, F. H., *Radiation Research*, **3**, 239–254 (1955).
3. Loftus, T. P., and Weaver, J. T., *Journal of Research of the National Bureau of Standards A*, **78**, 465–476 (1974).
4. Storm, E., and Israel, H. I., *Nuclear Data Tables A*, **7**, 565–681 (1970).
5. Berger, M. J., and Hubbell, J. H., Xcom: Photon cross sections on a personal computer, Tech. rep., NBSIR 87-3597, National Bureau of Standards (1987).
6. Cullen, D. E., Chen, M. H., Hubbell, J. H., Perkins, S. T., Plechaty, E. F., Rathkopf, J. A., and Scofield, J. H., Tables and graphs of photon-interaction cross sections from 10 ev to 100 gev derived from the llnl evaluated photon data library (epdl), Tech. rep., UCRL-50400 Vol 6, Rev 4, Parts A and B, Lawrence Livermore National Laboratory (1989).
7. Hubbell, J. H., Veigele, W. J., Briggs, E. A., Brown, R. T., Cromer, D. T., and Howerton, R. J., *Journal of Physical and Chemical Reference Data*, **4**, 471–538 (1936).
8. Cromer, D. T., *Journal of Chemical Physics*, **50**, 4857–4859 (1969).
9. Namito, Y., Ban, S., Hirayama, H., Nariyama, N., Nakashima, H., Nakaner, Y., Sakamoto, Y., Sasamoto, N., Asano, Y., and Tanaka, S., *Physical Review A*, **51**, 3036–99999 (1995).
10. Baro, J., Sempau, J., Fernandez-Varea, J. M., and Salvat, F., *Nuclear Instruments and Methods B*, **100**, 31–46 (1995).
11. Kawrakow, I., *Medical Physics*, **27**, 485–498 (2000).
12. Eisenberger, P., and Platzmann, P. M., *Physical Review A*, **2**, 415–423 (1970).
13. Ribberfors, R., *Physical Review B*, **12**, 2067–2074 (1975).
14. Biggs, F., Mendelsohn, L. B., and Mann, J. B., *Atomic Data and Nuclear Data Tables*, **16**, 201–309 (1975).
15. Berger, M. J., *Methods in Computational Physics*, **1**, 135–215 (1963).
16. Seltzer, S. M., *Applied Radiation and Isotopes*, **42**, 917–941 (1991).
17. Halbleib, J. A., and Mehlhorn, T. A., Its: The integrated tiger series of coupled electron/photon monte carlo codes, Tech. rep., SAND 84-0573, Sandia National Laboratory (1984).
18. Breismeister, J. F., Mcnp - a general monte carlo n-particle transport code, Tech. rep., LA-12625-M Version 4B, Los Alamos National Laboratory (1997).
19. Nelson, W. R., Hirayama, H., and Rogers, D. W. O., The egs4 code system, Tech. rep., 265, Stanford Linear Accelerator Center (1985).
20. Berger, M. J., Inokuti, M., Anderson, H. H., Bichsel, H., Dennis, J. A., Powers, D., Seltzer, S. M., and Turner, J. E., Stopping powers for electrons and positrons, Tech. rep., 37, International Commission on Radiation Units and Measurements (1984).
21. Bethe, H. A., *Annalen der Physik*, **5**, 325–400 (1930).
22. Pratt, R. H., Tseng, H. K., Lee, C. M., and Kissel, L. D., *Atomic Data and Nuclear Data Tables*, **20**, 175–209 (1977).
23. Moliere, G., *Z. Naturforsch.*, **2**, 133–145 (1947).
24. Goudsmit, S., and Sanderson, J. L., *Physical Review*, **57**, 24–29 (1940).
25. Kawrakow, I., *Nuclear Instruments and Methods B*, **134**, 325–336 (1998).
26. Berger, M. J., and Seltzer, S. M., Database of cross sections for the elastic scattering of electrons and positrons by atoms, Tech. rep., 6573, National Institute of Standards and Technology (2000).

Application of Radiation Track in Radiation Biophysics and Dosimetry

Hooshang Nikjoo[1] and Shuzo Uehara[2]

[1]*MRC Radiation & Genome Stability Unit, Harwell, OX11 0RD, UK*
[2]*School of Health Sciences, Kyushu University, Fukuoka 812, Japan.*

Abstract. This paper provides a brief summary of recent advances in track structure simulation of low energy alpha-particles. A description is given for biophysical modelling of the interaction of radiation in DNA in terms of energy deposition and complexity of DNA damage. Track simulations allow estimation of the molecular spectrum of DNA damage.

INTRODUCTION

Monte Carlo track structure method has been instrumental in advancing a greater understanding of the mechanism(s) of radiation damage and provides a more accurate description of the target for such damage. On the other hand, to describe the effect of radiation macroscopically, 'absorbed dose' is traditionally used in relating the imparted energy to subsequent radiobiological effects. Monte Carlo radiation transport has been widely used in computing the radiation dose and provides theoretical guidance for radiation dosimetry [1,2]. Radiation transport, either for track structure simulation, or for dosimetry purposes, relies on accurate modelling of various types of radiation interactions and their cross-section data in biologically relevant media. Historically, track structure simulation has been focused on water as a surrogate for soft tissue, inasmuch as soft tissue is over 80% water. In this work we describe our database for simulation of electron and ions in water. Using track structure simulation and models of DNA structure, we describe characteristics of molecular damage induced by ionizing radiation at the DNA level. To make biophysical models more predictive and useful in practice, there is a strong need for measurements of cross sections in composite materials of biological relevance.

TIME SCALE

Biological responses to ionizing radiation can be divided into early and late effects. Early effects are those which become apparent within milliseconds to days following irradiation. Late effects manifest themselves within weeks to years following exposures. Radiation injury begins with the physical processes of energy deposition in the medium leading to ionizations and excitations. The initial physical processes lead to molecular damage by direct interaction of the radiation with the target and indirectly through the generation of free radicals, oxidising agents and other molecular species. Molecular damage is expressed through alteration of biochemical processes and is amplified by biological mediators. Depending on the biological end-point examined, biological effects attributed to the radiation exposure may become evident within a femtosecond to years. In addition, damage may persist which is not generally visible or clinically detectable, leading to genetic instability. Although, initially 1 Gy of low LET radiation could cause many thousands of ionizations in the cell nucleus, most of these hits do not cause appreciable damage, leading to nearly 1000 single strand breaks and a few tens of double strand breaks. Normal human cells are very efficient in repairing the damage. In vitro experiments show very low frequency of genetic alterations to cells in the form of transformation and mutation of the order 10^{-5} - 10^{-6} per cell per Gy of radiation. Through biophysical modelling and epidemiological studies, advances have been made to elucidate the mechanism(s) of damage from initial molecular damage to late effects and cancer [2].

SIZE SCALE

Any attempt to understand the mechanism and effects of radiation damage requires a description of the radiation field at the resolution of the organisation of genome and the cell. DNA, the smallest unit of the genome, is composed of a linear array of deoxyribonucleic acids (~2 nm in nucleus of mammalian cell). A single nucleotide, the basic block of DNA, consists of three chemical parts: a sugar molecule; a phosphate and, a nitrogen containing base. Successive nucleotides are linked via a phosphodiester bond between the sugar and phosphate of adjacent nucleotides. The nitrogen containing bases are not involved in any covalent linkage other than their attachment to the sugar-phosphate backbone. It is the sequence of these nitrogen containing bases along the invariant sugar-phosphate backbone that constitutes the unique structural and functional individuality of each DNA molecule. A typical mammalian cell, of diameter ~10 µm, contains ~6 pg of DNA, highly condensed in the form of higher order molecular structures (nucleosome - diameter ~10 nm, chromatin fibre - diameter ~30 nm and chromosomes - diameter ~1 µm). The total mass of a nucleus is about 10^{-10} g which is a few percent of the total mass.

RADIATION TRACK SIMULATION

To date a large body of scientific literature has been generated which employ Monte Carlo track structure calculations for predicting the measurable parameters from biological experiments for understanding mechanism of damage in molecular radiation biology. Broadly, track structure codes can be classified into five categories [3]. First, target related methods based on a single parameter such as LET, and ionization yields. Second, descriptions based on track segment analysis and the distinction between track entities 'spur', 'blob' and 'short track'. Third, Monte Carlo codes for macroscopic description of dose distribution used e.g in radiological physics. Fourth, the amorphous track description based on average radial dose profiles around the charged particle, and fifth, the full molecular interaction by interaction track structure codes.

KURBUC - A CODE FOR ELECTRON TRACK STRUCTURE SIMULATION

The Monte Carlo track structure code KURBUC [4] simulates electron tracks in the energy range 10 eV to 10 MeV in water. The elastic scattering above 1 keV was calculated using the Rutherford formula taking into account the screening parameter. The effective atomic number and the screening parameter were chosen to fit the experimental data. Angular distribution for elastic scattering below 1 keV was obtained by direct sampling of experimental data. Inelastic cross sections for ionizations and excitations were compiled from different sources. In the range 10 eV - 10 keV, total ionization cross sections were obtained from the least-square fitting of various experimental data. Excitation cross sections were obtained from the data of Paretzke [1]. The calculation of the secondary electron spectrum was carried out using the method of Seltzer [3]. Angular distributions of secondary electrons were calculated using a combination of the experimental data and the kinematical relationships. Figure 1 presents the total ionization, excitation and elastic cross sections.

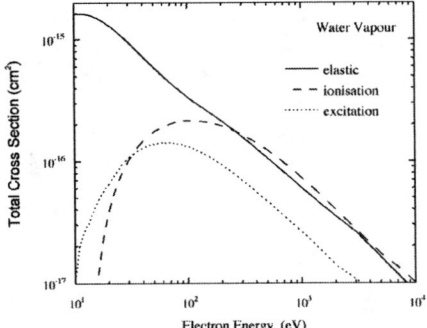
Figure 1. Total cross sections.

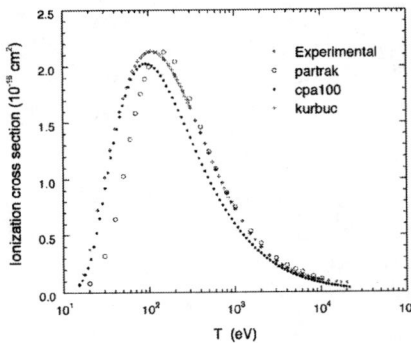
Figure 2. Total ionization cross-section.

IONIZATION CROSS SECTIONS

There are various theoretical treatments and measured experimental data of singly differential ionization cross sections for the ejection of an electron in collision with water molecule. Although there is close agreement between the experimental data over part of the energy spectrum, there is considerable variation in the range 50 eV to 1 keV. Figure 2 shows a comparison between the fitted experimental data and the ionization cross sections by model calculations in the codes *kurbuc* [4], *cpa100* [5] and *partrak* [6]. Detailed description of fitting parameters are given in reference [7].

EXCITATION CROSS SECTIONS

There are not many data sets of experimental excitation cross sections for electrons and these are mainly for water vapour. There are many modes of excitations for energies greater than 10eV. The code *kurbuc* [4] has adopted the data of Paretzke [1] for the 10 modes of excited states. Total excitation cross sections can be obtained by summing all the individual modes of excitations using the empirical formula of Berger and Wang [4]. Figure 3 shows comparison of model calculations by codes *kurbuc*, *cpa100*, *partrak* and *orec* [8].

ELASTIC SCATTERING

Experimental elastic scattering cross sections were least square fitted and compared with calculations by Moliere's formula which is adopted in the codes *kurbuc*. Theoretical cross sections used in the analysis were obtained by a linear combination of composite atoms, i.e. 2H + O. Fitted experimental data and model calculation by *kurbuc* are compared in figure 4.

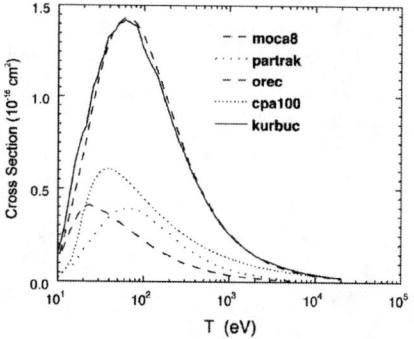

Figure 3. Total excitation cross-section.

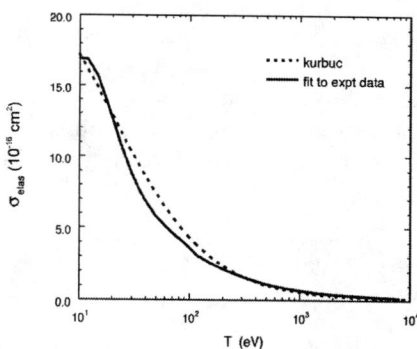

Figure 4. Total elastic cross section.

LEAHIST - A CODE FOR FULL SLOWING DOWN ALPHA-PARTICLE TRACK SIMULATION

This is the first of a new generation of Monte Carlo track structure codes simulating the full slowing down of particle track [9-10]. For ions heavier than protons, additional projectile charge states become important (Table 1). For alpha-particles, the single-electron transfer cross-sections σ_{21} are needed for He^{2+} interactions, σ_{10} and σ_{12} for He^+, and σ_{01} for He^0. The possibility of multiple-electron transfer (e.g., σ_{20} and σ_{02}) can also become important in an appropriate range of particle velocities. Therefore, six charge-changing cross sections are taken into account for a helium three-component system. Cross sections for ion and neutral particle impact were obtained from experimental data for water and where data were lacking the existing experimental data were fitted and extrapolated.

Table 1. Interactions of low energy helium ions with water.

Interactions	$He^{++}+H_2O \longrightarrow$	$He^{+}+H_2O \longrightarrow$	$He^{0}+H_2O \longrightarrow$
Elastic scattering	$He^{++}+H_2O$	$He^{+}+H_2O$	$He^{0}+H_2O$
Target ionization	$He^{++}+H_2O^{+}+e$	$He^{+}+H_2O^{+}+e$	$He^{0}+H_2O^{+}+e$
Target excitation	$He^{++}+H_2O^{*}$	$He^{+}+H_2O^{*}$	$He^{0}+H_2O^{*}$
2-electron capture	$He^{0}+H_2O^{++}$ (σ_{20})		
1-electron capture	$He^{+}+H_2O^{+}$ (σ_{21})	$He^{0}+H_2O^{+}$ (σ_{10})	
1-electron loss		$He^{++}+e+H_2O$ (σ_{12})	$He^{+}+e+H_2O$ (σ_{01})
2-electron loss			$He^{++}+e+e+H_2O$ (σ_{02})
1-electron capture*	$He^{+}+H_2O^{++}+e$	$He^{0}+H_2O^{++}+e$	
1-electron loss *		$He^{++}+e+H_2O^{+}+e$	$He^{+}+e+H_2O^{+}+e$

* and target ionization

Stopping Cross Sections

The reliability of basic cross section data appropriate to the macroscopic stopping of low-energy alpha-particles in water was examined. Analytical calculations of nuclear and electronic stopping cross sections were performed using the cross sections and mean energy loss per collision data. Figures 5 and 6 show the calculated stopping cross sections and ranges in water vapour in comparison with the ICRU data. The calculated stopping cross sections are consistent with published data to within a few % for energies between 1 keV/u and 2 MeV/u.

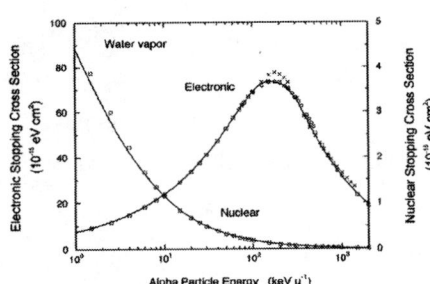

Figure 5. Electronic and nuclear stopping.

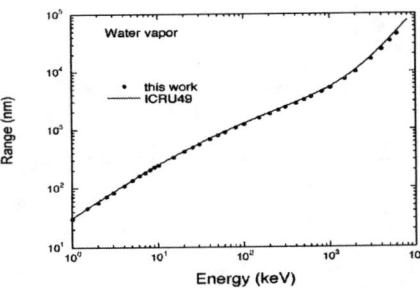

Figure 6. Electron range cross sections.

MODEL CALCULATIONS

Energy Deposition in Molecular targets

Relevant biological effects of radiation in mammals are due mostly to radiation damage to individual cells. It is also well established that the spatial and temporal distribution of radiation interactions within the cell or its nucleus has an important influence on biological effectiveness. Therefore these distributions must be considered if we are seeking either practical quantities for a comparison of different radiations in radiation protection or therapy, or if we are seeking a more fundamental understanding of the mechanism of action. Local energy deposition can readily be measured experimentally over the larger subcellular dimensions down to ~ 1 µm. Monte Carlo track structure techniques allow calculations of local energy depositions, DNA damage and many other properties, down to ~1nm. The conceptual framework of the method is to simulate the irradiation of biological experiments by generating a particle passing through a nucleus and locating the DNA hit and amount of energy deposited at the site of interaction. In this way, we have compiled an extensive data-base of energy deposition at DNA and higher order molecular structures by ultsasoft X-rays, electrons, protons (1 keV - 20 MeV), alpha-particles, and selected HZE particles in water [11]. These studies have been used in a variety of ways to seek quantities of energy deposition that correlate with the observed biological effectiveness of radiations of different qualities. Alternatively, these studies provide a detailed analysis in which energy is deposited inside a target. These energy depositions along with the distribution of reaction sites by reactive radical species with DNA have provided a means of obtaining spectrum of the initial DNA damage by ionizing radiations. Figures 7 and 8 show frequency distribution of energy deposition in a target corresponding in size to a volume of dimensions similar to a segment of (a) DNA and (b) chromatin fiber, in water irradiated with high- and low-LET radiations. The left ordinate gives the absolute frequency of deposition events greater than energy E deposited in the target volume when randomly irradiated with 1 Gy of the given radiation. The number of events in the target N, corresponding to the frequency of energy deposition for energies >E (eV) is obtained by N= $f(>E)/f(>0)*M$, where $f(>0)=1/z_F$ is the frequency of hits of any size and M is the total number hits. Relevant data can be obtained from reference [12].

DNA DAMAGE SIMULATION

An alternative way of using the detail provided by the Monte Carlo track structure approach is to examine the way in which energy is deposited by ionizations and excitations at sugar-phosphate moiety and the bases and relating it to an initial biological lesion which can be measured experimentally. Various parameters have been used to achieve such a relationship including 'ionization' and a 'quantity of energy' to induce a single strand break or double strand break by various authors [13]. In this work we use a quantity of energy, set at 17.5 eV, deposited by a single track at a sugar-phosphate site for induction of a single strand break. Using this criterion for the induction of a ssb the energy deposition patterns from charged particle tracks can be converted to distributions of single strand breaks along the hit section of the DNA. When two single strand breaks on opposite strands were separated by a distance less than ten base-pair it is assumed that a double strand break is produced. Figure 9 shows the relationship between the energy deposited and the probability of induction

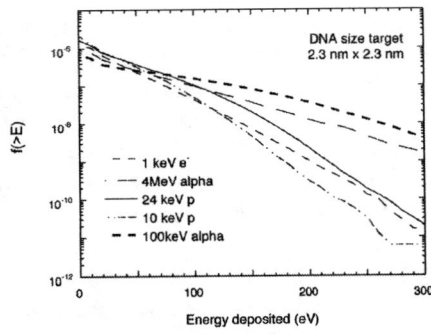

Figure 7. Frequency of energy depositions in DNA size targets.

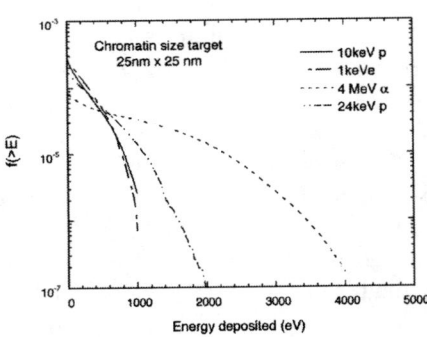

Figure 8. Frequency of energy deposition in chromatin.

of single and double strand breaks in DNA. The data shows a universal curve for all radiations of low and high LET independent of the type of radiation. This is expected as the complexity of damage depends on patterns of energy deposition rather than the source of radiation.

The threshold energy concept deals with the direct interaction of track with the target. Radical species such as e^-_{aq}, ˙H and ˙OH are also produced in the bulk water associated with DNA. The reaction of OH radicals with sugar produce sugar radicals. Not all sugar radicals lead to strand break. In general, about 20% of sugar radicals are converted to strand break. Therefore, in the cellular environment an activation probability of 0.13 for the induction of a single strand break by hydroxyl radical can be assumed.

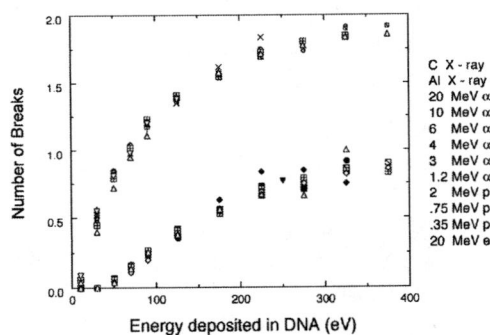

Figure 9. Relationship between energy deposited and probability of induction of DSB.

CLUSTERED DAMAGE IN DNA

There have long been indications that the biological consequences of ionizing radiations are determined by their clustering properties at the level of the DNA duplex. From the data base on the frequencies of energy deposition in volumes of biological dimensions it became possible to seek and correlate the size of energy deposition with particular biological effects. An early application of track scoring was made for X-rays of various energies indicating biologically critical properties are in the regions greater than 100 eV of energy deposition in volumes similar to that of DNA. In a similar manner the dominant feature associated with high-LET radiations was found to correspond to a class of clusters of energy depositions greater than about 300 eV in nucleosome size targets. The above approaches could not give information on the molecular nature of the DNA damage, but they indicated the need for more detailed simulations at the molecular level.

To provide relative distribution of damage in terms of the complexity of the damage, Monte Carlo calculated spectrum of DNA damage were analyzed for simple and complex strand breaks induced by electrons, protons and alpha particles of various energies. Complex strand breaks were defined as those segments of DNA containing more than one break associated with a single/double strand break on the same or the opposite strand. For low-LET radiations, track structure simulations have shown that a substantial proportion of the dose, approximately 30%, is always deposited via low energy secondary electrons that are efficient at producing clustered damage at DNA level. These clusters appear at sufficiently high frequency per unit dose to explain the observed yields, spectrum and spatial correlation of damage to DNA. For all radiations, the relative biological effectiveness for induction of single and double strand breaks is not dependent on the radiation quality, as seen in Figure 10. This is a reason the total initial DNA damage could not be correlated as a measure of the biological effectiveness of radiation. For the high LET radiations it is expected there will be fewer regions of damage to DNA per unit dose because of fewer tracks crossing the nucleus producing direct ionizations damage by radical species. Current calculations of complexity of DNA double strand breaks for a variety of radiations show a strong dependency on the ionization density of the track (Figure 11). These calculations show a substantial spectral shift toward greater complexity. For the higher LET alpha-particles, more than 70% of the double strand breaks estimated to have additional associated strand breaks within 10 base pairs. These calculations show protons produce more complex damage than alpha-particles. [14]

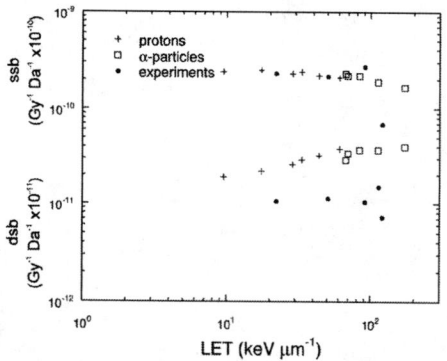
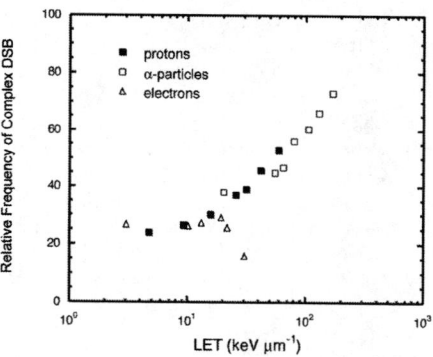

Figure 10. Relative biological-effectiveness (RBE) for double strand breaks vs LET.

Figure 11. Complexity of DNA double strand break vs LET.

CONCLUSIONS

Recent development in Monte Carlo track structure has enabled simulation of full slowing down of proton and alpha-particles in the energy range 1 to 8 MeV. However, there is a strong need for measurements of cross section for neutral and partially ionized particles. The Pathway to DNA damage is a highly complex subject as physical and physiological conditions affects the response of the cell to damage by ionizing radiation. The method adopted in these calculations is based on knowledge deduced from relevant biological experiment. In the first instance, it is assumed that the observed biological lesions are mainly mediated by the induction of double strand breaks. It is assumed that a double strand break is composed of two single strand breaks on opposite strand placed near each other and are induced by a single particle traversal. Calculations show that the majority of damaged sites are simple, consisting of a single strand break or base damage, but a substantial proportion, more than 20%, of the damages sites contain additional damage, and the contribution of hydroxyl radicals to the total yield of strand breakage is a function of activation probability. It is observed that for low LET radiations ~90% of total energy deposited in DNA are due to energy deposition events less than 60 eV but the largest dsb yield is due to energy depositions in the range 100-150 eV. Similarly, although the initial yield of strand breakage is nearly the same for all radiations, the complexity of strand breaks increase with increase in LET of the particle.

REFERENCES

1. Paretzke, H. G., *Kinetics of Nonhomogeneous Processes*, edited by G. R. Freeman, Wiley: New York, 1987, pp. 89-170.
2. Goodhead, D. T., "Relationship of Microdosimetric techniques..." in *The Dosimetry of Ionising Radiations*, edited by K. R. Kase, B. E. Bjarngard, F. H. Attix, vol. 2, Orlando, Academic Press, 1987, pp. 1-89.
3. Goodhead, D. T, Nikjoo, H., *Numero special de Radioprotection* **32**, C1-3 (1997).
4. Uehara, S., Nikjoo, H., Goodhead, D. T., *Phys. Med. Biol.* **38**, 1841 (1993).
5. Terrissol, M., *Int. J. Radiat. Biol.* **66**, 447-452 (1994).
6. Dingfelder, M., Inokuti, M., Paretzke, H. G., *Radiat. Phys. Chem.* **59**, 255 (2000).
7. Uehara, S., Nikjoo, H., Goodhead, D. T., *Radiat. Res.* **152**, 202 (1999).
8. Turner, J. E., Magee, J. L., Weight, H. A., Chatterjee, A., Hamm, R. N., Ritchie, R. H., *Radiat. Res.* **96**, 437 (1983).
9. Uehara, S., Toburen, L. H., Wilson, W. E., Goodhead, D. T., Nikjoo, H., *Radiat. Phys. Chem.* **59**, 1 (2000).
10. Uehara, S., Toburen, L. H., Nikjoo, H., *Int. J. Radiat. Biol* **77**, 139 (2001).
11. Cucinotta, F. A., Nikjoo, H., Goodhead, D. T., *Radiat. Res.* **153**, 459-468 (2000).
12. Copies of the Monographs on "Energy Depositions in cylindrical volumes" for electrons ultrasoft X-rays, protons and alpha-particles can be obtained from h.nikjoo@har.mrc.ac.uk
13. Nikjoo, H., O'Neill, P., Terrissil, M., Goodhead, D. T., *Radiat. Environ. Biophys.* **38**, 31-38 (1999).
14. Nikjoo, H., O'Neill, P., Wilson, W. E., Goodhead, D. T., *Radiat. Res.* **156**, 577-583 (2001).

Laboratory Data Needs and Applications for Assessing Radiation Effects in Biological Materials

L.H. Toburen* and J.L. Shinpaugh*

Department of Physics, East Carolina University, Greenville, NC 27858

Abstract. All types of ionizing radiation interact with material by producing atomic or molecular ions, excited states, and secondary electrons. Still, different types of radiation lead to quite different yields of biological damage. It is generally believed that the spatial distributions of ionization and excitation produced by the slowing down of charged particles, particularly electrons, govern the yields of bioactive molecular species. The assessment of these spatial patterns of ionization and excitation depend largely on the production cross sections for secondary electrons, the energies and angular correlations of their production, and the subsequent differential cross sections for their energy loss in the media of interest. The most thorough assessment of spatial patterns of energy deposition by charged particles is obtained using Monte Carlo simulation of charged particle track structure based on the available database of interaction cross sections. This step-by-step analysis of the interactions of charged particles, from their initial energies until stopped in the medium, provides detailed information on the spatial distribution of the products of ionization from which subsequent chemical and biochemical reactions can be assessed. Over the years a substantial database has been developed describing the interaction of fast, "bare" charged particles, e.g., electrons, protons, and alpha particles, with atomic and molecular targets. However, as charged particles slow they enter an energy regime where additional energy-loss channels open and the availability of data is often limited; interest in this region is also intensified because the biological effectiveness of these particles can increase as their energy decreases. This presentation will focus on the special needs for data involving low-energy electrons and ions, i.e., slowing protons, alpha particles and heavier ions, in a biological environment. A brief discussion of the availability of cross sections will be presented and areas of need for additional data will be discussed.

INTRODUCTION

Differences in biological effectiveness for different types of ionizing radiation are generally attributed to variations in spatial patterns of energy deposition produced when the radiation interacts with tissue. Although all types of ionizing radiation interact with tissue by the production of excitation and ionization resulting in ions and secondary electrons the spatial patterns of energy deposition vary greatly depending on the type of radiation in question. In particular, secondary electrons generated with spectra specific to the incident radiation produce additional ionization as they slow, leading to the formation of local regions of varying energy densities. This local density of energy deposition can have profound effects on the subsequent chemistry and thereby on the biological response to ionizing radiation.

Many different approaches have been developed to predict the variations in ionization density produced by the absorption of both directly and indirectly ionizing radiation. All of these approaches rely on understanding ionization and electron transport in biological material. Monte Carlo (MC) models of charged-particle track structure provide the most detailed event-by-event description of the energy deposition process. Such models provide detailed information on the spatial distributions of ionization and excitation events at dimensions comparable to critical sites in such important molecules as the nuclear cellular DNA and associated cellular structures (see, for example, Friedland et al. [1], Chatterjee and Holly [2], and Wilson and Toburen [3]). Although current models of charged particle track structure have been highly successful in supporting the assessment of mechanisms of radiation action, the potential to accurately assess the spatial patterns of energy deposition at the microscopic is being severely tested as biochemical tools seek to better define damage at increasingly detailed levels [4]. Additional data is sorely needed to account for the heterogeneity of tissue and to update the computational methodology for accurately assessing low-energy interactions in condensed phase media.

The simulation of charged particle tracks rely on a broad base of interaction cross sections. For high-energy charged particles (velocity significantly greater than that of the bound electron with which they interact) the primary interaction mechanisms are simple excitation and ionization of the target electron. Secondary electrons are generated with a wide range of energies and one must follow their interactions until they come to a stop in the media. As the ions slow, positive ions capture electrons and then must be treated in subsequent interactions as "dressed ions." To describe the full range of energies for the stopping of electrons and heavy charged particles one must have data for a wide range of interactions including:

- Excitation of the target by electrons, bare ions, and projectiles carrying bound electrons.
- Ionization of the target by electrons, bare ions and projectiles carrying bound electrons. In addition, these cross sections are needed differential in energy loss and the angle of emission of the secondary electron in order to determine spatial properties of the resulting particle track.
- Electron capture and loss cross sections for the incident projectile.
- Elastic and inelastic scattering cross sections for both the secondary electrons and the incident heavy particles.

For studies of importance in radiobiology, these interaction cross sections must reflect the heterogeneous molecular nature of the absorbing medium and be representative of the condensed phase of biological tissue. This latter requirement is perhaps the most difficult because measurements in condensed phase are particularly challenging - if at all possible, meaning that experimental cross sections are lacking.

The relative contribution of excitation, ionization, and charge transfer to the total stopping power for the stopping of protons in water vapor is illustrated in Fig. 1 using the data of Uehara et al. [5]. Note that the major contributions to stopping power are ionization by protons at high energies and ionization by hydrogen atoms at low energies. Charge transfer provides a sizeable contribution to the stopping power for energies less than about 100 keV and the estimate of excitation by protons never exceeds about 10% with the maximum contribution for energies from about 50 to 100 keV. Excitation by

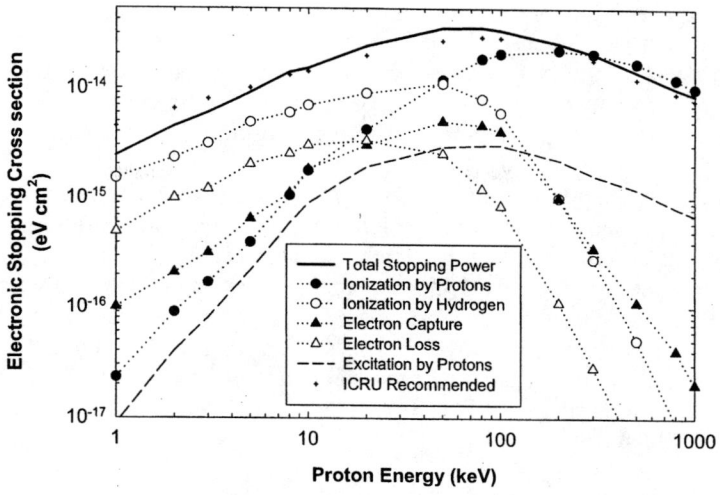

FIGURE 1. Contribution of excitation, ionization, and charge transfer to the total stopping power for protons traversing water vapor [5].

the neutral hydrogen fraction in the beam, formed from electron capture by the proton, is not shown here owing to lack of reliable data, but it is expected to be somewhat less than that estimated for protons owing to screening. In seeking data to develop models of charged particle track structure, an extensive data base exists for interaction cross sections for bare ions, but as electron capture becomes important there is little information on interactions of dressed ions. In the following a brief discussion of the data needs for modelling charged particle track structure in biological material will be provided.

DATA NEEDS

Excitation

Cross sections for excitation used in track structure models are generally obtained from an application of the analytic expression and parameters presented by Miller and Green [6]. These results are highly speculative considering the lack of data regarding excitation by charged particles. Fortunately the contribution of excitation to total energy loss is generally quite small, although for low-energy ions and neutrals, where the uncertainties are largest, the contribution can also be anticipated to be the largest. In developing the database for Monte Carlo calculations of the track structure of slow protons the most uncertain cross sections are often adjusted to provide consistency between calculated and measured stopping power [7]. Because of the high degree of accuracy afforded by both stopping power calculations and measurements, where accuracies of

1% are often quoted, this forced consistency of data is a relatively accurate means of refining uncertain cross sections when only one of the contributions to stopping power is in question, as is the case for proton energies greater than about 500 keV. Unfortunately, for low-energy ions uncertainties often exist in several of the energy loss components and forcing consistency with stopping power is less useful for adjusting uncertain cross sections.

It seems fair to characterize the state of our knowledge of excitation of biologically relevant media by charged particles, including dressed ions and neutrals, as essentially nil. First, few data exist for excitation of atoms and molecules, and we have little guidance as to how these meager data might be extrapolated to condensed phase material. Second, we have no data for excitation by screened nuclei, i.e., dressed ions. It is convenient to suggest that excitation by H^0 is the same as that for H^+ and that both can be scaled from electron impact, but we have no experimental confirmation of this scaling. What about other ions and neutrals and how do we treat condensed phase macromolecular systems? Much data is needed.

Charge Transfer

The state-of-knowledge for charge transfer cross sections needed for application in radiation biology was reviewed recently [8] and will be discussed only briefly here. For proton and neutral hydrogen impact we have a good understanding of electron capture and loss, respectively, in collisions with gas-phase atoms and molecules. For other bare and dressed ions the data are sparse. Even for interactions involving gas collisions, theory is severely limited, and there is no direct information on these processes in condensed-phase material. For proton impact we have been able to make some arguments regarding the relative lack of phase effects on electron capture and loss processes [8], but these arguments are of limited reliability owing to lack of data. In addition, as one looks to heavier ions the effects of excited states of the projectile cannot be ignored and there is little data on which to base estimates of these effects.

We are in need of data for all charge transfer channels for all dressed ions and neutrals in biologically relevant targets. Even for water vapor, data is limited and there is essentially no data for condensed-phase media.

Ionization

There have been considerable progress in the measurement of ionization cross sections for protons and helium ions with some data also available for heavier ions; these data have been discussed in several review articles (see, for example, Stolterfoht et al. [9], Toburen [10], and ICRU-55 [11]). The data for protons and atomic hydrogen impact cover a wide energy range with emphasis on molecular targets relevant to radiation biology. In particular, doubly differential ionization cross sections, differential in ejected electron energy and emission angle, have been measured from a few keV to a few MeV for protons in water vapor and hydrocarbon targets and from a few 10s to a few 100s of

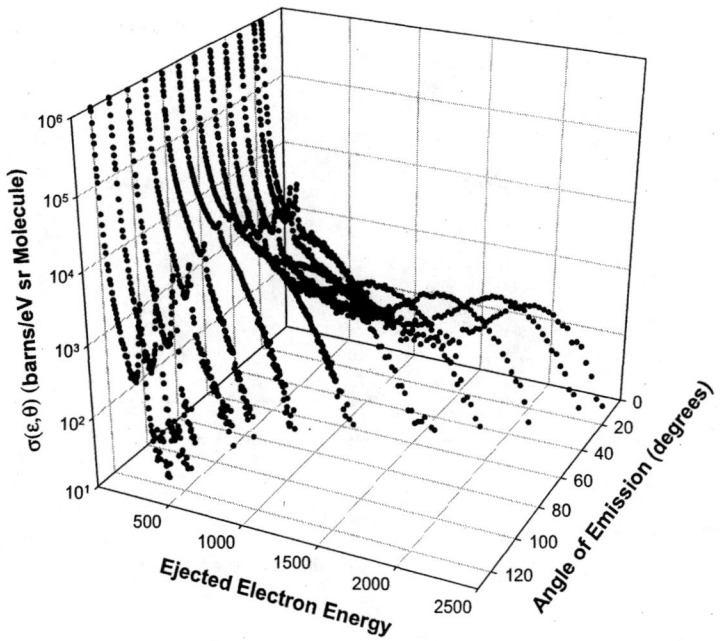

FIGURE 2. Electron emission spectra for ionization of CH_4 by 1-MeV protons [12].

keV for H^0 in H_2O.

The detail afforded by these measurements is illustrated in Fig. 2 where the doubly differential cross sections for ionization of CH_4 by 1 MeV protons are shown. These cross sections peak for emission of low-energy electrons resulting from "soft" or large impact parameter collisions and show the characteristic "binary encounter" ridge for ejection of high-energy electrons in small impact parameter Rutherford-like collisions. In fig. 2, excitation of the K-shell of carbon in the target is observed by the K-Auger spectrum superimposed on the continuum background at approximately 250 eV.

The data for ionization of molecular targets by protons is sufficiently extant that Rudd [13] has been able to develop an analytical model for the single differential cross sections, differential in ejected electron energy, with parameters obtained by fitting the experimental data. The reliability of his model is illustrated in Fig. 3 where the model calculations are compared to the experimental data of Lynch et al. [12] for ionization of methane. Agreement between experiment and model calculations is quite good throughout the ejected electron spectra shown, and has been tested and found reliable for proton energies from a few keV to several MeV. Differential cross sections for other bare ions can be scaled from this model by simply multiplying the proton cross sections by z^2 (z is the atomic number of the projectile) for equal ion and proton velocities, i.e., equal energies expressed in MeV/u.

The availability of differential ionization cross sections for dressed ion and neutral atom impact is very limited. Cross sections for low-energy electron emission in colli-

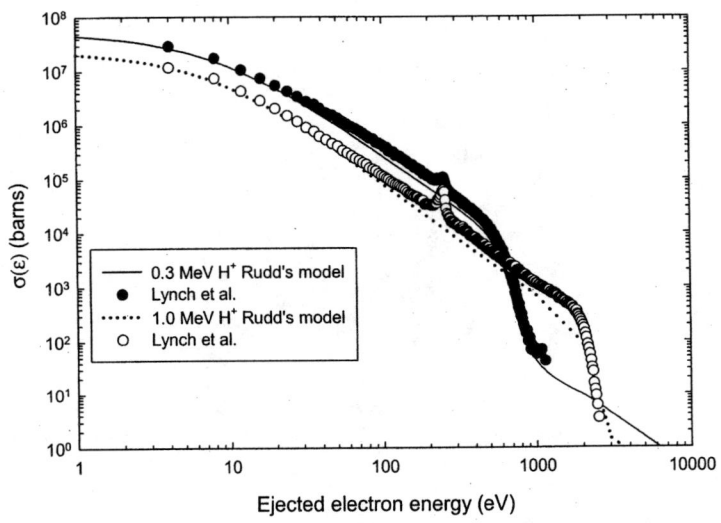

FIGURE 3. Cross sections for ionization of CH_4 by protons. Data are from [12].

sions involving ions carrying bound electrons are generally reduced relative to those for bare ions because of the effects of electronic screening. On the other hand, cross sections for "energetic" electron emission are nearly the same as bare ion cross sections because they arise from close collisions where screening is inefficient. In addition, the spectra show contributions from electrons stripped from the dressed ion in collisions with the target.

For H^0 impact, the ejected electron spectrum tends to very closely approximate the spectra derived from proton impact plus contributions from stripping an electron from H^0. The singly differential ejected electron spectrum for H^0 impact on H_2O measured by Bolorizadeh and Rudd [14] is shown in Fig. 4 along with the comparable proton spectrum [15]. The H^0-induced spectrum is seen to agree with the proton spectrum for both low- and high-energy ejected electrons. The effects of screening that reduce the cross sections for low-energy electrons is compensated by electrons from inelastic scattering of the projectile electrons by the target; this effect is illustated by the solid lines derived from inelastic scattering cross sections for electrons of comparable velocity to the incoming ion. The result is that through some accidental compensation the proton-induced cross sections give a relatively accurate representation of the neutral hydrogen cross section.

For heavier ions, the ability to scale proton data to dressed ion cross sections is more difficult. The spectra of electrons emitted in collisions of 0.3 MeV/u C^+ with CH_4 is shown in Fig. 5. In this figure there are several features that can be observed, these include a peak at about 150 eV for forward angles that results from electron stripping, a small peak that is superimposed on the electron continuum at high energies and small angles do to the Doppler shifted Auger electrons from inner shell ionization of the

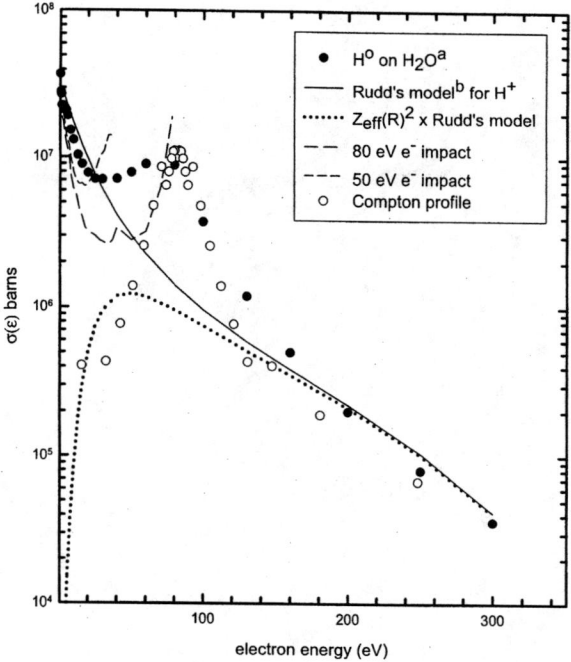

FIGURE 4. Cross sections for electron emission for collisions of H^0 and H^+ with H_2O. Experimental data are from [14] and [15].

projectile, an Auger spectrum from inner shell ionization of the carbon target, and a peak in each spectrum at low energies due to soft collisions. What is missing relative to the proton spectra shown in Fig. 2 is the binary encounter ridge for fast ejected electrons - it is still there but at a sufficiently low energy that it merges with the electrons from other processes making it less distinct.

The limited range of ionization data for heavy particles make modeling the cross sections in the manner conducted by Rudd for protons difficult. Still we can gain some understanding of the effects of bound electrons by comparing spectra to those for equal velocity protons, which is illustrated in Fig. 6 for 0.1 MeV/u ions. Note that the z^2 scaling of the proton data of Stolterfoht [16] is in quite good agreement with the C^+ data for high energy electron ejection, those electrons ejected in close collisions, but this scaling fails by a large amount for electron energies less that about 200 eV, i.e., contributions from distant collisions. We have tried the approach of using a screening function that depends on the distance of approach as estimated by the Massey criterion [17] and found that ions with energies greater than about 100 keV/u can be described reasonably well. Unfortunately there is inadequate data for ionization by heavy ions of various charge states and energies to allow a good test of the model.

For dressed ions we are therefore far short of an understanding that can lead us to the cross sections needed for track structure calculations. And, there are no data for

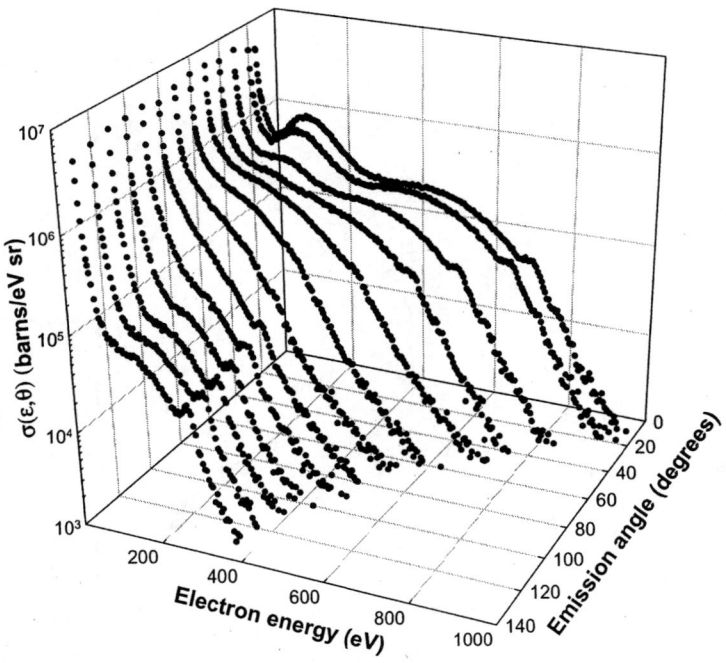

FIGURE 5. Electron emission spectra for collisions of 0.3 MeV/u C^+ with CH_4.

condensed phase single or double differential cross sections from condensed phase material. We can make guesses, and we can test codes against biological endpoints such as DNA strand break production, but we cannot currently estimate with confidence the cross sections for heavy ions as they slow and capture electrons. MORE data is needed.

Condensed Phase

Most, if not all, of the data discussed above have been measured for gas phase interactions. Little is know concerning the effects of condensed phase on interaction probabilities. Where calculations have been made for condensed phase, they are based on perturbation theories applicable to energy transfer by "fast" particles and they are largely untested by experimental results. It has only been in the past several years that laboratories are beginning to address the fate of charged particle interactions in condensed phase material. Of particular interest is work from Sanche and colleagues at the University of Sherbrooke on the interactions of slow electrons, those with energies from 1 eV to several 10s of eV, in such targets as water-ice, frozen hydrocarbons and biomolecules (see for example, Michaud and Sanche [18], Bass and Sanche [19], and Huels et al. [20]). Such data have the potential to complement existing data for high-

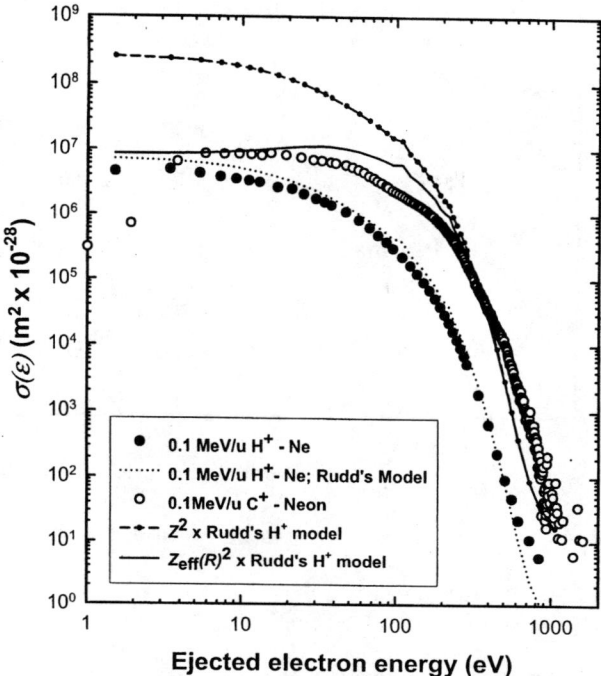

FIGURE 6. Electron emission spectra for 0.1 MeV/u ions incident on neon. Data for proton impact are from [16].

energy electrons where phase effects are not likely to be dominant. We show the data of Michaud and Sanche [18] for inelastic cross sections for low-energy electrons in Fig. 7 along with cross sections for ionization of water vapor by high-energy electrons [21]. We must await further measurements in the intervening energy range to gain information on the effects of phase on cross sections for the high-energy electron impact. It goes without saying that we have only touched the surface of this very important facet of understanding interactions of charged particles with biologically relevant material.

ACKNOWLEDGMENTS

Research funded in part by the Low Dose Radiation Research Program, Biological and Environmental Research (BER), U.S. Department of Energy, Grant DE-FG02-01ER63233.

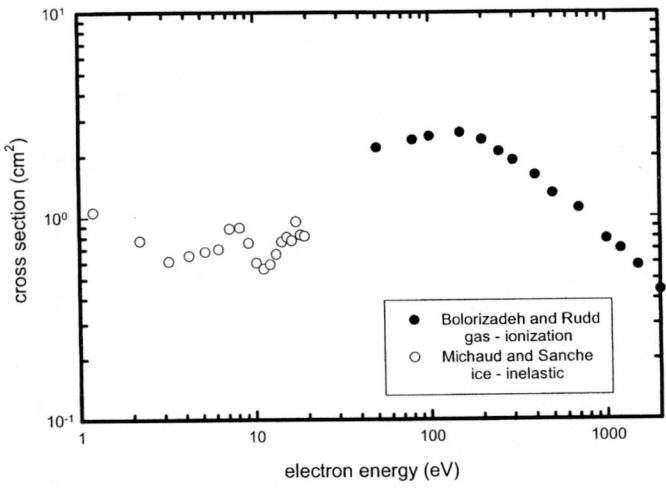

FIGURE 7. Cross sections for excitation of ice [18] and ionization of water vapor [21] by electrons.

REFERENCES

1. Friedland, W., Li, W. B., Jacob, P., and Paretzke, H. G., *Radiat. Res.*, **155**, 703–715 (2001).
2. Chatterjee, A., and Holly, W. R., *Int. J. Quantum Chemistry*, **39**, 709–727 (2001).
3. Wilson, W. E., and Toburen, L. H., "Biological response to inner-shell ionization," in *X-Ray and Inner-Shell Processes*, edited by T. A. Carlson, M. O. Krause, and S. T. Manson, AIP Conference Proceedings 215, American Institute of Physics, New York, 1990, pp. 878–888.
4. Sutherland, B. M., Sidorkina, P. V. B. O., and Laval, J., *Proc. Natl. Acad. Sci.*, **97**, 103–108 (2000).
5. Uehara, S., Toburen, L. H., Wilson, W. E., Goodhead, D. T., and Nikjoo, H., *Radiat. Phys. and Chem.*, **59**, 1–11 (2000).
6. Miller, J. H., and Green, A. E. S., *Radiat. Res.*, **54**, 343–363 (1974).
7. Uehara, S., Toburen, L. H., Wilson, W. E., Goodhead, D. T., and Nikjoo, H., *Int. J. Radiat. Biol.*, **77**, 139–154 (2001).
8. Toburen, L. H., *Radiat. Environ. Biophys.*, **37**, 221–233 (1998).
9. Stolterfoht, N., DuBois, R. D., and Riverola, R. D., *Electron production in heavy ion-atom collisions*, vol. 20 of *Springer Series on Atoms and Plasmas*, Springer-Verlag, Berlin, 1997.
10. Toburen, L. H., "Ionization by fast charged particles," in *Atomic and Molecular Data for Radiotherapy and Radiation Research*, IAEA-TECDOC-799, International Atomic Energy Agency, Vienna, 1995, pp. 51–162.
11. ICRU-55, Secondary electron spectra from charged particle interactions., Tech. Rep. 55, International Commission on Radiation Units and Measurements, Bethesda, MD (1995).
12. Lynch, D. J., Toburen, L. H., and Wilson, W. E., *J. Chem. Phys.*, **64**, 2616–2622 (1976).
13. Rudd, M. E., *Phys. Rev. A*, **38**, 6129–6137 (1988).
14. Bolorizadeh, M. A., and Rudd, M. E., *Phys. Rev. A*, **33**, 893–896 (1986).
15. Bolorizadeh, M. A., and Rudd, M. E., *Phys. Rev. A*, **33**, 888–892 (1986).
16. Stolterfoht, N., *Z. Physik*, **248**, 81–95 (1971).
17. Toburen, L. H., Stolterfolt, N., Ziem, P., and Schneider, D., *Phys. Rev. A*, **24**, 1741–1745 (1981).
18. Michaud, M., and Sanche, L., *Phys. Rev. A*, **36**, 4672–4683 (1987).
19. Bass, A. D., and Sanche, L., *Radiat. Environ. Biophys.*, **37**, 243–257 (1998).
20. Huels, M. A., Handorf, I., Illenberger, E., and Sanche, L., *J. Chem. Phys.*, **108**, 1309–1312 (1998).
21. Bolorizadeh, M. A., and Rudd, M. E., *Phys. Rev. A*, **33**, 882–887 (1986).

B. Lighting

Physical Aspects of Mercury-Free High Pressure Discharge Lamps

M. Born

Philips Research Laboratories, Weisshausstr.2, D-52066 Aachen, Germany

Abstract. This paper gives a summary of recent results about the replacement of mercury in high pressure discharge lamps by metallic zinc. Actually, this topic is of high relevance for the lighting industry due to the need of more environmentally friendly products. The work presented here is supported by the German government under contract no. 13N8072.

Pure zinc/argon discharges as well as lamps including zinc or mercury and metal halide additives are investigated. Experimental data are compared with model calculations of the energy balance involving the transport of heat and radiation. Since the excitation energies of relevant zinc transistions are lower than for mercury, axis temperatures of pure zinc lamps are about 300 K below the value of mercury arcs. In addition, the thermal conductivity of zinc including the contribution of radiation diffusion is larger than compared to mercury. From lamp voltage measurements it is found that the cross section for elastical electron scattering by zinc atoms is about the same as for mercury. When adding metal halides to a pure zinc discharge with argon as a starting gas, i.e. NaI, TlI, DyI$_3$, axis temperatures decrease to about 5100 K due to strong radiation cooling. In order to obtain sufficiently large lamp voltages, wall temperatures of more than 1300 K are adjusted by means of polycrystalline aluminaoxide (Al$_2$O$_3$) as a wall material. Electrical field strenghts of 6.0 V/mm and 8.6 V/mm are measured for metal halide lamps containing zinc or mercury, respectively. The light technical data of the discharges are very close, since mercury and zinc do not contribute significantly to the radiation in the visible range. Efficacies of up to 93 lm/W and 100 lm/W are found in metal halide lamps with zinc and mercury, respectively. Consequently, zinc turns out to be an attractive replacer for mercury in this type of lamp not only from an environmental point of view.

INTRODUCTION

Today, mercury is still part of the filling of high pressure discharge lamps (HID). In pure mercury lamps, the radiation of mercury in the visible yields low efficacies around 60 lm/W due to relatively high lying energy levels with moderate transition probabilities. For lighting applications, where efficacies of around 100 lm/W and good colour rendering properties are needed, metal halides (i.e. NaI, TlI, DyI$_3$) are added and dominate the visible spectra of the lamps. Mercury is still needed for elastical scattering of electrons in order to adjust a typical lamp voltage of about 80 ... 90 V. This value corresponds to a partial pressure of mercury of typically $p_{Hg} \approx 20\,bar$, which is about the total pressure in metal halide lamps. Thus, mercury takes the role of a so called buffer gas, which mainly affects the electrical properties of metal halide lamps. Unfortunately, mercury is an environmental hazard and therefore should be replaced by a non-toxic element [1]-[4], which has to fulfill the following requirements:

- large momentum transfer cross-section for elastical electron scattering
- large neutral particle density (or pressure)

- large excitation and ionisation energies

The first two items are related to the need of a sufficiently large resistance of the discharge in order to limit the discharge current at a given lamp power. Large excitation and ionisation energies are required since the buffer gas (mercury or the replacer) may not dominate the spectrum significantly. Besides these physical properties, chemical stability of the salt filling and the buffer gas must be guaranteed, e.g. the replacer may not significantly interact with the salt filling, the wall and electrode material. Finally, it should not be on the so called A-List of environmental hazards [5]. The above stated conditions can be matched by the usage of zinc instead of mercury, which is discussed in this paper.

EXPERIMENTAL

Two kinds of lamps are investigated: pure mercury and pure zinc discharges as well as metal halide lamps containing NaI, TlI, DyI_3 and mercury or zinc. All discharges are filled with argon as a starting gas. The lamps are made from polycrystalline aluminaoxide (PCA) which allows operation at mean wall temperatures of up to 1800 K. The electrical input power is varied between 60...90 W for the pure mercury or zinc discharges while the metal halide lamps are operated at a constant power of 75 W.

In figure 1 the interesting values of zinc and mercury pressures are indicated in the region between 10 and 100 bar. In the case of mercury, this region can be easily achieved due to its volatile character. Thus, mercury lamps are operated in the so called unsaturated region - the pressure is adjusted by the mass filled into the discharge vessel since mercury can be fully evaporized. In contrast, zinc lamps need to be operated in

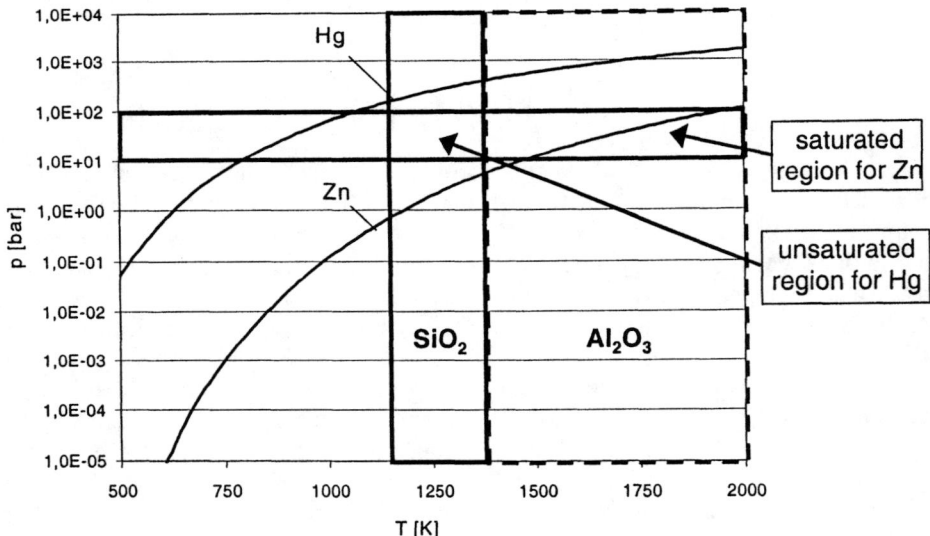

Figure 1. Evaporation of mercury and zinc and operation regions related to the wall materials quartz and PCA.

the so called saturated region due to the weaker evaporation behaviour. In this case, Zn is not fully evaporized due to its partial transport to colder parts of the lamp, i.e. to the end construction of the discharge vessel. In the indicated pressure region, sufficient evaporation of Zn is possible using polycrystalline Al_2O_3 (PCA) as a wall material.

MODELLING

As reported in [6] we solve the energy balance in cylindrical geometry since the lamps under investigation are of this symmetry. In addition, the contribution of convection can be neglected for pressures below ~30 bar [7].

$$\sigma(\rho)E^2 = u_{rad}(\rho) - \frac{1}{R^2\rho}\frac{\partial}{\partial\rho}\left[\rho\kappa(\rho)\frac{\partial T(\rho)}{\partial\rho}\right] \quad (1)$$

$$\rho = \frac{r}{R} \quad (2)$$

(σ: electrical conductivity, E: electrical field, κ: total thermal conductivity, T: plasma temperature, r: local radius, R: radius of discharge vessel).

In addition, we calculate the absolute local net emission u_{rad} by solving the two dimensional radiation transfer equation after [8]. Spectral line data and broadening parameters have been taken from [9]-[13]. Broadening parameters and the calculation of constants in the case of lack of experimental data are also given in [6] and [14]. For mercury and zinc, transistions according to figure 2 and figure 3 are considered.

The solution of eq. (1) is complicated by the fact that the temperature profile needed as an input for the solution of the radiation transfer equation is not known and thus has to be determined iteratively from the energy balance. The evaluation of the material functions κ and σ is described in more in detail in [14] and will be outlined only roughly in this paper.

The total thermal conductivity

$$\kappa = \kappa_{trans} + \kappa_{chem} + \kappa_{rad} \quad (3)$$

consists of a translational, chemical and radiative part, the latter being related to the radiation diffusion of non-optically thin lines which has been calculated after [6]. The electrical conductivity σ is related to electron-neutral collisions σ_{eo} and electron-ion collisions σ_{ei}

$$\frac{1}{\sigma(T)} = \frac{1}{\sigma_{e0}(T)} + \frac{1}{\sigma_{ei}(T)} \quad (4)$$

The second term on the right hand side of eq. (4) is small compared to the first and is evaluated after the Spitzer-Härm formula [15].

Electron-neutral collisions are combined from all species j with neutral densities n_j and effective collision frequencies $v_{j,eff}$

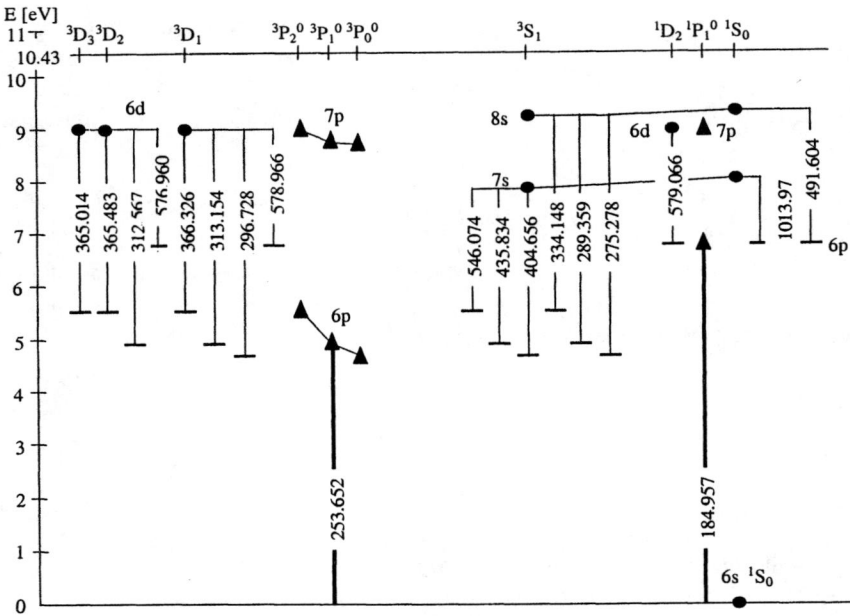

Figure 2. Scheme of mercury levels and transitions.

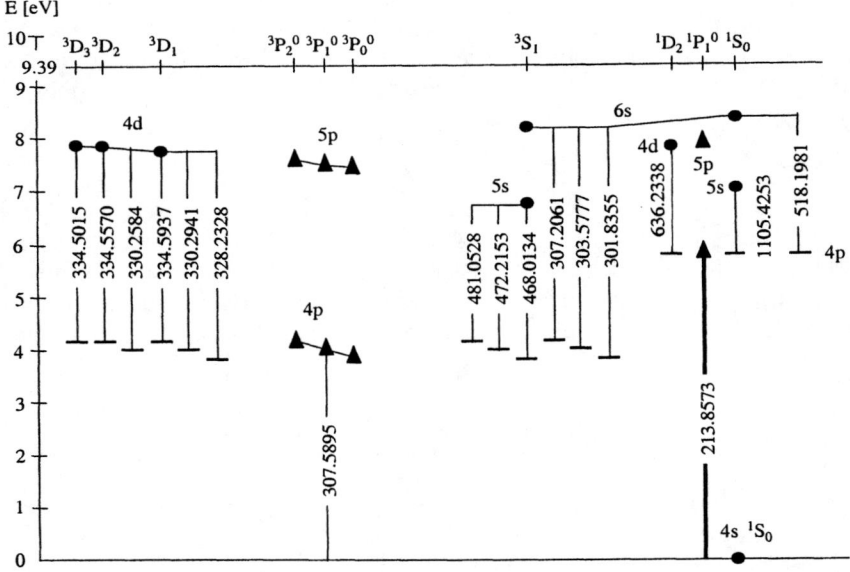

Figure 3. Scheme of zinc levels and transitions.

$$\sigma_{e0}(T) = \frac{e^2 n_e(T)}{m_e \sum_j \nu_{j,eff}(T)} \quad (5)$$

$$\nu_{j,eff}(T) = n_j(T) <q_j(T)\nu_e(T)> \quad (6)$$

$$\nu_e(T) = \sqrt{\frac{8kT}{\pi m_e}} \quad (7)$$

Data for the momentum transfer cross sections q_j are taken from Hg:[16], Na:[17], Tl:[18], Ar:[19]. While lamp voltage measurements of mercury discharges are in good agreement with calculations, this is not the case for zinc when taking the theoretical cross-sections from [20]. Experimental data in the low energy region below 1 eV is not known to the author. In order to obtain a good agreement between measurement and theory, the cross-section of Zn is modified by a factor of 0.42 in this work. Figure 4 displays the ratio of electrical conductivities of pure mercury and zinc discharges as a function of temperature and at equal pressures. The cross-section for Hg is taken from [16] whereas the one for Zn is taken from experimental data of this paper.

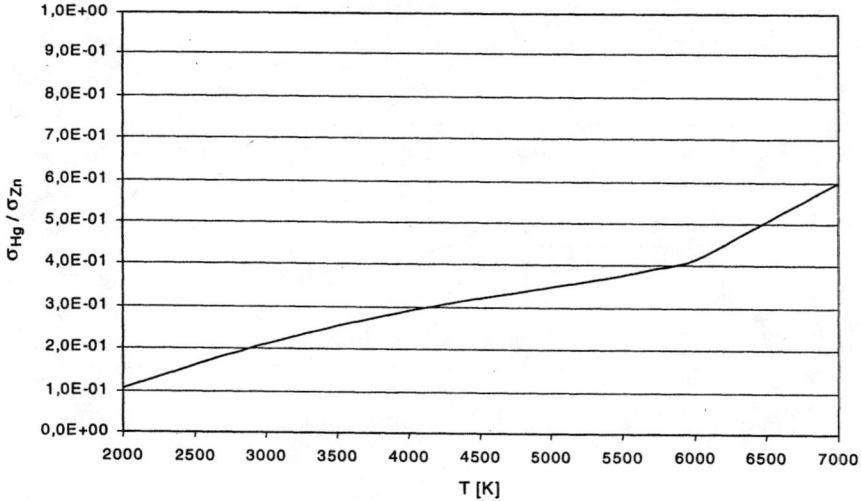

Figure 4. Ratio of electrical conductivities of pure mercury and zinc discharges at equal pressure.

RESULTS

Pure Mercury and Zinc Discharges

Fits of measured electrical field strengths E of pure mercury and zinc lamps are displayed in figure 5. The exponential dependency of E on partial pressure p is well described by the channel model of Elenbaas [21]

$$E \sim p^\gamma \tag{8}$$

$$\gamma \sim \frac{1}{4}\left(1 + \frac{E_{ion}}{E_{exc}}\right) \tag{9}$$

with E_{ion}: ionisation energy, E_{exc}: mean excitation energy. Since $E_{ion,Zn} = 9.4\,eV$, $E_{ion,Hg} = 10.4\,eV$, $E_{exc,Zn} = 7.4\,eV$, $E_{exc,Hg} = 7.8\,eV$ we find theoretical values of $\gamma_{t,Zn} = 0.568$ and $\gamma_{t,Hg} = 0.583$, which is close to the constants derived from the fits of measured data in figure 5: $\gamma_{m,Zn} = 0.556$ and $\gamma_{m,Hg} = 0.606$. As a result, electrical fields of pure mercury and zinc lamps are about the same at equal partial pressures.

The axis temperatures of zinc discharges are well below the data of mercury lamps due to larger transport of thermal heat and due to larger local net emission related to smaller mean excitation energy. For constant geometry and lamp power, we find a difference in plasma temperatures of $\Delta T \approx 300\,K$ (see figure 6). As for mercury lamps, axis temperatures and temperature profiles have been checked by comparing measured and calculated absolute power of several zinc transitions as indicated in figure 7 as well as from comparison of spectral line intensities.

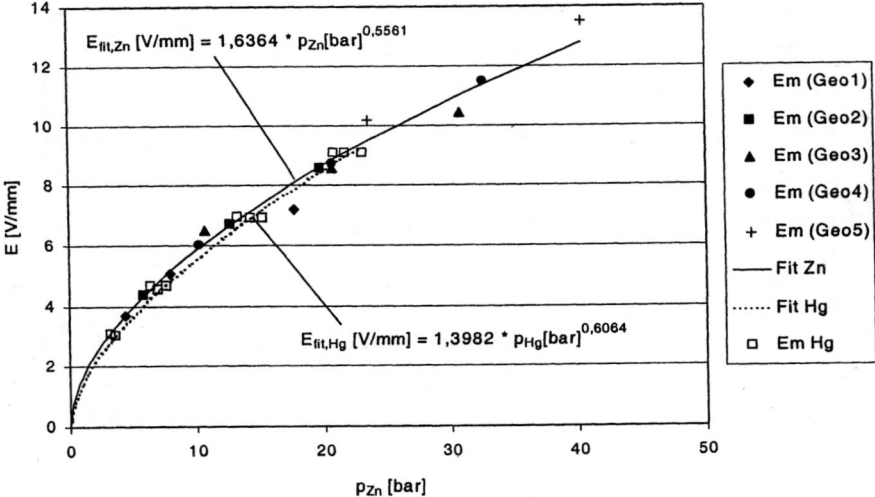

Figure 5. Fit of measured electrical field strengths of pure zinc (filled symbols) and mercury (open symbol) lamps as a function of the buffer gas pressure.

Mercury and Zinc Discharges with Metal Halides

The efficacy of zinc discharge lamps is poor due the lack of strong transition lines in the visible part of the spectrum (see figure 3). A typical value is $\eta_{Zn} \approx 15\,lm/W$ compared to $\eta_{Hg} \approx 60\,lm/W$. For lighting applications efficient radiators in the visible region are needed. Measured spectra of mercury and zinc containing metal halide lamps are displayed in figure 8 and figure 9. It is concluded that their light technical data are

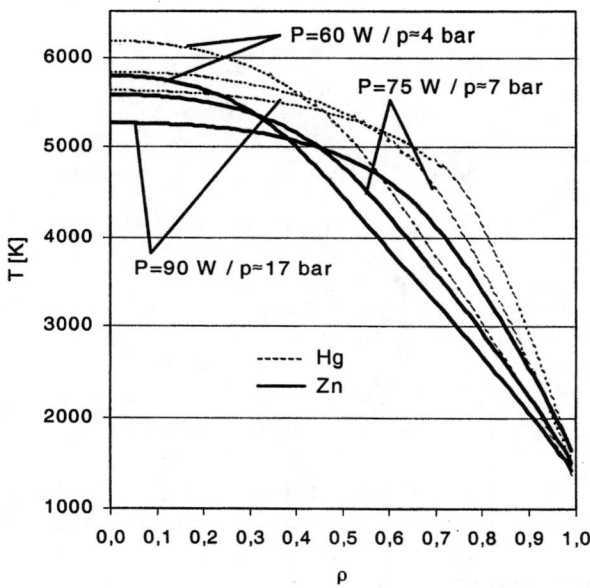

Figure 6. Calculated temperature profiles of pure zinc lamps in comparison to pure mercury lamps at different power levels and pressures.

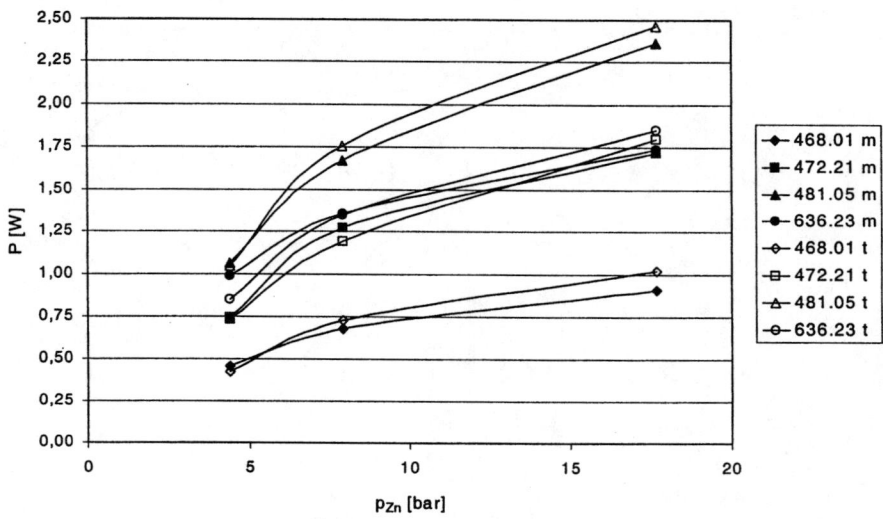

Figure 7. Measured (m) and calculated (t) absolute radiation power of several zinc transitions (see also figure 3). The increase of zinc pressure corresponds to a lamp power of 60, 75 and 90 W, respectively.

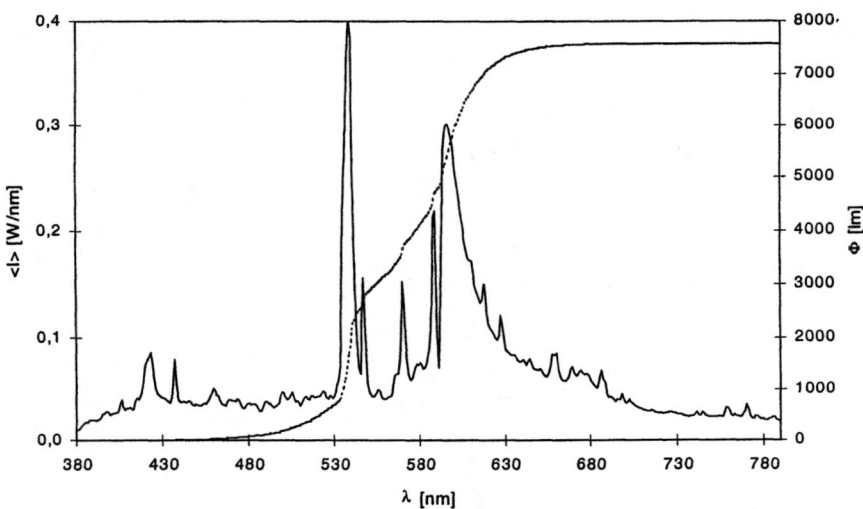

Figure 8. Measured spectrum and luminous flux of a mercury containing metal halide discharge at a lamp power of 75 W. Additives are NaI, TlI and DyI$_3$.

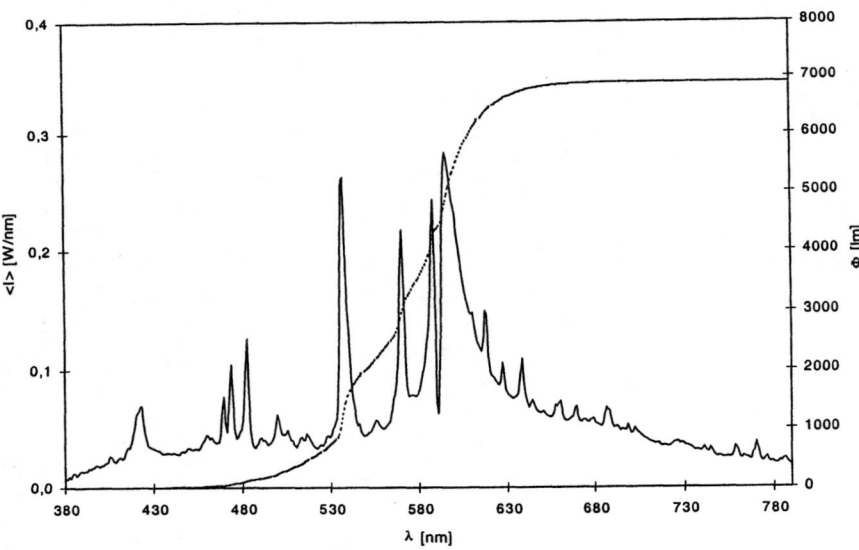

Figure 9. Measured spectrum and luminous flux of a zinc containing metal halide discharge at a lamp power of 75 W.

very close. Efficacies of $\eta_{MH,Hg} = 100$ lm/W and $\eta_{MH,Zn} = 93$ lm/W are measured in the case of using mercury or zinc, respectively.

Modelling of metal lamps is quite complex, since basic physical data, i.e. momentum transfer cross sections of neutrals (i.e. of iodine) and line broadening data, are hardly published. As a consequence, electrical conductivities are overestimated yielding too low calculated electrical field strengths. Missing experimental data of broadening parameters has been replaced by calculations [14]. An interesting fact is the difference in mean wall and coldest spot temperatures of mercury and zinc containing metal halide lamps of about 100 ... 150 K. This effect is caused on the one hand by larger thermal losses in case of zinc as shown in figure 10 and figure 11. On the other hand, radiation cooling is smaller for mercury metal halide lamps as shown in figure 12 and figure 13 which is a consequence of lower atomic densities of Na, Tl and Dy. From the latter two figures it is also concluded that the contribution of mercury or zinc to the total radiation power is small compared to the other metals.

The resulting temperature profiles are given in figure 14. In case of zinc enhanced radiation cooling leads to lower temperatures in the off-axis region of the plasma. In addition, the difference in coldest spot temperatures causes significantly different partial pressures of the buffer gas mercury ($p_{Hg} \approx 15\,bar$) and zinc ($p_{Zn} \approx 3...4\,bar$). However, the electrical field strength of metal halide lamps containing zinc is still sufficiently large due to a larger slope of the temperature profile close to the discharge axis compared to mercury. As mentioned before, calculated electrical field strengths are about 20% below measured data due to the lack of momentum transfer cross section data, i.e. of free iodine. For the metal halide lamps under investigation we find: $E_{meas,Hg} = 8.6\,V/mm$, $E_{calc,Hg} = 7.1\,V/mm$, $E_{calc,Zn} = 5.3\,V/mm$, $E_{meas,Zn} = 6.0\,V/mm$.

CONCLUSIONS

We have investigated the replacement of mercury in high pressure discharge lamps by metallic zinc. First, studies are carried out for pure mercury and pure zinc discharges in ceramic envelopes in order to determine the basic physical properties, i.e. lamp voltages and temperature profiles. From these investigations it is found that the momentum transfer cross section for elastical electron scattering by zinc atoms is about the same as for mercury atoms. This result is based upon a modification of the theoretical momentum transfer cross section of zinc by factor of 0.42 due to the lack of experimental data in the low energy region below 1 eV. Consequently, electrical field strengths of pure mercury and zinc discharges are nearly identical at equal partial pressures. In contrast to mercury, zinc lamps need to be operated in the saturated mode. Thus, the partial pressure of zinc depends strongly on the electrical input power. For zinc it is found that an increase of lamp power is accompanied by an increase of lamp voltage. In mercury lamps, an increase of the electrical input power is related to an increase of current, since mercury pressure and lamp voltage are nearly constant (unsaturated operation).

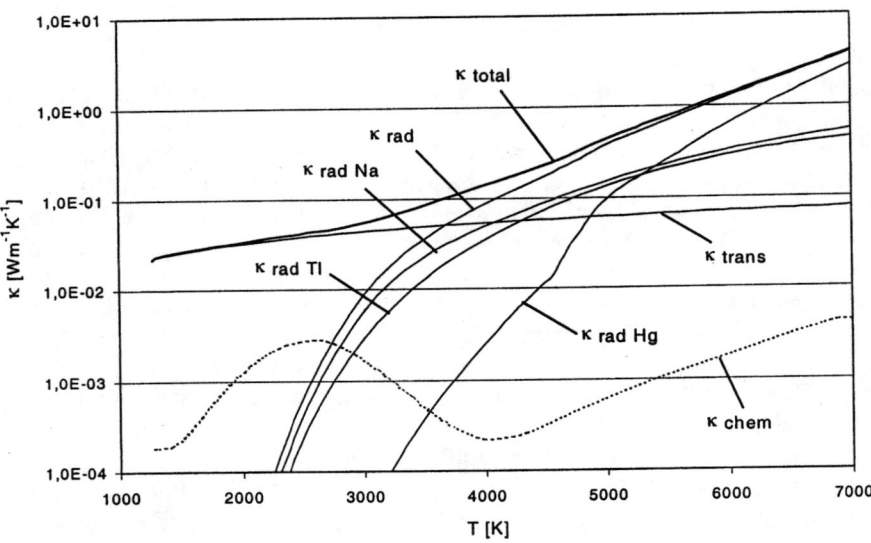

Figure 10. Calculated contributions to thermal conductivity for a mercury containing metal halide discharge at lamp power of 75 W (see eq. 3).

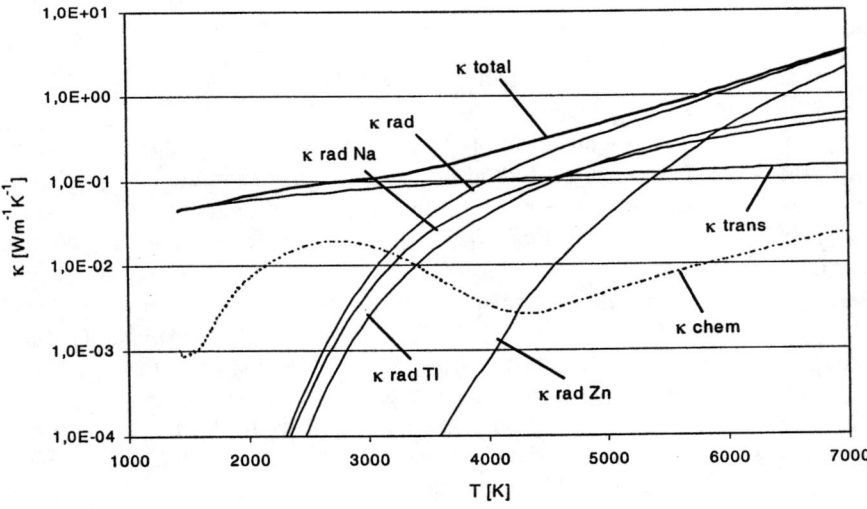

Figure 11. Calculated contributions to thermal conductivity for a zinc containing metal halide discharge at a lamp power of 75 W (see eq. 3).

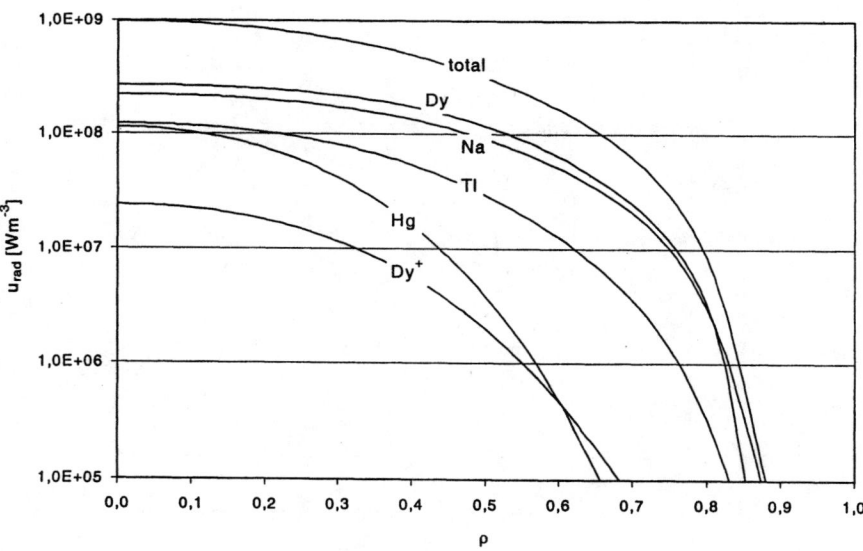

Figure 12. Power density of radiation of indicated species for a mercury containing metal halide discharge at a lamp power of 75 W.

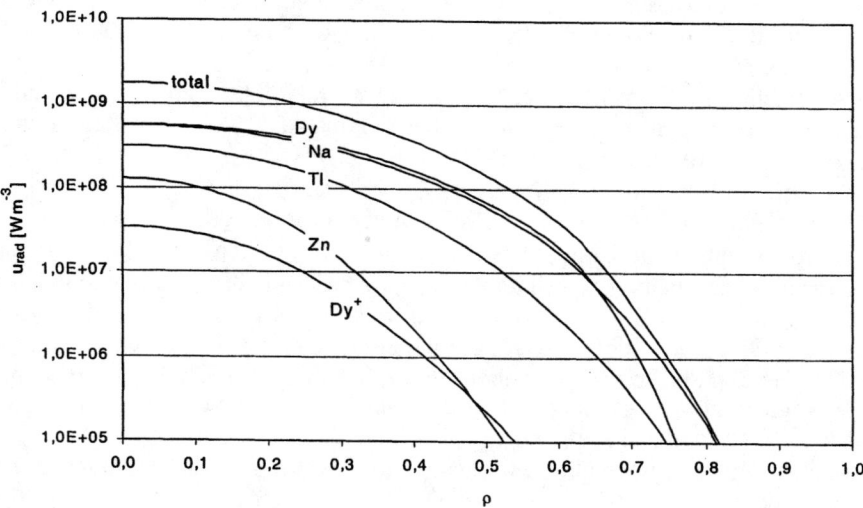

Figure 13. Power density of radiation of indicated species for a zinc containing metal halide discharge at a lamp power of 75 W.

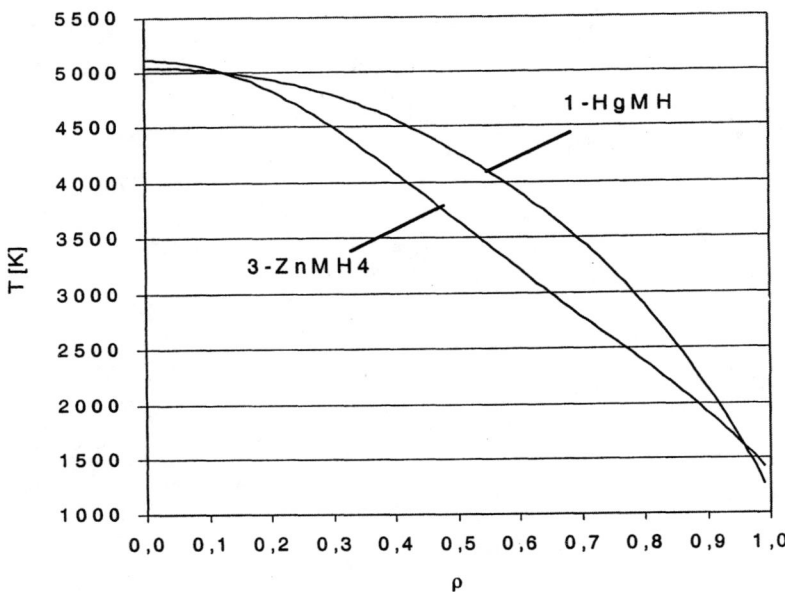

Figure 14. Calculated temperature profiles of metal halide discharges containing mercury or zinc at a lamp power of 75 W.

Since the energy levels of zinc are lower than for mercury, stronger radiation cooling in pure zinc discharges is observed. In addition, thermal losses of zinc lamps are larger. As a result, the temperature profiles of pure zinc lamps are typically 300 K below the ones for mercury.

When adding metal halides, i.e. NaI, TlI, DyI_3, contributions of mercury and zinc to the radiation power are small due to their comparatively large excitation energies. Consequently, light technical data of zinc and mercury containing metal halide lamps are about the same.

The axis temperatures decrease to a value of about 5100 K while the wall temperature of zinc containing metal halide lamps is typically 100...150 K larger than compared to the mercury case. The higher wall temperatures are related to larger thermal losses when using Zn instead of Hg. On the other hand, this results in larger densities of species and stronger radiation cooling. Due to a larger slope of the temperature profile, metal halide lamps containing zinc yield about 70% of the electrical field strengths of the ones containing mercury. Thus, lamp voltages of zinc containing metal halide lamps can be adjusted to typical values in the range of 70...90 V.

In this paper, experimental results clearly show that zinc would be a suitable replacement for mercury in metal halide discharge lamps, which are used in various lighting applications. It is emphasized that electrical and light technical data of the two lamp types are nearly identical.

REFERENCES

1. Ashurkov, S. G., Evolution of mercury-free HID lamps with radiant additives, Svetoteckhnika, No. 9, pp. 11-13 (1992).
2. Discharge Lamps and the Environment, European Lighting Companies Federation, 4th edition, (1997).
3. Muis, H., et al., Environmental Aspects of Lighting: A Product Oriented Approach, Final Report, Communicatie- En Adviesbureau over energie en millieu, Rotterdam, (1990).
4. Clear, R., and Berman, S., Environmental and Health Aspects of Lighting: Mercury, J. Illum. Soc., pp. 138 (1994).
5. Rouweler, G., Philips Lighting list of environmental relevant substances, Philips Report ULV-0088, Eindhoven (1997).
6. Born, M., Line broadening measurements and determination of the contribution of radiation diffusion to thermal conductivity in a high pressure zinc discharge, J. Phys. D: Appl. Phys. **32**, 2492 (1999).
7. Giese, H., Theoretische Untersuchungen zur Konvektion in Quecksilber Hochdruckgasentladungslampen, Dissertationsschrift, Verlag der Augustinus Buchhandlung, Aachen, Germany (1997).
8. Uhlenbusch, J., and Schramm, K., Zur Lösung der Strahlungstransportgleichung fAr den zylindrischen Lichtbogen, Bericht HMP 120, TH Aachen, (1968).
9. Amods, Atomic spectral lines, (1995), http://amods.kaeri.re.kr/spect/SPECT.html.
10. Skenderovic, H., and Vujnovic, V., A study of the line broadening constants obtained in a high-pressure mercury discharge. J. Quant. Spectrosc. Radiat. Transfer, **55**, 155 (1996).
11. Stormberg, H. P., Line broadening and radiative transport in high-pressure mercury discharges with NaI and TlI as additives, J. Appl. Phys. **51**, 1963 (1980).
12. Asselman, A., Aubes, M., Couris, S., and Salon, J., Stark broadening of the 690.7 nm mercury line in high pressure mercury discharges, J. Appl. Phys. **72**, 3341 (1992).
13. Elloumi, H., Aubes, M., Simonet, F., and Damelincourt, J. J., The determination of the neutral atom density in a high-pressure mercury discharge from the 253.7 nm self-reserved resonance line, J. Phys. D: Appl. Phys. **30**, 1893 (1997).
14. Born, M., Investigations on the replacement of mercury in high pressure discharge lamps by metallic zinc, J. Phys. D.: Appl. Phys. **34**, 909-924 (2001).
15. Seidel, J., and Wende, B., Plasmen, de Gruyter, Berlin-New York (1992).
16. Rockwood, S. D., Elastic and inelastic cross sections for electron-Hg scattering from Hg transport data, Phys. Rev. A **8**, 2348 (1973).
17. Norcross, D. W., Low energy elastic scattering of electrons by Li and Na, J. Phys. B: Atom. Molec. Phys. **4**, 1458 (1971).
18. Nakamura, Y., Electron drift velocity and momentum cross-section in mercury, sodium and thallium vapours: II. Theoretical, J. Phys. D: Appl. Phys. **11**, 337 (1978).
19. Frost, L. S., and Phelps, A. V., Momentum Transfer Cross Sections for Slow Electrons in He, Ar, Kr and Xe from Transport Coefficients, Phys. Rev. **136**, 1538 (1964).
20. Brown, S. C., Basic Data of Plasma Physics, M.I.T. Press, Cambridge, Massachusetts, 1959.
21. Elenbaas, W., The High Pressure Mercury Vapour Discharge, Publishing Company, Amsterdam, 1951.

The Physics of Fluorescent Lamps: Do We Understand the Atomic Processes?

Graeme G. Lister

OSRAM SYLVANIA, 71 Cherry Hill Drive, Beverly, MA 01915, USA

Abstract. Numerical models have provided insight into the operation of "standard" fluorescent lamps for more than 40 years. Recent developments in the lighting industry have led to products with much higher power loadings, for which modeling has been less successful in reproducing the experimental results. One of the potential weaknesses of the models is the absence of fundamental data to describe important phenomena in these "highly loaded lamps". The current state of our knowledge of available data is reviewed, together with an overview of the recently completed ALITE 1 project to reexamine the fundamental properties of fluorescent lamp operation.

INTRODUCTION

Fluorescent lamps are gas discharges in which mercury atoms are excited to produce ultra violet radiation, principally at 254 nm. The UV radiation is then converted to visible light by a set of phosphor coatings. This process is one of the most efficient ways of producing light, since about 25% of the electrical power applied to the lamps is converted to visible radiation.

The standard fluorescent lamp is a long cylindrical discharge filled with a mixture of a few mtorr of mercury and a rare gas, typically a few torr of argon. The rare gas serves two functions – firstly, it reduces sputtering at the electrodes, thus extending the life of the lamp, and secondly, it controls the diffusion of charged particles to the wall, to maintain the appropriate electron density.

Theoretical and numerical models of the positive column in fluorescent lamps have been developed for more than 40 years, since the pioneering work of Waymouth and Bitter [1] and Cayless [2]. Since that time, all major lighting companies have developed sophisticated numerical programs to describe the positive column [3-7]. There are also a number of other models reported in the published literature, principally from University researchers, which will be referenced, where appropriate, in the following text. These models have provided valuable insight into the performance of fluorescent lamps, but have yet to realize their full potential in the design and optimization of these lamps.

In the early 1960's, researchers at Philips conducted an extensive series of experiments in Hg-Ar discharges with an internal diameter (i.d.) of 36 mm, corresponding to the size of the standard fluorescent lamp at that time. They measured electric fields, electron densities and temperatures using Langmuir probes [8] and mercury excited state densities [9] and radiation balance using spectroscopic analysis

[10]. Until recently, this was the most detailed power and radiation balance study of fluorescent lamps, although a number of researchers have published extensive experimental measurements of different aspects of these discharges.

The standard operating conditions of fluorescent lamps at the time were 400 mA, 3 Torr argon pressure and 6 mTorr mercury vapor pressure. Despite the lack of fundamental data, numerical models reproduced the Philips' experimental measurements fairly well over a wide parameter range (100-800 mA discharge current, 0 to 30 Torr argon pressure and mercury vapor pressure 0.5 to 90 mTorr).

Recent developments in fluorescent lamps have led to increased power loading, either by reducing the diameter of the lamps (e.g. sub-miniature fluorescent lamps with 5 mm i.d., operating at 100 mA) or increasing the discharge current, (e.g. the 150 W ICETRON lamp from OSRAM SYLVANIA with 5 cm i.d. operating at 7 A). Further, in common with other inductively coupled fluorescent lamps, the absence of electrodes has enabled a reduction in rare gas pressure to 300 mtorr in ICETRON, leading to higher efficiency.

In general, models have done less well in reproducing experimental results in these highly loaded lamps, in particular overestimating the maintenance electric field and hence the electrical power in the lamp [11]. The most successful experimental and numerical studies of standard fluorescent lamps under highly loaded conditions (1-3A) is by Kreher and Stern [12], but their work is limited to argon pressures above 1 torr. The success of this model may in part be due to a more sophisticated treatment of electron kinetics than in the other positive column models reported in the literature.

THE ALITE 1 PROJECT

The need to obtain a better fundamental understanding of fluorescent lamps under highly loaded conditions was addressed by a consortium under the joint auspices of the Electric Power Research Institute (EPRI) and OSRAM SYLVANIA INC., in the framework of the first ALITE project (hereafter referred to as ALITE I). Participants were from University of Wisconsin (UW), Polytechnic University, New York (PU), Los Alamos National Laboratory (LANL) and the National Institute of Standards and Technology (NIST). The following projects were undertaken:
  a) Numerical and experimental studies of resonance radiation transport (UW/OSI)
  b) Experimental measurements (PU) and numerical calculations (LANL) of associative ionization cross sections in mercury.
  c) Calculations of electron impact and momentum transfer cross sections in mercury (LANL and Flinders University, South Australia).
  d) Power and radiation balance studies in a discharge with operating parameters appropriate to "electrodeless" fluorescent lamps (UW/OSI).
  e) Absolute radiometry measurements in mercury argon discharges (NIST).

Results from this program have been published or will be published in the near future. Results which have already been published and their impact on our understanding of fluorescent lamps form an integral part of this paper.

The new data obtained from ALITE I has been implemented into the positive column model GLOMAC [6]. The result of calculations to compare numerical predictions with experimental measurements has raised questions about our understanding of some of the fundamental physical processes in fluorescent lamps, even when operating at close to standard conditions. In order to resolve these discrepancies, it is necessary to use the models firstly to interpret the experimental results. The aim of the present paper is to discuss these issues in the light of the new ALITE I results.

PHYSICS OF HIGHLY LOADED FLUORESCENT LAMPS

The principal contribution to the physics of highly loaded lamps is the increase in electron density n_e, which tends to be almost linear with current. This increase manifests itself in a number of ways

a) The electron energy distribution function (EEDF) becomes closer to a Maxwellian distribution [13], due to the increased importance of electron-electron collisions, influencing both the "bulk" and "high energy" electrons.

b) "Two step" ionization and excitation into higher mercury states become more important, since rates for these processes are proportional to n_e^2.

c) *Radial cataphoresis* (or mercury depletion on the axis of the discharge) becomes more important. Cataphoresis is caused by mercury ions diffusing to the walls in the ambipolar field faster than the inward diffusion of mercury atoms from the walls and is proportional to the electron density. Cataphoresis has a strong effect on the production and trapping of UV radiation in the positive column [14].

d) Coulomb collisions play a more important role in reducing the electrical conductivity.

POSITIVE COLUMN MODELS

The goal of all positive column models is to calculate the efficiency of conversion of electrical power to useful radiation in fluorescent lamps as a function of discharge diameter, rare gas type and pressure, discharge current and mercury vapor pressure. In order to do so, it is necessary to solve a set of coupled equations:

Ohm's Law relates the discharge current density to the axial field and thus determines the V-I characteristics of the positive column.

Particle diffusion equations determine the densities of the relevant species (ground state and excited atoms, electrons and ions) in the positive column. In order to do this, it is necessary to have a description of the electron energy distribution function (EEDF).

A thermal conduction equation calculates the gas temperature.

A power balance equation partitions the electrical power supplied to the positive column into radiation, wall and volume losses.

Traditionally, there have been two approaches to modeling the positive column. The first [3-7,15] is to assume an analytic approximation for the EEDF, from which a set of rate and transport coefficients may be calculated. A set of rate equations (including the ambipolar diffusion for electrons) is then solved to compute particle densities, electric fields etc. The second approach [12,16] includes a direct solution of the Boltzmann equation to solve the electron kinetics directly, prior to computing rate and transport coefficients. Approximations to the EEDF have proved adequate for standard fluorescent lamp parameters, but must be used with caution for other parameter ranges. In any case, relatively complex Boltzmann equation models can now be run on desktop PCs in seconds and the case for using them is strong.

All models assume the positive column to be infinite in length, but particle densities may be introduced as radially averaged quantities [5,7,12,16] or the radial dependence is included explicitly in a set of differential equations [3,4,6,15]. The numerical methods required for these models are straightforward and easy to implement, but the amount of fundamental data required is formidable, and the validity of the models depends to a large extent on the accuracy of this data.

FUNDAMENTAL DATA

Electron Impact Excitation and Ionization Cross Sections

Electron impact excitation cross sections for mercury have been a subject of keen interest since the first fluorescent models were developed. Until recently, most models used cross sections for excitation from the ground state obtained from swarm data by Rockwood [17], augmented with estimates by Kenty [18] for electron excitations between states. Cross sections for ionization from excited states were estimated from a formula by Vriens and Smeets [19]. These so-called "two step" ionization processes become increasingly important as power loading in the lamp is increased.

The development of accurate quantum mechanical "close coupling" calculations of cross sections has been one of the exciting developments in atomic physics in recent years. Fursa and Bray [20] have shown excellent agreement of calculations with experimental measurements of excitation cross sections for barium and they were commissioned by the ALITE 1 team to calculate a relatively complete set of cross sections for mercury.

The first set of cross sections to be computed included excitation from the ground state to the $6^3P_{1,2,3}$ and 6^1P_1 states and the ionization cross section from these states. Results from modeling of an ICETRON-like positive column [13] using both the Rockwood [17] and Fursa and Bray [21] data have shown that the new cross sections predict a slightly higher (<10%) production of 254 nm, indicating that the older cross sections were reasonably accurate. The importance of cross sections for excitations between states are currently being analyzed.

Electron Momentum Cross Sections

The electron momentum transfer cross sections for Hg and rare gases are important for numerical models of the positive column, since they define the electrical conductivity of the discharge. Standard fluorescent operate at rare gas pressures around 3 Torr, and the cross sections for rare gases play the major role, although electron collisions with mercury atoms can be significant. There appears to be a consensus on the recommended cross section for argon, which is tabulated in the JILA database [22]. Many highly loaded electrodeless fluorescent lamps operate at rare gas pressures of a few hundred mTorr, and collisions with mercury atoms are as important as those with rare gases. Until recently, models have used the cross sections for mercury published by Rockwood [17], obtained from swarm experiments. More accurate measurements of these cross sections [23] have shown a resonance near 1 eV, leading to much larger cross sections than previously used in these energy region.

Chemi-Ionization Cross Sections

Chemi-ionization has been considered as an important process in the ionization balance of Hg rare gas discharges for some time [4]. Two distinct processes may be defined

Associative ionization $\quad Hg^* + Hg^{**} \rightarrow Hg_2^+ + e$
Penning ionization $\quad Hg^* + Hg^{**} \rightarrow Hg^+ + Hg + e$

There has been speculation that mercury molecular ions Hg_2^+ can form a major constituent of the ion population if the gas temperature is sufficiently large [4]. However, the cross section for dissociative recombination

$$Hg_2^+ + e \rightarrow Hg^* + Hg(^1S_0)$$

is also large [4] and the net increase of electron density by this process is computed to be relatively small in standard fluorescent lamps [6,15]. On the other hand, it has been shown numerically [6,15] that the Penning ionization involving two $Hg(^3P_2)$ atoms can, in principle, be an effective channel for electron production. Until recently, there were no experimental measurements for this process and a value of 100 $Å^2$ for this cross section was estimated [4]. Later, results of Boltzmann swarm calculations [24] predicted a value of 24 $Å^2$. More recent computations [25] and experiments [26] suggest this cross section may be significantly smaller than previous estimates, such that the role of this process in the ionization balance is greatly reduced. This result greatly influences the modeling of the ionization balance in fluorescent lamp discharges.

Radiation Transport

The total power per unit length from the radiative transition $k \rightarrow j$ reaching the walls

$$W_{rad}(\lambda_{kj}) = 2\pi\varepsilon_{jk}\beta_{jk}\int_0^R n_k(r)rdr \qquad (1)$$

where β_{jk} s^{-1} is the fundamental mode trapped decay rate for the transition $j \rightarrow k$ and ε_{jk} J is the energy difference between the excited states.

The analytic formula for β_{jk} for resonance lines as a function of the ground state density n_0, gas temperature T_g and the discharge radius R, developed by Lawler and Curry [27], has been corrected to include the influence of foreign gas broadening [28] and non-uniform distributions of ground state atoms, such as are produced by radial cataphoresis [14]. This formula can be directly applied to the 254 nm line in Hg-Ar discharges, provided the hyperfine structure of mercury is taken into account by dividing the effective mercury density by 5, the number of isotopes in natural mercury.

The treatment of the isotopic structure for the 185 nm radiation is somewhat more subtle, and a special formula, applicable only to this line, was developed from experimental measurements and Monte Carlo simulations [29]. This work confirmed and extended the earlier work of Post [30], with the advantage that all results could be reduced to a simple formula for inclusion in the numerical models.

DISCHARGE MODELING

Electrical Conductivity

The discharge current is related to the electric field in the positive column through

$$I = 2\pi E \int_0^R \sigma_e(r)rdr \qquad (2)$$

where σ_e is the electrical conductivity of the plasma, given by

$$\sigma_e = -\frac{n_e}{3N}\left(\frac{2e}{m_e}\right)^{1/2}\int_0^\infty \frac{\varepsilon}{q_t(\varepsilon)}\frac{\partial f_0}{\partial \varepsilon}d\varepsilon \qquad (3)$$

n_e is the electron density, N is the total gas density, $f_0(\varepsilon)$ is the isotropic component of the electron energy *probability* function for electron energy ε, normalized as $\int_0^\infty \varepsilon^{1/2} f_0(\varepsilon) d\varepsilon = 1$ and

$$q_t(\varepsilon) = q_{em}(\varepsilon) + q_{inel}(\varepsilon) + \frac{n_e}{N} \frac{q_e(\varepsilon)}{\gamma_E} \qquad (4)$$

is the total electron transport cross section [5], where $q_{em}(\varepsilon)$ is the total electron momentum transfer cross section, $q_{inel}(\varepsilon)$ is the total inelastic cross section, $\gamma_E = .582$ for a singly ionized gas [31] and $q_e(\varepsilon)$ is the Coulomb cross section [32]

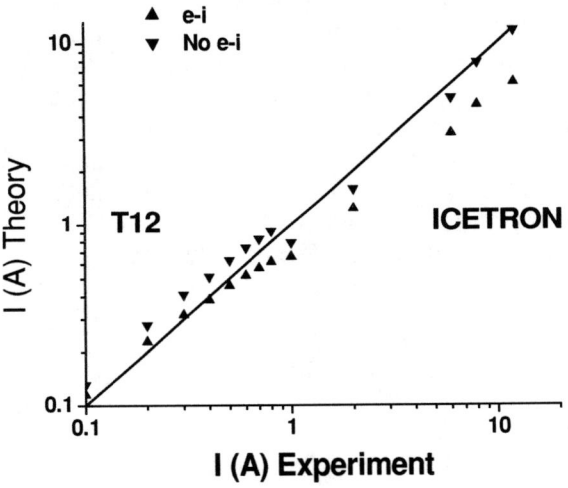

FIGURE 1. Comparison of discharge currents in T12 and ICETRON calculated from equation (1) with the measured values. Calculations used the experimental measurements of electron density and temperature for T12 from Verweij [8] and for ICETRON from Godyak *et al* [13], Electron momentum cross sections for Hg are from England and Elford [23] and for Ar from Phelps [22].

The influence of the updated mercury cross sections is to reduce the computed electrical conductivity. Further, the inclusion of Coulomb collisions in the total momentum transfer cross section reduces the conductivity even further. Coulomb collisions were neglected in the calculation of conductivity in many of the early codes and Zissis [15] has noted that it is difficult to reconcile numerical and experimental results when these collisions are included. To illustrate this point, Figure 1 shows the discharge current I, calculated from the Langmuir probe measurements of electric field and electron temperatures and densities as a function of the measured discharge current, for T12 (36 mm.i.d.) [8] and ICETRON [13]. Calculations in which Coulomb collisions have been neglected are included for comparison. A similar discrepancy was noted by Winkler et al [16b] and Kreher and Stern [12]. Their results show much higher electron densities (and hence electrical conductivity) than obtained from the probe measurements. This was at least partly attributed to inaccuracies in the measurements near the space potential [12]. The discrepancy is further increased when the new values of mercury cross sections are used.

It is important to stress the fundamental nature of this discrepancy. The measured current must be correct and the electric field can be estimated to within the uncertainties of measuring the electrode losses (a few per cent) from the voltage between the electrodes, even without the probe measurements. The electrical conductivity is relatively insensitive to the electron temperature – a change of 1000 K results in a 10% change in electrical conductivity. Further, the electrical conductivity is almost independent of the high energy electron tail in the EEDF. Thus, assuming the cross sections are correct, a much higher electron density is required to satisfy Ohm's law, indicating that a mechanism for ionization not presently included in the model is required. Results reported earlier in this paper indicate that chemi-ionization may not be as important as previously thought. Wani [7] has shown the importance of ladder like ionization of mercury and these process should be included in the codes, although data for cross sections remains inadequate.

Excited State Densities

As part of the ALITE 1 project, a series of experimental measurements of mercury excited state densities across a diameter in an ICETRON like discharge were performed using absorption spectroscopy. The results of measurements of the line densities of the $6^3P_{0,1,2}$ states at 40 °C (6.1 mtorr Hg pressure) are illustrated in Figure 2, together with numerical calculations using the GLOMAC code [6]. Calculated values are a factor of two to three higher than the measured values. Results using a fully independent model GLOW [5] have shown a similar discrepancy, hence the cause is unlikely to be numerical.

The measured column densities are considered to be very reliable, since they do not rely on any significant assumptions about the discharge. Errors should therefore be limited to the accuracy of the transition probabilities used in the interpretation ($\pm 10\%$), indicating that the description of the mechanism for depleting the metastable and resonance state densities used in the code is inadequate. As noted above, the cross sections for electron impact excitation out of these states, which are also important in describing "ladder like" ionization, are not well known. Further, mixing between atoms in excited states ("energy pooling") producing atoms in higher energy states, in a process similar to chemi-ionization, may also be playing a role and there are no available data for the cross sections for these processes. A further source for the discrepancy may be the description of the high energy tail of the EEDF, using a model due to Lagushenko [5].

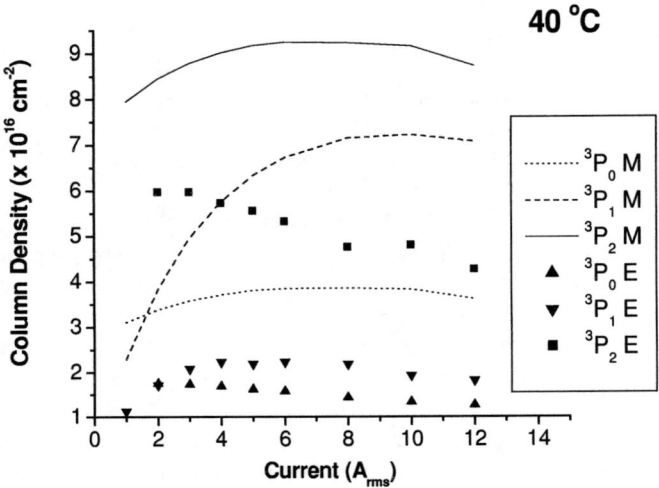

FIGURE 2. Measured and calculated 6 $^3P_{0,1,2}$ line densities: E- experiment [29], M- model [6].

CONCLUSIONS

As a result of the ALITE I project, a significant amount of new fundamental data has been obtained, which is applicable to a wide variety of fluorescent lamp parameters. In particular, this data is currently being applied to highly loaded fluorescent lamps, in an attempt to resolve discrepancies between modeling and experiment.

Of particular concern is the difficulty in reconciling the Langmuir probe measurements of electron density with the electrical conductivity of fluorescent lamps, even under modest power loading. We are re-examining the ionization balance in these discharges, particularly the role of "ladder like" ionization [7], since recent evidence indicates that chemi-ionization processes are less important than previously believed. In order to achieve this, more data on electron impact cross sections in mercury will be required.

Analysis of experimental measurements on mercury-argon discharges for a complete range of power loading in the light of new data, is in progress. The new data is allowing us to use the models to better interpret these results, thus providing the necessary bridge between modeling and experiment. The result of this exercise will be to give more insight into the important processes in highly loaded fluorescent lamps and provide the basis for the development of improved models to help optimize existing lamps for a variety of operating conditions and guide the development of new products.

ACKNOWLEDGMENTS

In preparing this paper, I am particularly grateful for valuable discussions with Jim Cohen, John Curry, Valery Godyak, Jim Lawler, Bob Piejak and Valery Sheverev. Support from the Electric Power Research Institute (EPRI) is also gratefully acknowledged.

REFERENCES

1. Waymouth, J.F., and Bitter, F., *J. Appl. Phys.* **27**, 122 (1956).
2. Cayless, M.A., in *Proc. 5th Int. Conf. on Ionization Phenomena in Gases*, Munich (1961).
3. Dakin, J.T., *J. Appl. Phys* **60**, 563 (1986).
4. Vriens, L., Keijser, R.A.J., and Ligthart, F.A.S., *J. Appl. Phys.* **49**, 3807 (1978).
5. Lagushenko, R., and Maya, J., *Adv. in Atomic, Molecular and Optical Phys.* **26**, 321 (1990).
6. Lister, G.G., and Coe, S.E., *Comput. Phys. Commun.* **75**, 160 (1993).
7. Wani, K., *J. Appl. Phys.* **75**, 4917 (1994).
8. Verweij, W., *Philips Res. Rep. Suppl.* **2**, 1 (1961).
9. Koedam, M., and Kruithof, A.A., *Physica* **28**, 80 (1962).
10. Koedam, M., Kruithof, A.A., and Riemens, J., *Physica* **29**, 565 (1963).
11. Fang, D.Y., and Huang, C.H., *J. Phys. D: Appl. Phys.* **21**, 1490 (1988).
12. Kreher, J., and Stern, W., *Contrib. Plasma. Phys.* **28**, 185 (1988).
13. Godyak, V.A., Piejak, R.A., and Alexandrovich, B.M., *Symp. Proc.: The 9th Int. Symp. on the Science and Technology of Light Sources*, Cornell Print & Digital Copy Services, 2001, pp. 157-158.
14. Curry, J.J., Lawler, J.E., and Lister, G.G., *J. Appl. Phys.* **86**, 731 (1999).
15. Zissis, G., Bénétruy, P., and Bernat, I., *Phys. Rev. A* **45**, 1135 (1992).
16. Winkler, R.B., Wilhelm, J., and Winkler, R., *Ann. Phys.* **40**, a) 90 (1983); b) *ibid* 119 (1983).
17. Rockwood, S.D., *Phys. Rev. A* **8**, 2348 (1973).
18. Kenty, C., *J. Appl. Phys.* **21**, 1309 (1950).
19. Vriens, L., and Smeets, A.H.M., *Phys. Rev. A* **22**, 940 (1980).
20. Fursa, D.V., and Bray, I., *Phys. Rev. A* **59**, 282 (1999).
21. Fursa, D.V., and Bray, I., to be published.
22. Phelps, A.V., unpublished, available at *ftp://jila.colorado.edu/collison.data*.
23. England, J.P., and Elford, M.T., *Aust. J. Phys.* **44**, 647 (1991).
24. Sawada, S., Sakai, S., and Tagashira, H., *J. Phys. D: Appl. Phys.* **22**, 282 (1989).
25. Cohen, J.S., Collins, L.A., and Martin, R.L., *Proc. XXII Int. Conf. On Photonic, Electronic and Atomic Collisions*, Santa Fe (2001).
26. Sheverev, V., Stepaniuk, V., and Lister, G.G., to be published.
27. Lawler, J.E., and Curry, J.J., *J. Phys. D: Appl. Phys.* **31**, 3225 (1998).
28. Lawler, J.E., Curry, J.J., and Lister, G.G., *J. Phys. D: Appl. Phys.* **33**, 252 (2000).
29. Menningen, K.L., and Lawler, J.E., *J. Appl. Phys.* **88**, 3190 (2000).
30. Post, H.A., *Phys. Rev. A* **33**, 2003 (1986).
31. Spitzer, L., and Härm, R., *Phys. Rev.* **89**, 977 (1953).
32. Spitzer, L., *Physics of Fully Ionized Gases* (second edition, Interscience, New York, 1961).
33. Curry, J.J., Lister, G.G., and Lawler, J.E., in preparation.

C. Etching, Plasma Displays, and Plasma Processing

Modeling of Moderate Pressure Microwave Plasmas Used for Diamond Deposition: Collisional Data Required for Process Simulation

K. Hassouni and A. Gicquel

LIMHP, CNRS-UPR 1311, Université Paris Nord, Avenue J. B. Clément, 93430 Villetaneuse, France

Abstract. Different aspects of the modeling of moderate pressure hydrogen and hydrogen methane plasma obtained under conditions relevant to diamond thin films deposition are discussed. First, a collisional-radiative model (CRM) of H_2 plasmas is presented and used in the frame of a quasi-homogenous plasma assumption to determine the main collisional phenomena that govern hydrogen discharges. The CRM was then used to build up more simplified thermo-chemical model that was used as a starting point for the development of a transport model for the more complex H_2/CH_4 plasmas. These have been investigated with a one dimensional (1D) that describes the plasma flow on the symmetry axis of a bell jar reactor. Discussion on the impact of the collisional data used in the different models is especially emphasized.

INTRODUCTION

Moderate pressure hydrogen-methane plasma obtained in microwave cavity systems under pressures ranging between 2500 and 20000 Pa present a significant interest in the field of Microwave Plasma Assisted Chemical Vapor Deposition (MPACVD) of diamond thin films for mechanical, optical and electronics applications. Two main microwave supplies were particularly investigated for this purpose. The first one works at 2.45 GHz with input powers up to 6 kW while the second makes use of 915 MHz microwave sources that can deliver up to 60 kW. The plasma deposition process is based on the possibility to produce active species such as H-atom and hydrocarbon radicals at relatively high density. These species are transported to a substrate surface where they undergo surface reactions that lead to the formation of diamond. There are two kinds of reactions involved in diamond formation. The first group includes the reactions of hydrocarbon species at the growing surface and leads to the formation of diamond and graphite, while the second group consists of etching reactions of graphite and diamond and is driven by H-atom. The formation of diamond is the net results of the competition between all these processes and although the reactions that produce graphite are faster than those resulting in diamond formation, the presence of H-atom insures a strong etching rate of graphite, the net result being the formation of diamond films. This shows the key role of H-atom in diamond deposition process. As far as hydrocarbon species are concerned, there is a quite large agreement in the literature on the key-role of CH_3-radical as the hydrocarbon species responsible for diamond formation [1].

The diamond growth rate and characteristics are governed on one hand by the substrate temperature and pretreatment which affect the kinetics of the surface reactions, and on the other hand by the discharge conditions (power density, feed gas composition) that define the densities of active species produced in the plasma bulk. The modeling studies performed on the investigated discharges are of two kinds. The first one makes use of self-consistent models and deals with the simulation of the plasma distribution in the deposition reactor. It is used to optimize the process in term of discharge homogeneity, power coupling efficiency and thermal management [2]. This optimization study may be performed on pure H_2 plasmas. The second kind of models addresses the energy dissipation channels and the chemistry in hydrogen-methane discharges [3]. The objective here is to investigate how the process parameters affect the species density and the discharge temperature. It is also to determine some estimates of the active species density at the substrate surface and to use these estimates in surface chemistry models that describe diamond growth.

The present paper discusses some modeling studies that belong to the second group of models and that have been performed on 2.45 GHz microwave cavity devices used for diamond deposition.

The next section describes the investigated device and gives the results of some experimental measurements that have been carried out on the plasma obtained in these devices. The third section discusses a CRM that was developed to describe moderate pressure H_2-plasma and used to investigate the different collisional phenomena involved in this kind of plasmas. This model was especially utilized to investigate the dissociation and ionization kinetics in the discharge and to develop a simple and accurate model that may be used in multidimensional transport codes. The fourth section presents a 1D transport model of H_2-CH_4 plasmas obtained under diamond deposition discharge conditions. The main phenomena that govern the density evolutions of the key-species for diamond deposition are also discussed. The last section gives the main conclusions that may be drawn from the different models.

MAIN CHARACTERISTICS OF THE CONSIDERED PLASMAS

The considered plasmas are obtained using a microwave cavity coupling system with a quartz bell jar reactor. The coupling system includes a microwave source, a transmission line and a field resonant system. The flow in the bell-jar reactor is monitored by an injection system, a gas flow control, a low pressure gauge and a vacuum pumping system. The cavity establishes a resonant electromagnetic mode structure that is axi-symmetric and the plasma source is typically operated so that the discharge volume is hemispherical and located just above the substrate holder (figure 1). The discharge conditions used in such devices correspond to pressure and input power values in the ranges 1000-12000 Pa and 600-2000 W, respectively. One of the most important controlling parameter for the discharges obtained in this system is the average microwave power density ($MWPD_{av}$). This parameter is changed by varying simultaneously the input power and the pressure in such a way to keep the volume of the plasma constant [4]. Its value ranges between 4 and 30 W/cm^3 in the considered discharges. Table I gives some values of the pressure, the input power and the corresponding values of $MWPD_{av}$ as well.

TABLE 1. Working pressure, input power and the resulting power density in the microwave discharge.

Input Power, MWP_{inp} (W)	Pressure (Pa)	Microwave power density, $MWPD_{av}$ (W/cm^3)
600	2500	9
1000	5200	15
1500	8400	22.5
2000	11000	30

Several measurements have been carried out on the plasma obtained under discharge conditions reported in Table I. Coherent Anti-Stokes Raman Scattering (CARS) and Optical Emission Spectroscopy have shown that the gas temperature, in the bulk of the plasma, ranges between 2200 and 3200 K [5]. The electron density in the investigated discharges was measured by Grotjohn et al. [6] who used microwave interferometry technique and values around 5.10^{11} cm^{-3} have been found.

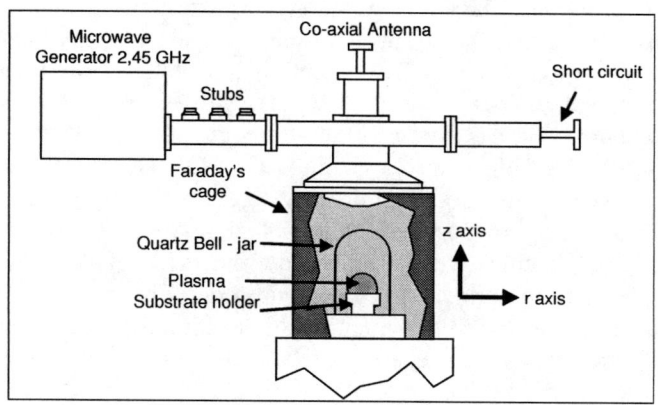

FIGURE 1. Schematic of the investigated device.

The gas temperature and electron density values determined experimentally show that the considered plasmas have an intermediate behavior with respect to the cold low pressure plasmas and the thermal atmospheric plasmas. The high values of the gas temperature achieved in these discharges may lead to an effective heavy species thermal chemistry, while the relatively low value of the measured electron density shows that moderate pressure H_2 plasmas should present a strong thermal non equilibrium.

The introduction of methane at concentration below 5% does not strongly affect the discharge in terms of gas temperature and plasma distribution. The H_2 and H emission line intensities show almost the same distribution as in the case of pure H_2 plasma. This does not mean however that the ionization and dissociation kinetics are similar for H_2 and H_2/CH_4 discharges.

COLLISIONAL RADIATIVE MODEL FOR H_2 PLASMAS

Before considering the discharge performed in H_2/CH_4 mixture and actually used for diamond deposition we first investigated pure hydrogen plasmas obtained under discharge conditions corresponding to those used for diamond deposition. The objective was to develop a simple and accurate enough collisional model that may be used as a starting point for the investigation of the more complex H_2/CH_4 plasmas.

The ionization degree of the considered plasmas is less than 10^{-4} and the charged species densities are much smaller than those of H_2 and H which represent the major chemical species in the investigated hydrogen plasmas. Consequently, the internal energy modes were considered only for H_2 and H and the CRM developed for H_2 discharges distinguishes the following four classes of reactions:

(1) The vibrational, dissociation and dissociative attachment kinetics of H_2,
(2) The kinetics of electronically excited states and of ionization for H_2,
(3) The kinetics of electronically excited states and of ionization for H-atom,
(4) Charged species chemistry.

The details of the CRM model are discussed in reference [7] and we will limit ourselves in the present paper to describe the main peculiarities of this model. The vibrational kinetics take into account the direct and step-wise excitation through electron-impact processes. The direct vibrational excitation of the first 5 excited levels from the v=0 has been extensively investigated and there is a significant amount of cross section data in the literature. The vibrational excitation from upper vibrational levels has been much less considered and we used in this work the recent theoretical values published in [8]. The step-wise vibrational excitation through the excitation and radiative decay of H_2 singlet states ($B^1\Sigma$, $C^1\Pi$, $D^1\Sigma$) has been also extensively investigated and we used the cross section data suggested by Hiskes [9]. De-excitation of H_2 vibrational levels through second kind collisions were also considered and the corresponding cross sections were determined from those of the reverse processes using the detailed balance principle. The second group of processes involved in the vibrational kinetics is related to the vibrational relaxation through vibration-vibration (v-v) and vibration-translation energy exchanges. For these last processes we distinguished the energy exchange due to H_2-H_2 collisions (v-t) and those due to H-H_2 collisions (v-T). The reaction rate constants of v-v and v-t processes were obtained from SSH theory [10]. The v-T collisions presents a reactive character and may lead to a multi-quantum de-excitation, the SSH theory is therefore no longer valid in this case and we used the rate constant values published in [11]. To take into account the dissociation processes through the vibrational pumping mechanism we introduced a fictive fully dissociative vibrational levels which may be obtained by v-v, v-t or v-T process involving the upper vibrational levels. The dissociative attachment processes that leads to the formation of H^- from v=4-10 levels were also taken into account using the cross section data published in [11].

The details of the processes taken into account in the kinetic model of H_2 electronically excited states may be found in reference [7]. This model is based on the works performed in [12,13] and takes separately into account the different triplet and singlet states with Rydberg numbers less than 4. Higher Rydberg states are lumped. The electronically excited states of H_2 are produced through electron impact processes and some of them are linked through radiative de-excitation and collisional quenching. Several quenching processes of excited states leads to the dissociation of H_2 molecule and represents an important source of atoms. There is a significant lack of data on the collisional and radiative processes that links the different excited states of H_2. The consequence of such lack of knowledge remains however quite limited in the case of moderate pressure H_2 plasmas since the kinetics of electronically excited states is mainly governed by the production through electron impact excitation on the ground state (the different vibrational levels have to be considered) and the consumption through collisional quenching and radiative processes. The information that is really needed is linked to the product and the cross section data of the quenching reactions. Such information are of prime interest for an accurate estimation of the H-atom density and a satisfactory description of the vibrational kinetics since the production of vibrational levels through radiative de-excitation of hydrogen singlet states could be strongly limited if the collisional quenching of these states becomes significant.

Starting from the studies carried out in [14,15], a CRM was also developed for H-atom. The transitions between the different electronic states of H-atom through collisions with electron and heavy species as well as radiative processes were considered. Ionization and radiative recombination involving the different states were also considered. The major processes that were introduced in the present work are those one describing the ionization and dissociation through the quenching of H-atom excited states, and especially H (n=2) and H (n=3), by H_2. As will be discussed latter, the processes H (n=2-3)+ H_2 => H_3^+ represents a major ionization channels that enable to sustain the moderate pressure H_2 plasmas used under diamond deposition conditions.

The last group of reactions involves ion conversion between H^+, H_2^+ and H_3^+, mutual neutralization of these ions with the negative ion H^- and some ion radiative (for H^+) or dissociative (for H_2^+ and H_3^+) recombination processes [7].

A schematic of the different species and reaction groups involved in the whole collisional radiative model is given in figure 2.

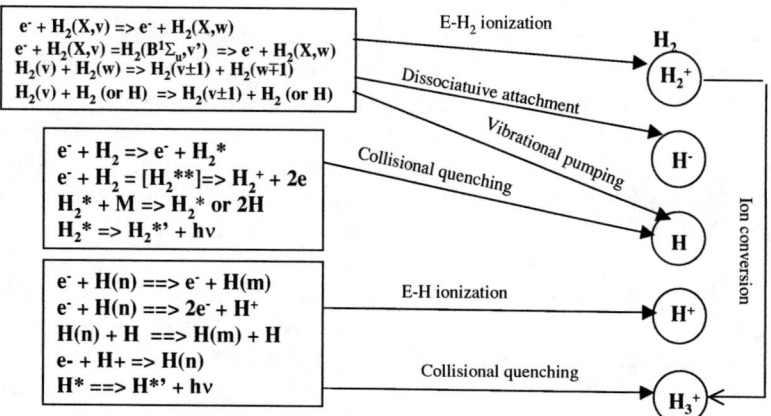

FIGURE 2. Schematic of the different processes considered in the model.

65

This full model was used to estimate some plasma characteristics in the frame of quasi-homogenous plasma assumption where we consider that the discharge involves two parts: a large homogenous plasma bulk and a thin diffusion boundary layer in the vicinity of the substrate (figure 3a). The active species may diffuse through this boundary layer and recombine or de-excite at the substrate surface. The numerical model is based on the coupled solution of the balance equations of the chemical species, including H_2 vibrational levels and H_2 and H electronically excited states, the Boltzmann equation that yields the electron energy distribution and a total energy equations to determine the gas temperature [7]. A flow sheet that describes the numerical model and the solution procedure is given in figure 3b.

The calculated electron energy distribution distinguishes two energy groups. The first one corresponds to low energy electrons, i.e. typically less than 8 eV, and involves the major part of electrons, while the second involves the higher energy electrons that govern the high energy-threshold dissociation and ionization processes. The eedf may be therefore characterized by two temperatures: T_{e-l} for the low-energy electrons and T_{e-h} for the high-energy electrons. The electron average energy, $<\varepsilon_e>$ is related to the temperature of the low-energy electrons that represent the major fraction of these species. Figure 4 shows for example the variation of electron and gas temperatures as function of the absorbed microwave power density ($MWPD_{av}$) in the investigated discharges. The increase of $MWPD_{av}$ leads to a strong increase of the gas and vibration temperatures that vary from 1600 K up to 3300 K and a significant decrease of the electron average energy that varies from 3 to 1.5 eV. Figure 5a shows that such variations have an important consequence on the dissociation kinetics and the weights of the different dissociation channels in the whole production of H-atom. This is governed by electron-impact processes at low power density and by heavy species collisions involving upper vibrational levels at high power density. A more detailed analysis of the dissociation kinetics shows that an accurate estimation of the dissociation yield requires the knowledge of the electron impact dissociation cross sections from the first three vibrational levels at low power density and the dissociation from v=10-14 levels through H_2-H_2 or H-H_2 collisions at high power density.

The analysis of the ionization kinetics showed that the major ionization channel is the quenching of the first excited state (n=2) of H by H_2 (figure 5b). This reaction has been suggested as a possible dissociation and ionization channel in [16] and reconsidered in [17]. The conclusion drawn in these two publications would suggest a collision cross section around 15-20 Å^2 for the production of H_3^+ by H_2-H (2s) collisions. Such values of the quenching collision cross section makes the quenching process much more efficient than the ionization by electron impact on the different vibrational and electronic states of H_2.

Further analysis of the results obtained from the full model showed that a model assuming a thermal equilibrium for the vibrational mode of H_2 and takes into account the 7 ground state species, H_2, H, H^+, H_2^+, H_3^+ and e⁻ as well as the first two excited state of H-atom would be sufficient for an accurate description of the ionization, the dissociation and the dissociative attachment kinetics in the investigated plasmas. Such simplification is made possible by the weak coupling between the kinetics of

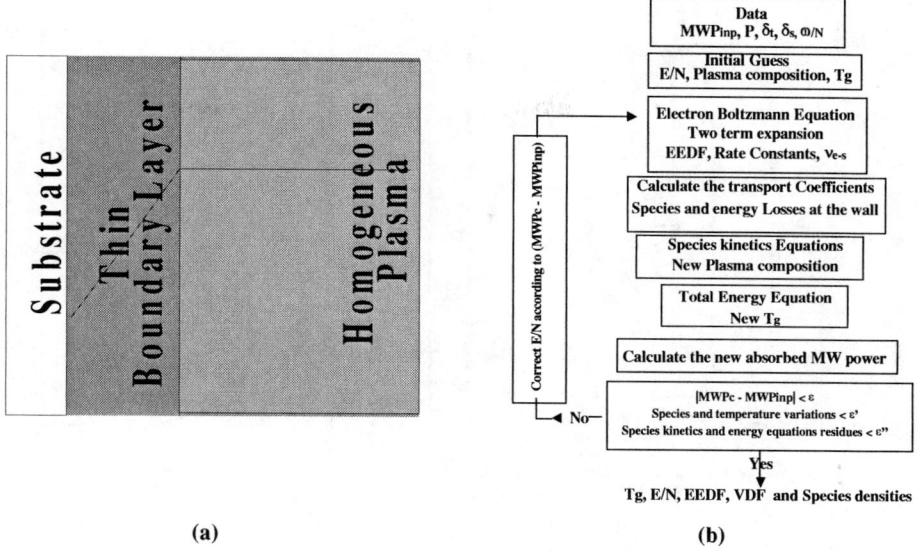

FIGURE 3. Principle of the quasi-homogenous plasma model (a) and computation procedure used in the quasi-homogenous model.

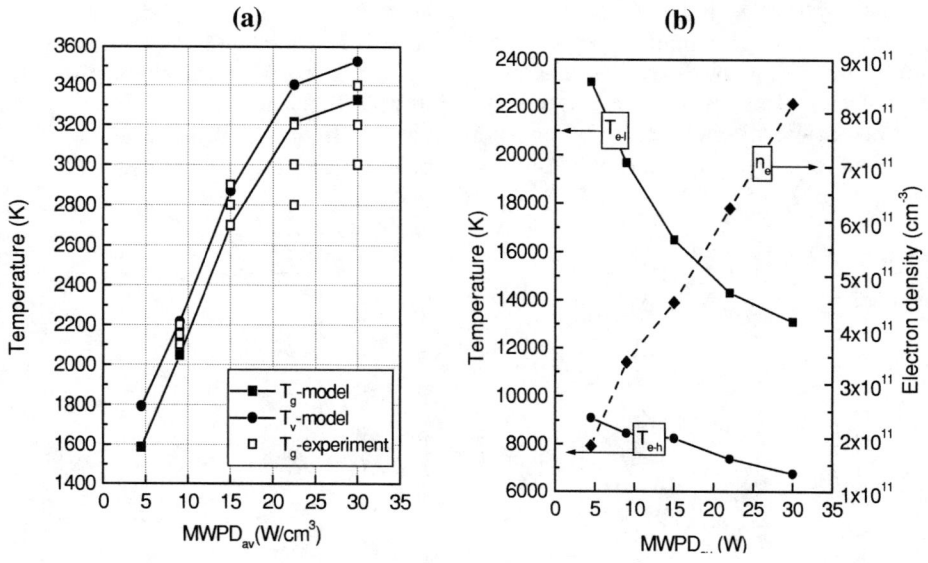

FIGURE 4. Calculated gas and vibration temperatures (a) and calculated electron density and low-energy ($T_{e\text{-}l}$) and high energy ($T_{e\text{-}h}$) electron temperatures.

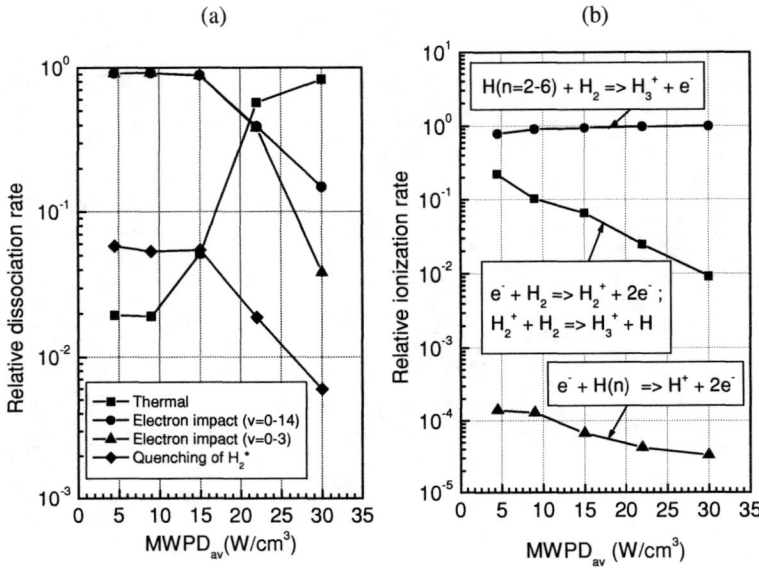

FIGURE 5. Weights of the different dissociation (a) and ionization (b) channels.

H(n=2-3) states and the remaining excited states of H-atom. In particular, neglecting n>3 excited states does not alter the estimation of H (n=2-3) states which show the major participation to the ionization kinetics through quenching processes. A model involving these species and taking into account 30 chemical reactions was developed and its results were compared with those of the full model described in this section. Some of these comparisons are shown in figure 6. A good agreement is obtained for the prediction of the H-atom dissociation and the electron density.

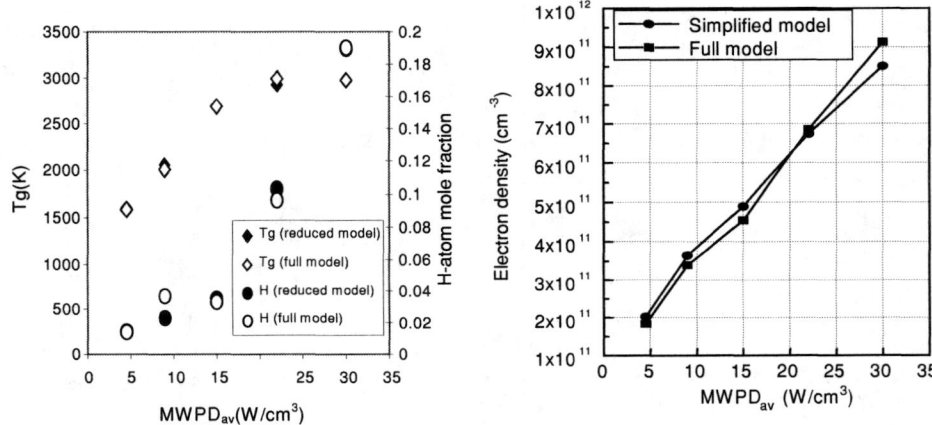

FIGURE 6. Comparison between the H-atom mole fractions and gas temperatures (a), and electron densities obtained from the full and the simplified models

ONE-DIMENSIONAL MODEL OF MODERATE PRESSURE H_2/CH_4 PLASMAS

The simplified collisional model described in the last section was used as a starting point to develop a transport model for H_2/CH_4 plasmas. With more than 95% hydrogen content in the feed gas, the hydrogen to carbon ratio is high enough to prevent the formation of hydrocarbons with more than two carbons [18]. Therefore the following hydrocarbon species with one and two carbons, i.e. CH_4, CH_3, CH_2, CH_2^*, CH, C, C_2H_6, C_2H_5, C_2H_4, C_2H_3, C_2H_2, C_2H and C_2, were taken into account. Besides the model takes into account the major ions that may form under the considered discharges conditions, i.e. electron density around 10^{11}-10^{12} cm^{-3} and T_e=1-3 eV. These are: CH_4^+, CH_5^+, $C_2H_6^+$, $C_2H_5^+$, $C_2H_4^+$, $C_2H_3^+$ and $C_2H_2^+$. These hydrocarbon ions may be produced either through electron-impact collision on neutral molecules, i.e. CH_4^+, CH_6^+ and $C_2H_2^+$ or ion conversion processes involving a proton transfer from H_3^+ or a hydrocarbon ion to a hydrocarbon molecule, i.e. CH_5^+, $C_2H_5^+$, $C_2H_3^+$ [19].

The chemical model used to describe the hydrogen-methane plasma involves three groups of reactions. The first one corresponds to pure H_2 plasmas. The second set of reactions describes the neutral chemistry in C/H system with a small C/H ratio. These reactions have been extensively studied in the frame of the research work in combustion field [20]. There is a quite large agreement on most of the rate constants of the processes belonging to this group. The only exception is may be the three-body recombination processes, the rate of which are pressure dependent and are accurately known only for the high and low pressure limits. The estimation of these rate constants at a moderate pressure requires the use of pressure correction relations that still need investigation and improvement. Accurate determination of these pressure-correction terms is of prime importance for a better prediction of the reacting boundary layer flows that are usually encountered in MPACVD processes as will be discussed in a next section. The last group of reactions is that one corresponding to the ionic species kinetics. It involves electron-impact process, the cross sections of which are reviewed in reference [21]. Cross section or rate coefficient data of ion conversion and charge transfer between the different hydrocarbon ions and between H_3^+ and hydrocarbons do not exist for all the hydrocarbon species. A better knowledge of these data are really needed since for example ion conversions involving H_3^+ are very fast processes that determine the nature of the major ions in H_2/CH_4 plasma which is generally a hydrocarbon ion even when the methane amount in the feed gas is as small as 0.1% [3]. There is also a lack of data with respect to the dissociative recombination of hydrocarbon ions which represents the major ion loss mechanism in H_2/CH_4 discharges. The data for these reactions are only known in the case of CH_5^+ and $C_2H_5^+$ [22], and we assumed that ions with same number of carbons have the same dissociative recombination rate constants when no data are available.

The collision model summarized above leads to the description of H_2/CH_4 in term of 32 species and 104 reactions. This kinetic model was used to describe the coupled phenomena of transport and chemistry on the axis of the plasma reactor from the reactor inlet to the substrate surface (figure 1). Since the considered geometry has a cylindrical symmetry, the 2D transport equations that describe the plasma flow in the

whole reactor may be reduced on the symmetry axis to a system of 1D equations that governs the species and energy transport on this axis. The flow rates used in the present reactor are quite low and the inertial convection flow may be neglected. Also, the density remains low enough to neglect the natural convection phenomena. Consequently, the species and energy were assumed to be transported only by diffusion, which results in the following flow governing equations:
- Continuity equations for species:

$$\frac{d}{dz}\left[\rho \frac{M_s}{M} D_s \left(\frac{\theta_s}{T_g}\frac{dT_g}{dz} - \frac{dx_s}{dz}\right)\right] - W_s = 0 \qquad (1)$$

Where D_s, M_s, x_s and W_s are the mole fraction, the diffusion coefficient, the molar mass and the net production rate of the sth species. ρ and M are the average mass density and molar mass of the plasma. θ_s is the ratio of the Fick's diffusion coefficient to the thermal diffusion coefficients (Soret's effect). The diffusion coefficients are calculated from the collision integrals given by Yos [23] for hydrogen and using Lennard-Jones interaction potential for hydrocarbons. Due to the relatively high pressure, ambipolar diffusion assumption was used in the case of charged species. The net production rate is calculated on the basis of the collisional model described above, which requires to determine the reaction rate constants and therefore the gas and the electron average energy temperatures. Although the eedf is not Maxwellian, the electron-heavy species rate constants are functions of the only electron average energy, which is linked to the temperature of the low-energy Maxwellian group $T_e = 2/3 <\varepsilon_e>/k$.

- Electron energy equation

The conservation equation of the electron average energy may be written:

$$\frac{d}{dz}[J_e h_e] + \frac{d}{dz}\left(-\frac{2}{3k}\lambda_e \frac{d<\varepsilon_e>}{dz}\right) - MWPD(z) + Q_{e\text{-}t} + Q_{e\text{-}v} + Q_{e\text{-}x} = 0 \qquad (2)$$

Where $<\varepsilon_e>$, h_e, λ_e, J_e are the average energy, the enthalpy, the Spitzer conductivity and the diffusion flux of electrons, respectively. MWPD is the absorbed power density while, $Q_{e\text{-}v}$, $Q_{e\text{-}x}$, $Q_{e\text{-}t}$ are the power loss through e-v processes, chemistry and electron-translation energy transfer.

- Total energy equation:

$$\frac{d}{dz}\left(-\lambda_g \frac{dT_g}{dz} - \lambda_e \frac{dT_e}{dz}\right) + \sum_s \frac{d}{dz}[J_s h_s] - MWPD(z) + Q_{rad} = 0 \qquad (3)$$

Where T_g, λ_g are the gas temperature and the thermal conductivity (vibration + translation + rotation) of the gas. J_s and h_s respectively denote the diffusion flux and the enthalpy of the sth species. Q_{rad} is the rate of energy loss through radiation.

The solution of the transport equations (1)-(3) requires the knowledge of the axial profile of the microwave power density. This may be estimated by two possible methods. The first one is based on the use of a self consistent model where the plasma species and energy equations are coupled to the electromagnetic field equations that need to include for the investigated discharges the full Maxwell equations. This method requires however to consider a 2D geometry, since the Maxwell equations cannot be reduced to describe the field on the symmetry axis. It was used in reference [2] to describe 2D H_2 plasma and will not be discussed in the present paper. The second method, described in [3] is based on the use of experimental measurements. A small amount of argon was introduced in the discharge and the axial profile of the 750 nm argon emission line was used to determine the power density. This is performed by coupling to the transport equations an additional equation that links the axial profiles of the power density and the 750 nm argon emission line intensity. The full set of equations is then solved so that the calculated and measured profiles of a given emission line intensity coincide. The model therefore requires the measured axial profile of the 750 nm emission line as an input and yields the axial profiles of the species densities, the gas and electron temperatures and the power density.

Typical axial profiles of the argon line intensity and the resulting power density obtained from this model are presented in figure 7. This shows that the power deposition in the investigated discharge is strongly non-uniform and a preferential power deposition takes place near the substrate surface. This result is in good agreement with the one obtained from the self-consistent model described in reference [2]. Figure 8 shows the axial profiles of the measured and calculated H-atom mole fractions (8a) and gas temperatures (8b). Good agreements between the model and the experiment is obtained for both these quantities.

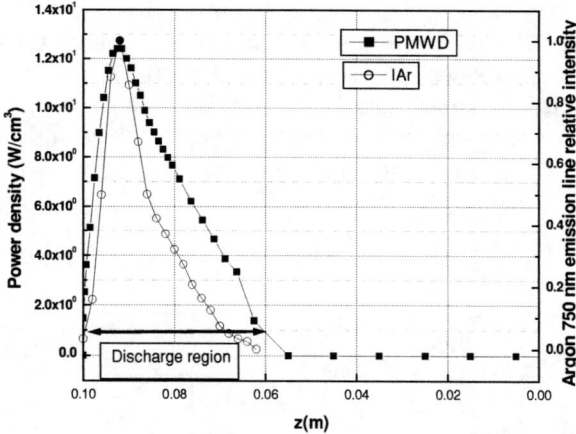

FIGURE 7. Axial profiles of the 750 nm argon line intensity and the resulting (calculated) absorbed power density.

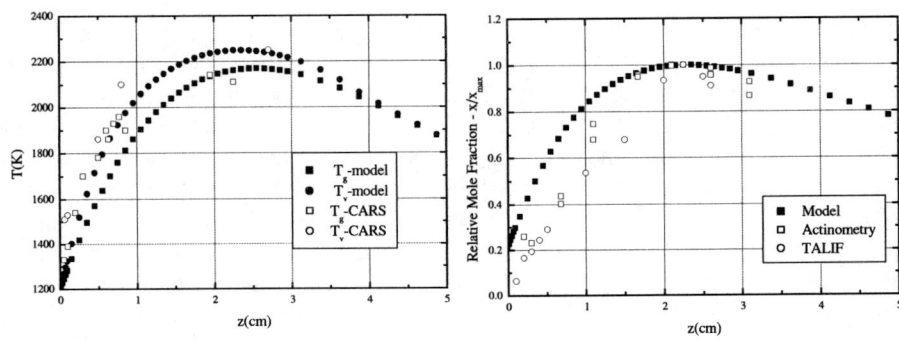

FIGURE 8: Axial profiles of calculated and measured gas temperature (a) and H-atom mole fraction (b).

Figure 9 presents the axial profiles of the neutral hydrocarbon species for an average power density of 15 W/cm^3. It shows that methane is quantitatively converted to acetylene as soon as the gas enter the reactor (outside the discharge zone). Acetylene becomes therefore the major species in the discharge, and especially in the high temperature region. In fact, the neutral hydrocarbon densities are mainly governed by chemistry (transport is negligible) and a partial equilibrium exists between these species. The boundary layer between the plasma and the substrate is characterized by very steep density gradients correlated to the strong decrease of the gas temperature which results in the back-conversion of acetylene to methane. Although this back conversion is kinetically limited and acetylene remains the major hydrocarbon species, the density of some species vary by several orders of magnitude near the substrate. These density variations especially govern the methyl radical concentration at the substrate surface which is, along with atomic hydrogen, responsible for diamond growth. The three body recombination processes of hydrocarbon species play a key-role in the hydrocarbon chemistry that takes place in the boundary layer. The predicted amount of CH$_3$ at the growing surface, and therefore the predicted diamond growth rate, strongly depend on the rate constant of these processes.

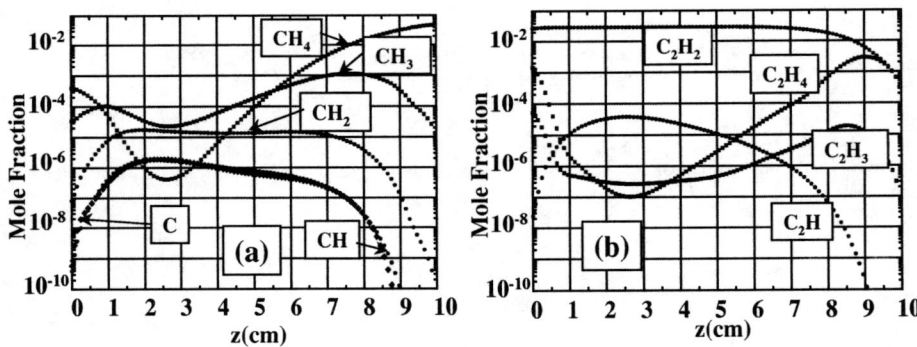

FIGURE 9. Axial profiles of the densities of some hydrocarbon species: (a) CH$_y$ and (b) C$_2$H$_y$.

The axial profiles of the major ionic species are shown in figure 10. The ionization may proceed by ion conversion involving a hydrocarbon molecule and H_3^+ ion, which explains the strong decrease of H_3^+ density even for very small amount of CH_4 in the feed gas. There is also a significant ionization through electron impact on acetylene in the bulk of the plasma. A very strong ion conversion kinetics takes place in the boundary layer between the different hydrocarbon ions and species, e.g. $CH_4 + C_2H_2^+ \Rightarrow C_2H_3^+ + CH_3$. This conversion is mainly driven by the change in the neutral density discussed in the last paragraph. For example, the strong increase of CH_4 in the boundary layer leads to an enhanced production of $C_2H_5^+$ through ion conversion. The ion loss is mainly governed by dissociative recombination processes, which yield a very strong decrease of the ionic species at the discharge edges. This is very different from what is obtained in hydrogen discharge where the major ion H_3^+ has smaller dissociative recombination constant and can diffuse along with the electrons outside the discharge leading to a small power absorption outside the visible plasma volume as was clearly shown in reference [2].

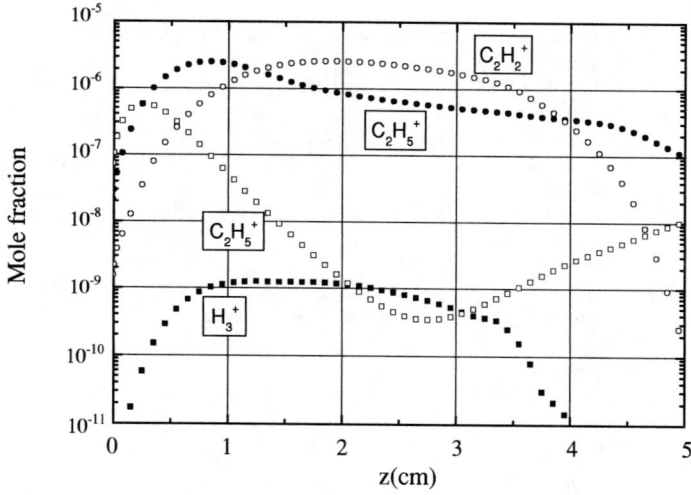

FIGURE 10. Axial profiles of the major hydrocarbon ions.

CONCLUDING REMARKS

There is certainly a lack of information on a large number of collisional data required to describe the details of the chemistry that governs the discharges used for thin film deposition. We tried, however, in this work and in the case of H_2/CH_4 plasmas used for diamond deposition to show that some data are much more needed than others, and also that the amount of data that are really required for an accurate simulation of the process of interest, although significant, remains reasonable. The analysis made in this work showed that further investigation of the product formed by the quenching of the n=2-3 excited H-atom are required to verify the validity of the ionization channel discussed in this paper. Further studies are also needed for hydrocarbon ion conversion reactions that determine the nature of the major ions in these plasmas. The cross sections or rate constants of the dissociative recombination of some hydrocarbon ions still need complementary measurements or theoretical estimations. The neutral hydrocarbon chemistry also need further investigation especially with respect to the pressure dependence of the three body recombination processes that may play a key-role in the chemistry that governs the boundary layer region and therefore the deposition process.

REFERENCES

1. Gicquel, A., Anger, E., Ravet, M.F., Fabre, D., Scatena, G., and Wang, Z.Z., Diamond and Related Materials **2**, 417 (1993).
2. Hassouni, K., Grotjohn, T. A., and Gicquel, A., Journal of Applied Physics **86**(1), 134 (1999).
3. Hassouni, K., Leroy, O., Farhat, S., Gicquel, A., Plasma Chemistry and Plasma Processing **18** (3), 325 (1998).
4. Gicquel, A., Hassouni, K., Farhat, S., Breton, Y., Scott, C. D., Lefebvre, M., and Péalat, M., Diamond and Related Materials **3**, 581 (1994).
5. Gicquel, A., Hassouni, K., Breton, Y., Chenevier, M., Cubertafon, J., Diamond and Related Materials **5**, 366 (1996).
6. Grotjohn, T.A., Vikharev, A., and Asmussen, J., to be published.
7. Hassouni, K., Gicquel, A., Capitelli, M., and Lourreiro, J., Plasma Sources Sciences and Technology **8**(3), 494 (1999).
8. Celiberto, R., Lammana, U. T., and Capitelli, M., Phys. Rev. A **50**(6), 4778 (1994).
9. Hiskes, J.R., J. Appl. Phys. **70**(7), 3409 (1991).
10. Loureiro, J., and Ferreira, C.M., J. Phys. D: Appl. Phys. **22**, 1680 (1989).
11. Gorse, C., Capitelli, M., Bacal, M., Bretagne, J., and Lagana, A., Chem. Phys. **117**, 177 (1987).
12. Sawada, K., and Fujimoto, T., J. Appl. Phys. **78**(5), 2913 (1995).
13. Miles, W.T., Thompson, R., and Green, A.E.S., J. Appl. Phys. **43**(2), 678 (1972).
14. Kunc. J.A., Phys. Fluids **30**(7), 2255 (1987).
15. Vriens, L., and Smeets, A.H.M., Phys. Rev. A **22**(3), 940 (1980).
16. Glass-Maujean, M., Phys. Rev. Letters **62**(2), 144 (1989).
17. Terazawa, N., Ukai, M., Kouchi, N., Kameta, K., and Hatano, Y., J. Chem. Phys. **99**(3), 1637 (1993).
18. Yu, B. W., and Girshick, S. L., J. Appl. Phys. **75** (8), 3914 (1994).
19. Tahara, H., Minami, K. I., Murai, A., Yasui, T., Yoshikawa, T., Jpn. J. Appl. Phys. **34**, 1972 (1995).
20. Bowman, C. T., Hanson, R. K., Davidson, D. E., Gardiner, W. C., Jr., Lissianski, V., Smith, G. P., Golden, D. M., Frenklach, M., and Goldenberg, M., http://www.me.berkley.edu/gri_mech/ .
21. Tawara, H., Itakawa, Y., Nishimura, H., Tanaka, H., and Nakamura, Y., NIFS-DATA Research Report **6** (1990).
22. Tachibana, K., Nishida, M., Harima, H,. and Urano, Y., J. Phys. D: Appl. Phys. **17**, 1727 (1984).
23. Yos, J. M., Technical Memorandum RAD, TM-63-7, AVCO-RAD, 1963, Wilmington, MA.
24. Hassouni, K., Capitelli, M., Farhat, S., Scott, C. D., and Gicquel, A., Surface Coating and Technology **97**(1), 391 (1997).

Ultraviolet Production Efficiency of AC-PDPs and Ways to Increase It

Keizo Suzuki[1], Norihiro Uemura[1], Shirun Ho[2], and Masatoshi Shiiki[1]

[1]*Hitachi Research Laboratory, c/o Central Research Laboratory, Hitachi, Ltd.,*
1-280 Higashi-Koigakubo, Kokubunji-shi, Tokyo 185-8601, Japan
[2]*Advanced Research Laboratory, c/o Central Research Laboratory, Hitachi, Ltd.,*
1-280 Higashi-Koigakubo, Kokubunji-shi, Tokyo 185-8601, Japan

Abstract. The mechanism of VUV production in an AC-PDP was theoretically studied, and ways to increase VUV production efficiency were discussed. It was then shown that precise and dynamic control of the non-uniform and non-steady discharge-radiation process will become even more important in the future development of new AC-PDP technologies.

1. INTRODUCTION

The plasma display panel (PDP) is expected to serve as a new information window in the IT era because of its superior properties, namely, a flat, thin shape and a digital, clear display. Much effort is being put into expanding the PDP market, and the key technologies involved in these efforts aim to increase luminous efficiency and decrease device cost. A bright picture with low power consumption requires high luminous efficiency, which is expected to reach a value comparable to that of a CRT (cathode-ray tube) display (approximately 5 lm/W) in the near future. In this presentation, vacuum ultraviolet (VUV) production efficiency, which is a significant parameter determining the luminous efficiency, and the ways to increase it will be discussed in terms of discharge and gas-reaction characteristics. The main part of this presentation follows a paper already presented by the authors in Japanese [1].

2. DISPLAY PRINCIPLE IN AN AC-PDP AND VUV PRODUCTION EFFICIENCY

Two fundamental methods, one for AC-PDP (alternating-current PDP) and one for DC-PDP (direct-current PDP), have been developed. Of these two, AC-PDP, in which the electrodes are covered by insulating layers, is more commonly applied in commercial devices because it provides long-life operation and has a simple structure. In our work, accordingly, we mainly focus on the AC-PDP method.

Figure 1 shows the AC-PDP structure. Pairs of the sustain-discharge electrodes perpendicularly cross the address electrodes to form a matrix, whose intersecting points are the display cells. The bright and dark cells are selected by means of the address discharge between the two crossing electrodes. The walls of the bright cells have electric charge formed during the address discharge on the insulating film surface covering the sustain-discharge electrodes. Sustain discharge (display discharge) occurs only in the bright cells under the effect of the wall voltage induced by the wall charge. A picture is displayed by visible red, green, and blue (R, G, and B) lights emitted by phosphors excited by the VUV photons produced in the sustain discharge. Thus, essential features of the AC-PDP are the nonlinear function of discharge and the memory function of the wall charge, which make possible the bright cell selection and the selection maintenance.

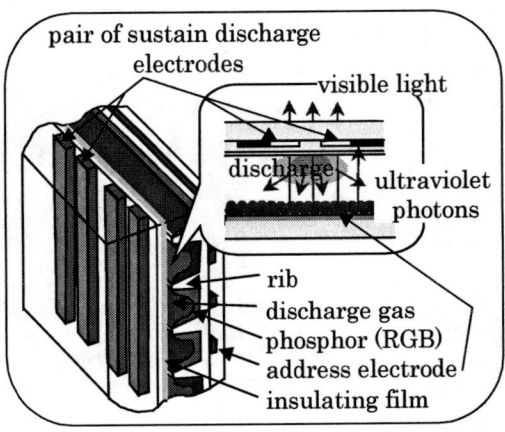

FIGURE 1. AC-PDP structure (reference [1]).

A mixture of neon and xenon is usually used for the discharge gas in the cells. The VUV photons, whose wavelengths are mainly 147 nm (photon energy: 8.44 eV) and 173 nm (photon energy: 7.17 eV), are radiated by excited xenon atoms (Xe^3P_1) and excited Xe_2 molecules ($Xe_2^*(^3\Sigma_u^+)$, $Xe_2^*(^1\Sigma_u^+)$, $Xe_2^*(1_u)$, and $Xe_2^*(O_u^+)$) produced in the sustain discharge.

To increase display brightness without increasing power consumption, luminous efficiency must be increased. Current value of AC-PDP luminous efficiency, along with fluorescent lamp luminous efficiency, is shown in Table 1. Luminous efficiency, η, is usually defined as [visible light output power of the display]/[input power to make and maintain the discharge]. The unit of η thus becomes lm/W if the visible light output power is expressed in lumens, but it has no units if the visible light output power is expressed in watts.

According to the display principle previously outlined, luminous efficiency η is given by the product of the efficiency elements: VUV production efficiency η_v, VUV transportation efficiency η_t, phosphor emission efficiency η_p, and visible light utilization efficiency η_u, as shown in Table 1 (in comparison with a fluorescent lamp). There have been no reports on measured efficiency elements for an AC-PDP; the values given in Table 1 are those estimated by the author using published information [2], [3]. The fluorescent-lamp efficiency elements are already given in a published paper [2].

Table 1 shows that the VUV production efficiency, η_v, will make the largest contribution to increasing PDP luminous efficiency. That is, a revolutionary technology to increase η_v must be developed -- along with the continuous improvement in other efficiency elements -- in order to achieve a luminous efficiency comparable to that of CRT display devices.

TABLE 1. Luminous Efficiency of an AC-PDP and a Fluorescent Lamp (reference [1]).

		AC-PDP	Fluorescent Lamp
Luminous efficiency $\eta = \eta_v \times \eta_t \times \eta_p \times \eta_u$		2.5x10^{-3} (1.2 lm/W)	0.20 (60 lm/W)
Efficiency elements	η_v VUV production efficiency	0.06	0.6
	η_t VUV transportation efficiency	0.40	0.9
	η_p phosphor emission efficiency	0.25	0.4
	η_u visible light utilization efficiency	0.40	0.9

3. BASIC MECHANISM OF VUV PRODUCTION IN AN AC-PDP AND WAYS TO INCREASE VUV PRODUCTION EFFICIENCY

Figure 2 shows a schematic representation of VUV production and the related energy flow. The energy flow ratios are those estimated by the author using simulation results [3], [4]. Almost 40% of the input power into the plasma is transformed into electron kinetic energy, a small part (3 to 6%) of which is used to form the excited xenon atoms (Xe*) and excited Xe$_2$ molecules (Xe$_2^*$) that can radiate VUV photons. The rest of the electron kinetic energy (94 to 97%, including the input power taken up as ion kinetic energy) is all consumed to heat the discharge gas and the surrounding cell walls. In addition, some of the excited xenon atoms and Xe$_2$ molecules radiate VUV photons, and the others lose their excitation energy without producing radiation.

According to the above energy process, VUV production efficiency η_v is defined and expressed as

$$\eta_v \equiv W_v / W_p, \tag{1}$$
$$= \eta_{ep} \eta_{xe} \eta_{vx}, \tag{2}$$
$$\eta_{ep} = W_e / W_p, \tag{3}$$
$$\eta_{xe} = W_x / W_e, \tag{4}$$
$$\eta_{vx} = W_v / W_x, \tag{5}$$

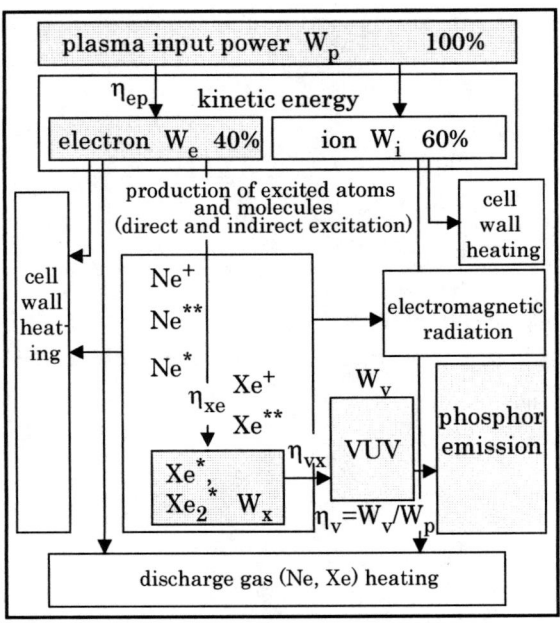

FIGURE 2. VUV production process in an AC-PDP (reference [1]).

where

W_p : input power into a unit-volume plasma, called plasma input power [Wm^{-3}],

W_e : input power into a unit-volume plasma transformed into electron kinetic energy, called electron kinetic energy input power [Wm^{-3}],

W_x : input power into a unit-volume plasma used to form excited xenon atoms and Xe$_2$ molecules which can radiate VUV photons, called VUV excitation power [Wm^{-3}],

W_v : VUV production power of a unit-volume plasma, called VUV production power [Wm^{-3}],

η_{ep}: efficiency of transforming plasma input power into electron-kinetic energy, called power-input efficiency,

η_{xe}: efficiency of the electron-kinetic energy for forming the excited xenon atoms and Xe$_2$ molecules that radiate VUV photons, called excitation efficiency, and

η_{vx}: efficiency of the excited xenon atoms and Xe$_2$ molecules for radiating VUV photons, called radiation efficiency.

Equation (2) shows that VUV production efficiency, η_v, can be increased by (a) increasing power-input efficiency η_{ep}, by (b) increasing excitation efficiency η_{xe}, or by (c) increasing radiation efficiency η_{vx}. Method (b) is especially useful to produce a satisfactory VUV production efficiency, η_v.

Figure 3 is a schematic diagram showing the process by which electron kinetic energy is transformed into the excitation energy of the excited xenon atoms and Xe_2 molecules that then radiate VUV photons. The discharge gas is supposed to be a mixture of neon and xenon. Ions and excited atoms of neon and xenon are formed through the collisions between accelerated electrons and ground-state neutral atoms. These excited atoms include "Xe^*" radiating VUV photons. This process for forming

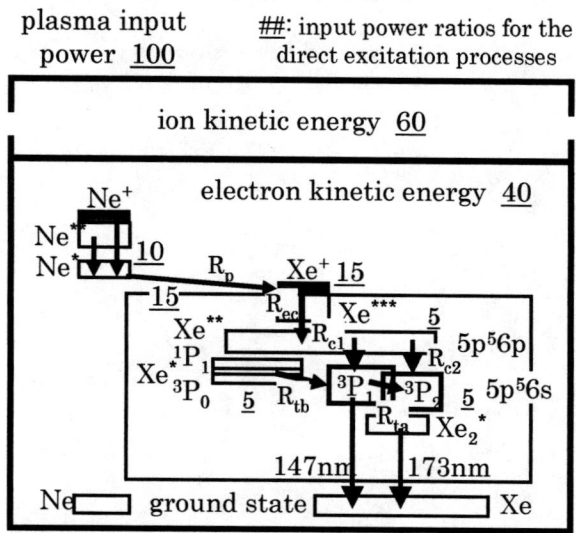

FIGURE 3. Direct and indirect excitation processes for forming Xe^* and Xe_2^* (reference [1]). R_p: Penning reaction, R_{ec}: electron-capture reaction, R_{ta}: transformation reaction from $^3P_1$ to $^3P_2$, R_{tb}: transformation reaction from $^1P_1$ or $^3P_0$ to $^3P_1$ or $^3P_2$, R_{c1}: cascade reaction to $^3P_1$, R_{c2}: cascade reaction to $^3P_2$.

"Xe^*" is called direct excitation. Input power ratios for these direct-excitation processes were calculated by simulation and are shown in the figure as underlined numbers. A discharge gas of ordinary xenon partial pressure ratio is supposed. After the direct excitation process, excited xenon atoms and Xe_2 molecules radiating VUV photons are newly formed through radiative and collisional decays from higher to lower excitation levels. This process for forming VUV radiating particles is called indirect excitation. The dominant reaction channels for indirect excitation are indicated in the figure by arrows. According to the above processes, excitation efficiency η_{xe} is given by

$$\eta_{xe} = \eta_{xed} + \eta_{xei}, \tag{6}$$

where

η_{xed}: efficiency of the electron-kinetic energy for forming the excited xenon atoms and Xe_2 molecules that radiate VUV photons through direct excitation, called direct excitation efficiency, and

η_{xei}: efficiency of the electron-kinetic energy for forming the excited xenon atoms and Xe_2 molecules that radiate VUV photons through indirect excitation, called indirect excitation efficiency.

Equation (6) shows that excitation efficiency η_{xe} can be increased by (b1) increasing direct excitation efficiency η_{xed} or by (b2) increasing indirect excitation efficiency η_{xei}.

4. WAYS TO INCREASE DIRECT EXCITATION EFFICIENCY

Under the assumption of a uniform and Maxwell-energy-distribution plasma, direct production efficiency for xenon excited atoms of level u through electron-xenon collision is given by

$$\eta_{ud} = \eta_{ep}\eta_{ued}, \quad (7)$$

$$\eta_{ued} = \left[\frac{2^{3/2}\alpha_{Xe}}{(\pi m_e)^{1/2}(kT_e)^{1/2}}\varepsilon_u^2 \tilde{\sigma}_u \exp(-\varepsilon_u/(kT_e))\right] /$$

$$\left[\frac{2^{3/2}}{(\pi m_e)^{1/2}(kT_e)^{1/2}}\sum_{i=1}^{I}\alpha_i\varepsilon_{ia}^2 J_i\tilde{\sigma}_{ia}\exp(-\varepsilon_{ia}/(kT_e)) + \frac{2^{7/2}m_e^{1/2}(kT_e)^{3/2}}{\pi^{1/2}}\sum_{i=1}^{I}\frac{\alpha_i\tilde{\sigma}_{iel}}{m_i}\right], \quad (8)$$

where

η_{ud}: efficiency of plasma input power for forming the excited xenon atoms of energy level u through direct excitation, called direct production efficiency of level-u xenon,

η_{ued}: efficiency of electron-kinetic energy for forming the excited xenon atoms of energy level u through direct excitation, called direct excitation efficiency of level-u xenon.

Parameters α_{Xe}, T_e, k are xenon partial pressure ratio in the discharge gas, electron temperature of the discharge, and Boltzmann's constant. The meanings of other parameters in equation (8) are given conventionally in reference [5].

In Equation (8), all the parameters except α_{Xe} and T_e are natural constants, and are automatically determined when excitation level u is set. This means that direct excitation efficiency η_{ued} is practically a function of α_{Xe} and T_e. Calculation results, in the case that the excitation level u is supposed to be Xe^3P_1, are shown in Fig. 4. In the calculation, η_{ep} was assumed to be 0.4 [3], [4]. Electron temperature in an ordinary discharge ranges from 2 eV to 10 eV. Thus, Fig. 4 shows that the direct production efficiency η_{ud} of Xe^3P_1 increases with increasing xenon partial pressure ratio α_{Xe} and decreasing electron temperature kT_e. Since the other levels of excited xenon atoms and excited Xe_2 molecules have similar relationships with these parameters, direct excitation efficiency η_{xed} can be increased by (b1-1) increasing xenon partial pressure ratio α_{Xe} or by (b1-2) decreasing electron temperature kT_e.

Method (b1-1) (increasing α_{Xe}) has already been confirmed by the experimental results; that is, the luminous efficiency of an AC-PDP increases with increasing xenon partial pressure ratio in the discharge gas [6], [7]. It has also been reported that luminous efficiency also increases with increasing discharge gas pressure [6], [7] and by using rf discharge [8]. These findings are considered to be due to the above-mentioned method (b1-2), decrease in electron temperature, kT_e, caused by the small electron mean free path under a high gas pressure and by the weak electric field in an rf discharge. In addition, the increase of power-input efficiency η_{ep} is also considered to contribute to the luminous efficiency increase in the rf-discharge case.

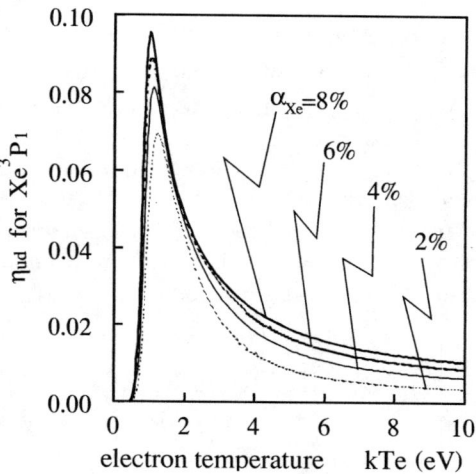

FIGURE 4. Direct production efficiency η_{ud} for Xe^3P_1 (reference [1]).

5. WAYS TO INCREASE INDIRECT EXCITATION EFFICIENCY

In the following, we suppose the discharge gas is a neon-xenon mixture as in Fig. 3. The excitation levels that mainly contribute to VUV radiation are Xe^3P_1 (excitation energy: 8.45 eV) and Xe^3P_2 (8.32 eV), which exist in the lowest energy-level group, $Xe5p^56s$. The VUV radiation from the other excitation levels, such as Xe^1P_1 (9.57 eV) and Xe^3P_0 (9.45 eV), in $Xe5p^56s$ is weak.[9] Indirect excitation efficiency can thus be increased by increasing the energy flow from the higher energy levels to the Xe^3P_1 and Xe^3P_2 levels. This can be done in the following three ways:

- (b2-1) Acceleration of the transformation reaction in the $Xe5p^56s$ level group, as shown by R_{ta} and R_{tb} in Fig. 3. In particular, reaction R_{tb} from Xe^3P_0 to Xe^3P_1 and Xe^3P_2 is expected to make a large contribution to increasing indirect excitation efficiency.
- (b2-2) Acceleration of the cascade reaction from the higher energy-level groups to the Xe^3P_1 and Xe^3P_2 levels, as shown by R_{c1} and R_{c2} in Fig. 3. In particular,

the cascade reaction from the lowest upper-energy-level group, $Xe5p^56p$, will make a large contribution.
- (b2-3) Acceleration of the electron-capture reaction of Xe^+ ions, as shown by R_{ec} in Fig. 3. This reaction is expressed as $Xe^+ + e + A \rightarrow Xe^* + A$.

The effectiveness of the first two ways can be assessed by study of the discharge gas. Namely, it has been reported that luminous efficiency is increased by addition of helium gas to the neon-xenon discharge gas [9]. This increased efficiency is a result of the effect that the addition of such a low-mass atom like helium to the discharge gas enhances the collisional decay of the excited xenon atoms and accelerates cascade reaction R_{cl} and transformation reaction R_{ta}.

6. DISCHARGE-RADIATION SIMULATOR AND ITS APPLICATION

Discharge-radiation processes occurring in an AC-PDP can be well reproduced by a two-dimensional simulator using a drift-diffusion model and a local-field approximation. This is because the physical phenomena in an AC-PDP are mainly gas-phase reactions in a rare-gas condition, so the whole process can be described with a simple physical model. Figure 5 compares the experimental and simulation results on the luminous-efficiency dependence on discharge gas pressure [4]. The discharge-gas composition is (Ne+5%Xe). It is clear from the figure that the simulation results reproduce the experimental results well.

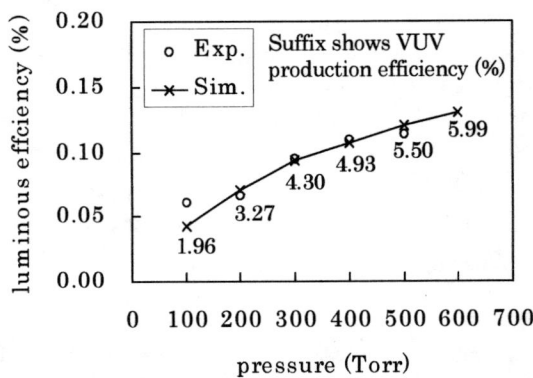

FIGURE 5. Experimental and simulation results on luminous efficiency dependence on discharge gas pressure (reference [1]).

The discharge-radiation simulator can be used to understand the physical processes in a real AC-PDP. It is also expected to be used as a design tool for developing AC-PDPs.

7. NECESSITY OF PRECISE AND DYNAMIC CONTROL OF THE DISCHARGE-RADIATION PROCESS

The ways to increase VUV production efficiency η_v were discussed in this paper. They are briefly summarized here as (a) increase of power-input efficiency η_{ep},

(b) increase of excitation efficiency η_{xe}, and (c) increase of radiation efficiency η_{vx}. Strategy (b) can be split into two methods: (b1) increase of direct excitation efficiency η_{xed}, and (b2) increase of indirect excitation efficiency η_{xei}. Furthermore, the first method (b1) can be split into (b1-1) increase of xenon partial pressure ratio α_{Xe} and (b1-2) decrease of electron temperature kT_e. Similarly the second method (b2) can be split into (b2-1) acceleration of the transformation reaction in the $Xe5p^56s$ level group, (b2-2) acceleration of the cascade reaction from the higher energy-level groups to the VUV radiation levels, and (b2-3) acceleration of the electron capture reaction of Xe^+ ions.

The methods described above except (b1-1) are expressed using the physical parameters of the discharge-radiation process. However, these methods have to be expressed in terms of operation (or design) parameters such as cell-structure constants, input-voltage pulse shape, and gas composition. Furthermore, at the same time, fundamental requirements for a practical technique, such as operation reliability, long lifetime, low operation voltage, and low production cost, must be satisfied. The trade off that appears between these two sets of requirements makes it difficult to develop practical technologies for improving VUV production efficiency.

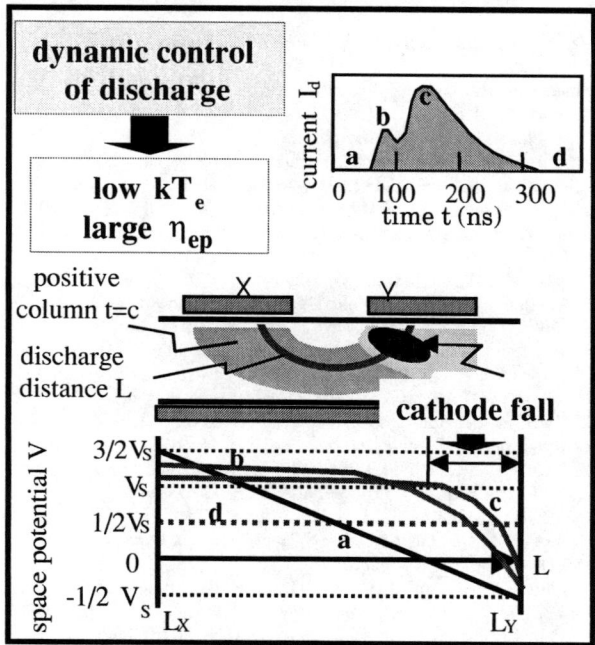

FIGURE 6. Dynamic control of discharge-radiation process in an AC-PDP (reference [1]).

An AC-PDP is an electronic device whose performance parameters, such as luminous efficiency, luminosity, power consumption, life time, picture quality, and production cost, are determined by the operation parameters described above. Between these operation and performance parameters, plasma parameters such as electron temperature and electron density exist. We thus have to study the relationship between these operation parameters, plasma parameters and performance parameters in order to resolve the trade off between the requirements described above.

The essential feature of an AC-PDP is non-uniform and non-steady characteristics of the discharge. The spatial changes in plasma parameters such as electron temperature and electron density are steep in the cathode-fall region as shown in Fig. 6. Moreover, the electrical field and discharge current in the whole discharge space change dramatically during 100 to 300 ns. Therefore, precise and dynamic control of the discharge-radiation process becomes even more important in the future development of new AC-PDP technologies.

ACKNOWLEDGMENTS

The authors wish to express their thanks to Dr. Y. Hatano of Kyushu University and Dr. N. Kouchi of the Tokyo Institute of Technology for their fruitful discussions.

REFERENCES

1. Suzuki, K., Uemura, N., Kirin, K., and Shiiki, M., *Monthly Display*, Vol. 7, No. 5, pp. 48-53 (May 2001, published in Japan).
2. Doyeux, H., "Luminance and Luminous Efficiency of PDPs," in *Proceedings of The 2nd International Display Workshops IDW'95*, Vol. **1**, PDP-7, pp. 53-56 (1995).
3. Meunier, J., Belenguer, Ph., and Boeuf, J.P., *J. Appl. Phys.*, Vol. **78**, pp. 731-745 (1995).
4. Ho, S., Saji, M., Ihara, S., Shiiki, M., Suzuki, K., Yuhara, A., Yokoyama, A., Ishigaki, M., Sato, R., Kouchi, N., and Hatano, Y., "Numerical analysis of discharge voltage and light emission efficiency in AC-PDPs," in *Proceedings of The Fifth International Display Workshops IDW'98*, PDP1-2, pp. 479-482 (1998).
5. Suzuki, K., Kawanami, Y., Ho, S., Uemura, N., Yajima, Y., Kouchi, N., and Hatano, Y., *J. Appl. Phys.*, Vol. **88**, pp. 5605-5611 (2000).
6. (a) Yoshioka, T., Miyakoshi, A., Okigawa, A., Mizobata, E., and Toki, K., "A high luminance and high luminous efficiency AC-PDP using high Xe-content gas mixtures," in *Proceedings of The Seventh International Display Workshops IDW'00*, PDP1-1, pp. 611-614 (2000). (b) Koshio, C., Taniguchi, H., Amemiya, K., Saegusa, N., Komaki, T., and Sato, Y., "New High Luminance 50-inch AC PDPs with an Improved Panel Structure using "T"-shaped Electrodes and "Waffle"-structured Ribs," in *Proceedings of The 8th International Display Workshops IDW'01*, PDP1-2, pp. 781-784 (2001).
7. (a) Oversluizen, G., Zwart, S. de, Heusden, S. van, and Dekker, T., "Dependence of PDP efficacy on the gas pressure," in *Proceedings of The Seventh International Display Workshops IDW'00*, PDP2-1, pp. 631-634 (2000). (b) Oversluizen, G., Zwart, S. de, Dekker, T., Juestel, T., and Heusden, S. van, "Characteristics of a high Xe-concentration 4-inch PDP," in *Proceedings of The 8th International Display Workshops IDW'01*, PDP4-3, pp. 833-836 (2001). (c) Gillies, M.F., Oversluizen, G., Dekker, T., Heusden, S. van, and Zwart, S. de, "Spectroscopic Study of a Xe-Ne Plasma Display Panel," in *Proceedings of The 8th International Display Workshops IDW'01*, PDP4-4, pp. 837-840 (2001).
8. Kang, J., Kim, O.D., Jeon, W.G., Song, J.W., Park, J., Lim, J.R., and Boeuf, J.P., "Panel Performance of RF PDP," in *Proceedings of The Seventh International Display Workshops IDW'00*, PDP2-4, pp. 643-646 (2000).
9. Uemura, N., Yajima, Y., Kawanami, Y., Suzuki, K., Kouchi, N., and Hatano, Y., "Kinetic Model of the VUV Production in AC-PDPs as Studied by Time-resolved Emission Spectroscopy," in *Proceedings of The Seventh International Display Workshops IDW'00*, PDP2-3, pp. 639-642 (2000).

Atomic and Molecular Data Needs in Thermal Plasmas

P. Fauchais, V. Rat, J. Aubreton and M.F. Elchinger

Laboratoire Science Des Procedes Ceramiques Et Traitements De Surface (Spcts) Umr 6638 Cnrs - Universite De Limoges (France)

Abstract. This paper is related to an overview of data needs to calculate thermodynamic and transport properties of thermal plasmas in local thermodynamic equilibrium (LTE) and in non-LTE (when the electron kinetic temperature is assumed to be different of that of heavy species). Methods of calculation of plasma composition are first presented at equilibrium as well as in non-equilibrium conditions with the data needs for partition functions and kinetic reaction rates. Then a particular attention is paid to transport coefficients in non-equilibrium thermal plasmas with the data needs: interaction potentials, collision cross sections.... The comparison between the non-equilibrium simplified theory of transport properties of Devoto and Bonnefoi and that recently developed by Rat et al is made. The influence of the plasma composition on transport coefficients is also presented. Finally, two-temperature combined diffusion coefficients are shown in an argon-hydrogen plasma highlighting the mass conservation. Such calculations show that, besides the data uncertainties, the calculation method plays a critical role in the values of plasma thermodynamic and transport properties, especially in non-equilibrium conditions.

1. INTRODUCTION

Thermal plasmas are widely used in various application fields such as spraying, cutting, welding, surface treatment, metal purification, smelting, extractive metallurgy, ultrafine particles, waste treatments... [1]. They are produced either by direct current (d.c.) plasma torches (blown plasmas) or transferred arcs, or by Radio-Frequency R.F. plasma torches. D.c. plasma torches with stick or button type hot cathodes or cold well type cathodes are characterized by high temperatures 8000-14000 K and velocities 500-2500 m/s at their nozzle exit, and high radial temperature gradients: up to 5000 K/mm. Power levels range between 1 and 6000 kW. In transferred arcs with hot cathodes, temperatures close to the cathode tip can reach 25000 K, whereas at the anode, they are between 9000-14000 K. The power level P depends on the cathode material since, for a thoriated tungsten cathode, $1 < P < 8000$ kW, while, for a graphite cathode, $P < 100$ MW. For R.F. plasmas temperatures are below 10000 K and velocities below about 150 m/s for power levels at the maximum of 150 kW.

Generally, in these plasma processes local thermodynamic equilibrium is not satisfied in regions of strong temperature gradients and/or when a cold gas or a liquid is injected. The ratio of the electron kinetic temperature T_e to that of heavy species

T_h, θ, can vary between 1 and 3 in regions where $T_h < 8000-9000$ K. The main used plasma forming gases are binary mixtures: Ar-H_2, Ar-N_2, N_2-H_2, air, or ternary mixtures Ar-He-H_2 and thus below 9000 K they are mainly consisted of molecular species.

The understanding of plasma processes is growing with the development of plasma modelling (with 2D or 3D plasma flow codes) which consists in solving the conservation equations (mass, momentum and energy) with the appropriate boundary conditions. These models require the knowledge of composition, thermodynamic and transport properties of plasmas which themselves depend on the equilibrium constants, kinetic reaction rates, interaction potentials, collision cross sections.

To illustrate these data needs, the second section of this paper presents how the plasma composition at chemical equilibrium but not necessarily in LTE conditions is calculated while the third one is dedicated to the calculation methods of transport properties of thermal plasmas.

2. PLASMA COMPOSITION

Plasma composition is a prerequisite since it allows one to characterize the plasma with for example the Debye length, the dissociation and ionisation degrees... Moreover, the thermodynamic properties of the plasma, such as internal energy, enthalpy or entropy are completely determined by the plasma composition [2]. Finally, transport properties also strongly depend on plasma composition.

Chemical equilibrium assumes that local gradients in the plasma are weak enough so that the difference between the creation and destruction rates of a species vanishes. This assumption is closely linked to the plasma process used and the chemical reactions considered.

Assuming chemical equilibrium and LTE, plasma composition can be obtained from the minimization of the Gibbs free energy using the "Steepest descent algorithm" of White et a [3] with a low-time consuming calculation, or from the well-known method of equilibrium constants related to the reactions chosen. These methods require the knowledge of chemical species partition functions which are available data for the major species. For minor species data are not necessarily available or are known with a poor precision (two to three orders of magnitude) [4]. For example a molar fraction of 10^{-4} or 10^{-8} of phosgen COF does not change at all the enthalpy but the former is lethal!

The stationary kinetic calculation [5], neglecting diffusion and convection, often appears to be the most reliable method, since it takes into account many reaction paths. It consists in determining kinetic coefficients of direct and reverse reactions, respectively k_d and k_r. The former must be found in the literature whereas the latter is deduced from the former assuming micro-reversibility of processes introducing the equilibrium constant K_p of the reaction considered. However, a lack of data regarding the kinetic coefficients is noted and plasma composition can only be obtained for rather simple plasma mixtures. At LTE, all methods provide same results.

However, when non-LTE is considered for which the kinetic electron temperature is assumed to be different from that of heavy species, strong disagreements are observed between different methods of calculation of plasma composition (Gibbs free minimization, equilibrium constant methods, stationary kinetic calculation) [6-8, and references therein].

Moreover, for example in the dissociative recombination reaction $AB^+ + e \Leftrightarrow A + B$, the direct kinetic coefficient should be calculated at T_e, whereas the reverse kinetic coefficient at T_h. However, the equilibrium constant of the reaction depends on only one temperature and, at chemical equilibrium, is equal to k_d/k_r. Thus, in order to ensure micro-reversibility, a temperature T^* can be introduced as function of species fluxes ensuring a smooth transition between T_e and T_h. This intermediate temperature T^* is defined [7,9] by:

$$T^* = T_e - (T_e - T_h) \exp(-R) \quad \text{where} \quad R = \frac{\overline{n_e v_e}}{\sum_i n_i v_i} \quad (1)$$

R has been defined as the ratio of the electron flux to that of neutral species: T^* tends towards T_h when n_e is low, and towards T_e when the densities of neutral species are low.

Moreover, non-equilibrium thermal plasmas require the definition of a non-equilibrium parameter $\theta = T_e/T_h$ which should be function of the electron number density n_e since equilibrium is reached as soon as $n_e > 10^{23}$ m^{-3}. In accordance with experimental measurements, θ_e has been defined as follows:

$$\theta = 1 + A \ln\left(\frac{n_e}{n_e^{max}}\right) \quad (2)$$

n_e^{max} being the electron density over which equilibrium can be assumed, that is $n_e^{max} = 10^{23}$ m^{-3}. The constant A is fitted to experiments [7,9].

Tanaka et al [10] also propose an effective excitation temperature, depending on T_e and T_h, which allows one to calculate the plasma composition. For example, figure 1 shows the electron temperature dependence of number densities, calculated with a stationary kinetic calculation using the temperature of Tanaka et al, of an atmospheric argon-hydrogen plasma (50mol%) for $\theta = 1.6$ [11,12]. It has to be noted that, in this calculation, θ has been fixed and does not depend on n_e as defined in equation (2).

First, in figure 1, a strong discontinuity is observed at $T_e = 11000$ K. This can be explained regarding the equilibrium constants of reactions. For example for the reaction $A + B \Leftrightarrow C + D$, the equilibrium constant is calculated as $Kp(\text{theoritical}) = Q_C Q_D \exp(-E_R/kT)/Q_A Q_B$ and it should be checked that:

$$Kp = n_C n_D / n_A n_B = Kp(\text{theoritical}) \quad (3)$$

However, it is observed that below the discontinuity temperature, this relationship is satisfied for charge transfer reactions and dissociative recombination reactions mainly through H_2, these reactions limiting ionizations of Ar and H. It is to be noted that, when the number density of molecular hydrogen becomes low (10^{21} m^{-3}) as temperature increases, the previous ionization limiting reactions do not hold anymore, inducing an electron avalanche. Equation (3) is then satisfied mainly for ionization reactions of Ar and H as described by standard Saha equations.

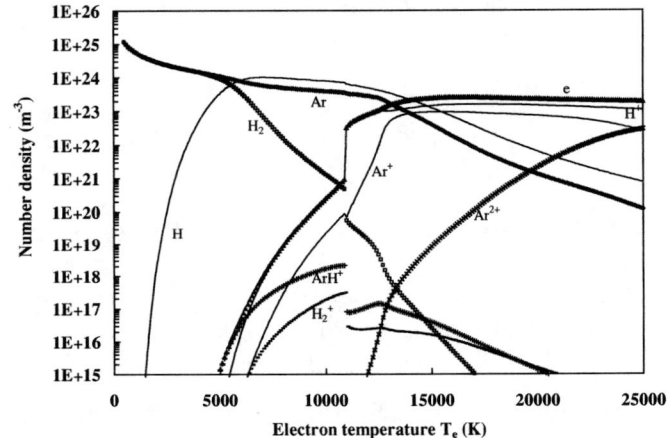

FIGURE 1. Evolution with the electron temperature of the non-equilibrium composition of an Ar – H_2 (50% mol) mixture at atmospheric pressure for $\theta=1.6$ obtained using the stationary kinetic calculation.

3. TRANSPORT PROPERTIES

3.1 Calculation

Transport properties are obtained from the solution of the Boltzmann equation of each species (let N be the total number of species in the plasma). The Boltzmann equation is solved using the well-known Chapman-Enskog method assuming that the distribution function is a Maxwellian perturbed by a first-order perturbation function. It introduces gradients, responsible of transport phenomena, and is developed in the form of a series of Sonine polynomials to obtain a linear form of the Boltzmann equation for each species present in the mixture. These N equations generate a matrix which can be solved (where the unknowns are the coefficients of the previous linear combinations) to determine the first-order perturbation function. Thus, from the knowledge of the distribution function of each species, transport coefficients can be defined and are also calculated from a matrix inversion, depending on the approximation order chosen for the expansion of Sonine polynomials. The matrix elements are functions of collision integrals which describe the type of interaction between two colliding species. Collision integrals therefore represent the second data base, after that for the plasma composition which is necessary, to calculate transport coefficients. The accuracy of collision integrals strongly influences the accuracy of transport coefficients as shown by [2,13]. Collision integrals are numerically calculated from data found in the literature as differential cross sections measured by beam experiments, quantum phase shifts and interaction potentials obtained by quantum calculations. At LTE, transport coefficients have been widely studied and calculated for pure or binary plasma forming gases.

When deviations to LTE occur, that is when the kinetic temperature of electrons T_e is considered to be different from of that of heavy species T_h, Devoto [14] was the first to propose an approach which allows the calculation of transport coefficients. His method was improved later by Bonnefoi [15]. Devoto and Bonnefoi assumed it is possible to neglect collision terms between electrons and heavy species in the Boltzmann equation. This procedure was aimed at reducing the calculations since, for example, for the electron translational thermal conductivity, the third approximation order is required to obtain an accurate value whereas the second approximation order for heavy species is sufficient.

However, recently, Rat et al [16] have questioned the Bonnefoi's approach when calculating at equilibrium combined diffusion coefficients [17] from the simplified theory of transport properties [15]. They have shown that the mass conservation is not satisfied in the plasma using the simplified approach of Bonnefoi. Consequently, Rat et al [18] have proposed to maintain a coupling between electrons and heavy species in the calculation of transport coefficients. Transport coefficients can then be calculated in non-equilibrium thermal plasmas keeping the mass balance.

3.2 Data for Transport Coefficients

Transport coefficients are calculated once collision integrals have been determined. The calculation of the latter consists in successive numerical integrations of the transport cross sections to obtain the collision integral $\Omega_{ij}^{(\ell,s)}$ for one pair of species i-j, where ℓ and s are indices which depends on the approximation order chosen for transport coefficients. For example, in the approach proposed by Rat et al [18], if ξ is the order of approximation, all combinations of collision integrals between ℓ and s satisfying $0<s\leq 2\xi-1$, $0<\ell\leq\xi$ and $\ell\leq s$ have to be calculated for diffusion coefficients and thermal conductivity. For viscosity, these conditions are $0<s\leq 2\xi$, $0<\ell\leq\xi+1$ and $\ell\leq s$. It means that, to compute the translational thermal conductivity of electrons to the third approximation, 12 different collision integrals have to be calculated for one pair of species which represents, for a argon plasmas with only 3 species (electrons, Ar, Ar^+), 72 different collisions integrals! However, this data base is limited by the lack of data concerning potential interactions or differential cross sections which at the moment are limited to the plasma forming gases (see section1). However in plasma processes are often injected gases such as C_nH_m, C_nF_m, MCl_n, MH_n (M standing for a metal) for which data are scarce especially for interactions of the type C-O-F, C-H-O-N...

3.3 Results

Figure 2 shows the temperature dependence of combined thermal diffusion coefficients derived [16] at equilibrium from the simplified theory of transport properties of Bonnefoi [15] for an atmospheric argon-nitrogen mixture (50wt%). It clearly highlights for temperatures above 10000 K that the mass conservation is not satisfied since $\overline{D}^T_{ArN_2} \neq -\overline{D}^T_{N_2Ar}$ due to the decoupling of electrons and heavy species in the calculations of transport coefficients.

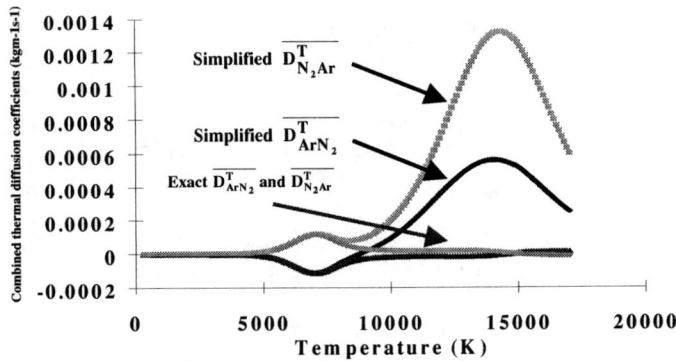

Figure 2. Temperature dependence of combined thermal diffusion coefficients of an Ar-N$_2$ (50wt%) plasma at atmospheric pressure calculated from the simplified transport properties theory of Bonnefoi [15].

Figure 3 shows a comparison of electrical conductivity plotted as a function of the kinetic electron temperature T_e, and calculated either with the simplified theory of transport properties [15] or that of Rat et al [18]. The calculation has been performed for an argon-hydrogen mixture (50mol%) at atmospheric pressure for different values of $\theta = T_e/T_h$, where T_h is the heavy species temperature. First, a good agreement is found at equilibrium. Second, discrepancies are observed, between the two approaches, reaching more than 20%, for $\theta = 1.6$, at $T_e = 25000$ K. Similar behaviors of the electron translational thermal conductivity can be shown.

Figure 4 depicts the electron temperature dependence of electrical conductivity of an argon-hydrogen mixture (50mol%) at atmospheric pressure for $\theta = 1.6$ calculated with a constant equilibrium method (KpC) and a stationary kinetic calculation (KinC). It shows the influence of the plasma composition since a discontinuity, invoked in section 2, occurs around 11000 K in the electrical conductivity obtained from the stationary kinetic calculation. Moreover, in the latter, an ionisation delay is also observed due the charge transfer reactions and the dissociative recombination reactions.

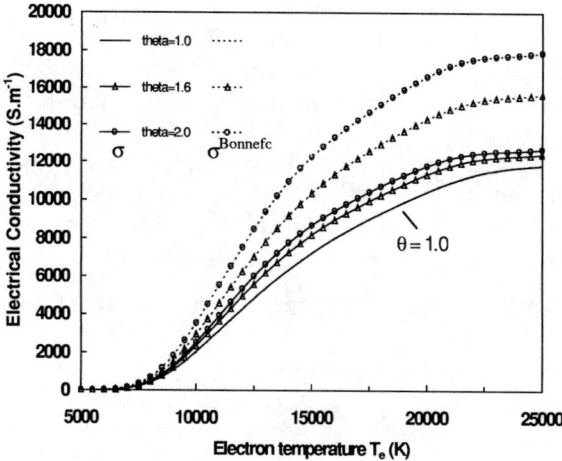

Figure 3. Evolution with the electron temperature of the electrical conductivity calculated at atmospheric pressure either from 18] or the simplified theory of transport properties of Bonnefoi [15] in an Ar-H_2 (50mol%) plasma for different values of $\theta = T_e/T_h$. Both approaches give the same results at $\theta = 1.0$.

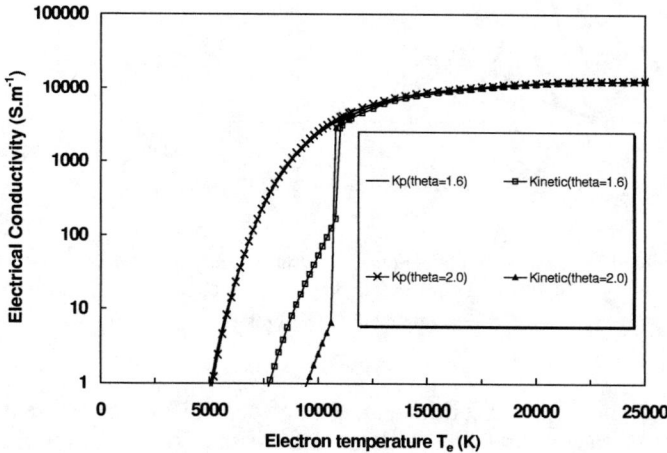

Figure 4: Evolution with the electron kinetic temperature, at atmospheric pressure, of the electrical conductivity of an $Ar - H_2$ (50mol%) mixture using compositions calculated by the stationary kinetic calculation (Kinetic) and the equilibrium constant method (Kp) for $\theta = 1.6$ and $\theta = 2.0$.

Figure 5 shows the electron temperature dependence of the total thermal conductivity of an argon-hydrogen mixture (50mol%) using the compositions calculated at atmospheric pressure by the KpC and KinC methods for $\theta = 1.6$ and $\theta = 2.0$. The total thermal conductivity includes translational, internal and reaction contributions [12]. For $\theta = 1.6$, a good agreement is observed between the dissociation peaks for both the KinC and KpC methods. However, for $\theta = 2.0$, the maximum value of the KpC dissociation peak is higher than that of the KinC. Moreover, for a temperature range of 8000-11000 K, the KinC method leads to an ionization delay of argon atoms, $\kappa_{tot}(KinC) < \kappa_{tot}(KpC)$, but at temperatures above the avalanche phenomena, n_{Ar^+} increases faster (see figure 1), giving $\kappa_{tot}(KinC) > \kappa_{tot}(KpC)$.

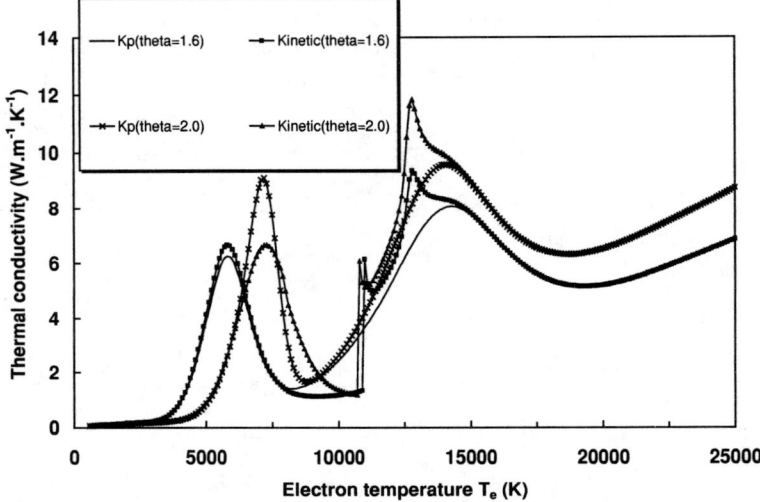

Figure 5: Evolution with the electron temperature at atmospheric pressure of the total thermal conductivity of an $Ar - H_2$ (50mol%) mixture using compositions calculated by the stationary kinetic calculation (Kinetic) and the equilibrium constant method (Kp) for $\theta = 1.6$ and $\theta = 2.0$.

Finally, figure 6 shows the electron temperature dependence of non-equilibrium combined thermal diffusion coefficients derived from the non-equilibrium diffusion velocities defined in [18] in an argon-hydrogen mixture (50mol%) for $\theta = 1.6$ and calculated from the KinC method. This figure shows that the mass balance is satisfied whatever the temperature.

4. CONCLUSION

Modelling of thermal plasmas processes strongly depends on data bases related to plasma composition and transport coefficients. The chemical equilibrium plasma

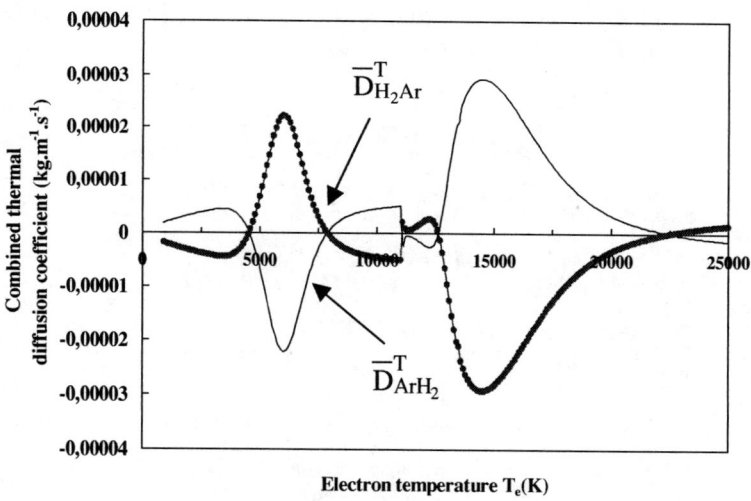

Figure 6: Evolution with the electron temperature at atmospheric pressure of combined diffusion coefficients of an $Ar-H_2$ (50mol%) mixture using a plasma composition calculated by the stationary kinetic method for $\theta = 1.6$. The mass conservation is satisfied since $\overline{D}^T_{ArH_2} = -\overline{D}^T_{H_2Ar}$ whatever the temperature.

composition can be calculated at equilibrium or in non-LTE either with the stationary kinetic method, for which data are not always available, mass action laws using equilibrium constants or Gibbs free minimization for which only data related to minor species are known with a poor precision. However, in non-LTE, besides the lack of data, the different methods of calculation lead to different results [7]. Moreover, in order to take into account dissociation recombination reactions such as $AB^+ + e \Leftrightarrow A + B$ for example, in the stationary kinetic method, an intermediate temperature has to be introduced maintaining micro-reversibility of processes and depending on the electron T_e and heavy species T_h kinetic temperatures [10,11]. An example is given for an argon-hydrogen mixture at atmospheric pressure for $\theta = T_e/T_h = 1.6$. It exhibits a strong discontinuity around $T_e = 11000$ K which is explained by the key role of charge transfer and dissociative recombination reactions. Non-equilibrium transport coefficients are calculated from the solving of the Boltzmman equation keeping the coupling between electrons and heavy species [18]. Collision integrals are a prerequisite to calculate transport coefficients and strongly influence results of transport coefficients. However at the moment data to calculate them are limited to the usual plasma forming gases. It is also shown that transport coefficients can also be affected by the plasma composition, demonstrating the importance of the choice of the plasma composition. Finally, two-temperature combined thermal diffusion coefficients have been calculated according to a new theory [18] and results, for an argon-hydrogen mixture, show that the mass conservation is fully satisfied contrarily to those obtained with the simplified theory of transport properties of Devoto or Bonnefoi [14,15].

REFERENCES

1. Fauchais, P., and Vardelle, A., *IEEE Trans. Plasma Sci* **25**, 1258-1280 (1997).
2. Boulos, M.I., Fauchais, P., and Pfender, E., *Thermal Plasmas: Fundamentals and Application*, Vol. 1, Plenum, New York (1994).
3. White, W.B., Johnson, S.M., and Dantzig, G.B., *J. Chem. Phys.* **28**, 751-755 (1958).
4. Fauchais, P., Elchinger, M.F., and Aubreton, J., *High Temp. Material Processes* **4**, 21-42 (2000).
5. Richley, E., and Tuma, D.T., *J. Appl. Phys.* **53**, 8537-8542 (1982).
6. Chen, Xi, and Han, Peng, *J. Phys. D: Appl. Phys.* **32**, 1711-1718 (1999).
7. Rat, V., André, P., Aubreton, J., Elchinger, M.F., Fauchais, P., and Lefort, A., *J.Phys. D: Applied Phys.* **34**, 1-14 (2001).
8. Giordano, D., and Capitelli, M., *Phys. Rev. E* **65**, 16401 (2002).
9. André, P., Aubreton, J., Elchinger, M.F., Fauchais, P., and Lefort, A., *Plasma Chem. Plasma Process.* **21**, 83-106 (2001).
10. Tanaka, Y., Yokomizu, Y., Ishikawa, M., and Matsumura, T., *IEEE Trans. Plasma Sci.* **25**, 991-995 (1997).
11. Rat, V., André, P., Aubreton, J., Elchinger, M.F., Fauchais, P., and Lefort, A., Two-temperature transport coefficients in argon-hydrogen plasmas Part I: Elastic processes and collisions integrals, *Plasma Chem. Plasma Process.*, in press.
12. Rat, V., André, P., Aubreton, J., Elchinger, M.F., Fauchais, P., and Lefort, A., Two-temperature transport coefficients in argon-hydrogen plasmas Part II: Inelastic processes and influence of composition, *Plasma Chem. Plasma Process.*, in press.
13. Murphy, A.B., *Plasma Chem. Plasma Process.* **20**, 279-297 (2000).
14. Devoto, R.S., Ph.D. Thesis, Stanford University (1965).
15. Bonnefoi, C., State Thesis, University of Limoges, France (1983) (in French).
16. Rat, V., Aubreton, J., Elchinger, M.F., and Fauchais, P., *Plasma Chem. Plasma Process.* **21**, 355-364 (2001).
17. Murphy, A.B., *Phys. Rev. E* **48**, 3594-3603 (1993).
18. Rat, V., André, P., Aubreton, J., Elchinger, M.F., Fauchais, P., and Lefort, A., *Phys. Rev. E* **64**, 26409(1-20) (2001).
19. Rat, V., Aubreton, J., Elchinger, M.F., Fauchais, P., and Murphy, A.B., *Diffusion in a two-temperature thermal plasmas*, to be published.

New Gas Chemistry for High-Performance SiO$_2$ Patterning in Sub-0.1 μm ULSIs

Seiji Samukawa

Institute of Fluid Science, Tohoku University
2-1-1 KatahiraAoba-Ku Sendai, 980-8577, Japan

Abstract. SiO$_2$ etching is done by using fluorocarbon gases to deposit a fluoropolymer on the underlying silicon. This deposit enhances the etching selectivity of SiO$_2$ over silicon or silicon nitride. CF$_2$ radicals are used as the main gas precursor for polymer deposition. In a conventional gas plasma, however, the CF$_2$ radicals and other radicals (high-molecular-weight-radicals: C$_x$F$_y$) lead to polymerization. This condition causes microloading and etching-stop in high-aspect contact-hole patterning due to the sidewall polymerization during SiO$_2$ etching processes. Conversely, by using new fluorocarbon gas chemistries (C$_2$F$_4$/CF$_3$I), we achieved selective radical generation of CF$_2$ and eliminated high-molecular-weight radicals. Under this condition, microloading-free and etching-stop-free high-aspect-ratio contact-hole patterning of SiO$_2$ was accomplished. Thus, the higher molecular weight radicals play an important role in the sidewall polymerization in contact holes because these radicals have a higher sticking coefficient than CF$_2$ radicals. Selective generation of CF$_2$ radicals and suppression of C$_x$F$_y$ radicals are thus necessary to eliminate microloading and etching-stop in the formation of high-aspect-ratio contact holes.

INTRODUCTION

High-density plasmas, such as electron-cyclotron resonance (ECR) plasma and inductive coupled plasma (ICP) are used in etching processes to fabricate ultralarge-scale integrated circuits (ULSIs) because they efficiently provide the radicals and ions required for etching. In the patterning of high-aspect-ratio contact holes,[1-4] however, these plasmas have serious problems related to the low selectivity of the underlying silicon or silicon nitride, "microloading," and "etch-stop" (the interruption of etching). Generally, SiO$_2$ etching is done by using fluorocarbon gases to deposit a fluoropolymer on the underlying silicon. This deposit enhances the etching selectivity of SiO$_2$ over silicon or silicon nitride. In particular, CF$_2$ radicals are used as the main gas precursor for polymer deposition. CF$_3^+$ ions are the dominant etchant for SiO$_2$ films because their etching yield for Si and SiO$_2$ is larger than those of CF$^+$, CF$_2^+$, and C$_2$F$_4^+$.[5,6] Moreover, CF$_3^+$ ions are efficiently generated from CF$_3$ radicals because of their lower ionization thresholds (CF$_3$+e$\rightarrow$CF$_3^+$: 10.3 eV). Plasmas made from gases with a low molecular weight (CF$_4$, CHF$_3$, and C$_2$F$_6$) generate fewer CF$_2$ and CF$_3$ radicals, as well as a large amount of F atoms, because of the higher degree of dissociation caused by high-energy electrons. As a result, a lesser degree of polymerization reduces the etching selectivity. Conversely, gases with a higher molecular weight (C$_3$F$_8$, C$_4$F$_8$) result in more polymerization with a larger number of CF$_2$ radicals because the number of high-energy electrons is relatively reduced by the large cross-sections of the electron collisions in terms of momentum transfer and vibrational excitation at around 5 eV.[7] However, smaller amounts of ions (CF_3^+) have also been observed in these plasmas. Particularly, in high-aspect-ratio contact holes of less than 0.25 ?m in diameter, the polymer deposition rate is higher than the SiO_2 etching rate. This condition causes aspect-ratio-dependent etching and etching-stop. It is thus very difficult to use conventional gas chemistry to control the balance between the radical flux (CF_2 radicals) and ion flux (CF_3^+ ions) during SiO_2 contact etching in fluorocarbon plasma. Although Ar dilution (over 80%) of C_4F_8 is widely used to improve the control of polymerization in high-aspect contact holes, it causes problems such as a decreased etching rate and low etching selectivity. To solve these problems, we proposed a new gas chemistry (C_2F_4/CF_3I)[8-11] to generate selected CF_2 radicals and CF_3^+ ions in the plasma. This gas chemistry could enable SiO_2 etching that is free of both microloading and etching-stop.

In this paper, we report the effects of high-molecular-weight radicals on microloading and etching-stop in a conventional high-molecular-weight gas plasma. We also clarify the effectiveness of selective CF_2 radical generation for microloading-free sub-0.1-μm SiO_2 patterning by using CF_3I and C_2F_4 in an ultrahigh-frequency (UHF) plasma.

EXPERIMENTAL

A new non-perfluorocarbon gas chemistry[8-11] for generating selected radicals in a UHF plasma has been developed. Our UHF plasma source[12-18], which has a spokewise antenna, is shown in Fig. 1. We found that CF_3I and C_2F_4 plasmas are good sources of CF_3 (CF_3^+) and CF_2 radicals in a UHF plasma (Fig. 2). This is because their gas chemistries have weaker C-I and C=C bonds when combined with UHF plasmas. The flux of CF_2 and CF_3 radicals can be controlled by changing the ratio of these gases in the plasma. By applying ultrahigh-frequency (500 MHz) power to a high-density plasma, non-Maxwellian electron-energy distribution functions (EEDFs) can be produced.[17,18] The UHF plasma appears to exhibit a so-called bi-Maxwellian EEDF. The majority of the electrons (i.e. bulk electrons) are relatively cold and can be represented by a Maxwellian distribution with a relatively low electron temperature. The high-energy tail of the distribution is enhanced and this enhancement can be described by the presence of the second (much less populated) pool of electrons with higher (with respect to that of bulk electrons) electron temperature. These high-energy electrons can achieve a sufficient level of ionization to produce a higher density plasma ($> 10^{11} cm^{-3}$).

The lowest energy channels for the generation of CF_3 and CF_2 radicals from conventional C_2F_6 and C_4F_8 sources that do not involve complicated, unlikely bond rearrangements (for CF_2 formation from C_4F_8) are

$$C_2F_6 + e \rightarrow 2CF_3 + e \qquad \Delta H = 4.17 \text{ eV}, \qquad (1)$$

$$C_4F_8 + e \rightarrow CF_2 + \cdot C_3F_6 \cdot + e \qquad \Delta H = 4.08 \text{ eV}, \qquad (2)$$

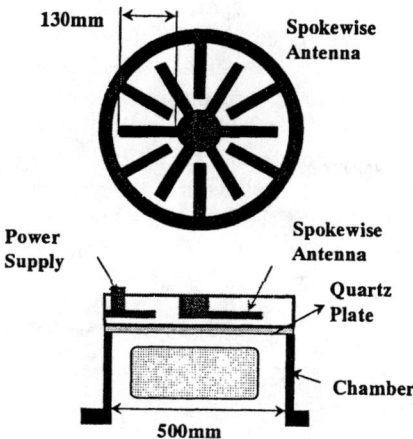

FIGURE 1. Schematic illustration of developed ultrahigh-frequency (UHF) plasma source with a spokewise antenna.

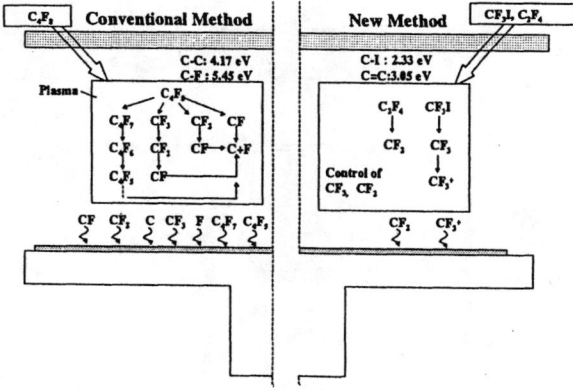

FIGURE 2. Concept of new radical control method for selective generation of CF_2, CF_3 radicals and CF_3^+ ions from CF_3I and C_2F_4 gases in UHF plasma.

where ΔH represents the thermodynamic heat of reaction[19] (also the bond strength for these reactions and those below) and $\bullet C_3F_6 \bullet$ is a di-radical species. With the new compounds explored here, CF_3 and CF_2 can be formed by the reactions

$$CF_3I + e \rightarrow CF_3 + I + e \qquad \Delta H = 2.33 eV, \qquad (3)$$
$$C_2F_4 + e \rightarrow 2CF_2 + e \qquad \Delta H = 3.05 eV, \qquad (4)$$

Because the peak electron density is at an energy of about 2-3 eV (electron temperature T_e: 2-3 eV) in the UHF fluorocarbon gas plasma, a C-I bond of about 2.33 eV and a C=C bond of about 3.05 eV are efficiently dissociated in CF_3I and C_2F_4 in reactions (3) and (4), whereas the rupturing of C-F bonds (5.45 eV) and the formation of CF_3 and CF_2 by reactions (1) and (2) require energies above the peak electron energy in the UHF plasma.[19] In fact, the reported threshold for CF_2 formation from electron impact dissociation of C_4F_8 is quite high (10.5 eV).[20] The UHF plasma and new source gases (CF_3I and C_2F_4) are thus an ideal combination for selective generation of CF_2 and CF_3 radicals in the plasma. Conversely, in the case of a conventional high-density plasma (ICP, ECR plasma), high dissociation of the C-C and C-F bonds of C_2F_4 and CF_3I gas molecules is caused by the high-energy electrons. Even under this condition, however, we believe that the new gas chemistry results in a higher density of CF_2 radicals and CF_3^+ ions and that the concentration ratios of CF_2 radicals and CF_3^+ ions are more controllable in comparison with conventional higher-molecular-weight gases.

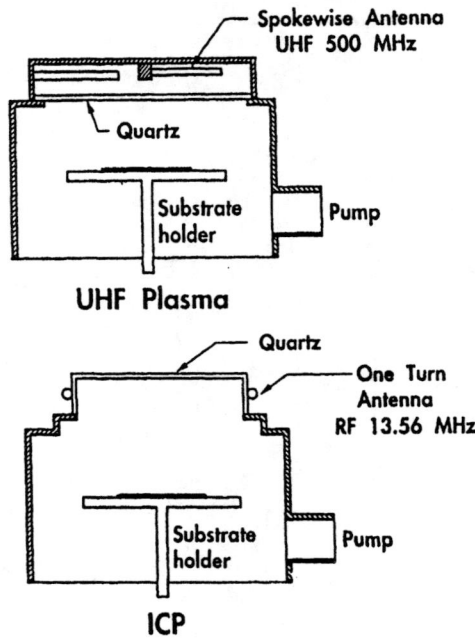

FIGURE 3. Schematic illustrations of UHF spokewise antenna and inductively coupled plasma (ICP) one-turn loop antenna.

In this experiment, the reactor consisted of a 40-cm-i.d. anodized aluminum chamber, a helium-cooled clamped-wafer chuck, a 2200-l/s turbomolecular pump, and a plasma source, which was either a spoked antenna for UHF plasma or a one-turn loop antenna for ICP.[17, 18] The two source configurations are shown in Fig. 3. When the reactor was configured for the UHF plasma, there was a 3-cm-thick 42.5-cm-diameter quartz window at the top of the chamber directly below the antenna. When the reactor was configured for the ICP, a flat-topped 10-cm-high 30-cm-diameter quartz bell jar was placed in the chamber. The one-turn antenna encircled this bell jar. We analyzed radicals and ions for our new gas chemistry and conventional gas chemistry in the UHF plasma and ICP. Infrared-diode laser absorption spectroscopy (IR-LAS)[21-23] and optical emission spectroscopy (OES) were used to measure the densities of CF, CF_2, and CF_3 radicals and F atoms. The IR-LAS laser beam passed through the plasma region 16 times during the radical measurements. The CF, CF_2, and CF_3 radical densities were measured 3 cm above the substrate. The absorption lines used were 1308.5 cm^{-1} for the CF radicals, 1132.7 cm^{-1} for the CF_2 radicals, and 1262.1 cm^{-1} for the CF_3 radicals. We used the optical emission intensities for CF (202.4 nm), CF_2 (251.9 nm), and F (703.7 nm). The relative amounts of ion species, such as CF_3^+ and Ar^+ ions, were measured with a mass spectrometer. The plasma density and electron temperature in these plasmas were measured with a Langmuir probe (Irie Co., Japan) located at the center of the plasma reactor at the height at which the OES and IR-LAS measurements were done. To keep the tip surface clean, the probe tip (with an area of 0.0942 cm^2) and the reactor walls were heated to over $100^{\circ}C$. The reproducibility of the measurements was confirmed by repeating the measurements several times. Six-inch-diameter substrates were used on a 20-cm-diameter electrode. The rf bias (1 MHz) on the substrate was fixed at 500 W (1.6 w/cm^2) when the SiO_2 etching was done. The gas flow rate was 100 sccm, the pressure was 5 mTorr, and the UHF and ICP power was 1 kW. The polymer deposition rate and the polymer surface roughness on the silicon substrate were measured without rf-bias voltage. The surface roughness of the deposited polymer was measured with an atomic force microscope (AFM: HITACHI WA-200).

RESULTS AND DISCUSSIONS

Relationship between Generation of Reactive Species and Polymerization in UHF Plasma or ICP

The measured radical densities in the CF_3I and C_2F_4 plasmas show that they are efficient sources of CF_3 (CF_3^+) and CF_2 radicals compared with conventional C_4F_8 and C_2F_6 plasmas.[11] This is easily understood from the differences in the bond strengths of reactions (1) – (4). We also found that the density ratios of CF_2 and CF_3 could be independently controlled by changing the gas-flow ratio of the CF_3I and C_2F_4 mixture (Fig. 4) and that the polymer deposition rate can be accurately controlled by changing the density of CF_2 radicals (Fig. 5). Since polymer deposition occurred when using only the CF_2 radicals, the higher molecular weight radicals can be eliminated from the CF_3I/C_2F_4 mixture plasma. The densities of the CF_2 radicals and the CF_3^+ and Ar^+ ions were also measured in the C_4F_8/Ar plasmas (Fig. 6). By increasing the Ar dilution in the C_4F_8 and Ar mixture, the CF_2 radical density was drastically decreased and the total amount of CF_3^+ and Ar^+ ions was relatively increased so that the polymer deposition rate decreased (Fig. 7). However, the polymer deposition rate did not correspond to the CF_2 radical density in the C_4F_8/Ar plasma. This result suggested that other radicals, such as the higher molecular weight radicals (C_xF_y),[3,24] play an important role in the polymerization in the C_4F_8/Ar plasma. Under these conditions (CF_3I/C_2F_4, C_4F_8/Ar), the surface roughness of the deposited polymer was measured with an AFM, as shown in Fig. 8. A fine high-density polymer was deposited on the silicon substrate in the C_2F_4/CF_3I mixture plasma, whereas a porous low-density polymer was deposited in the C_4F_8/Ar mixture plasma, when the polymer thickness was 3000 A. This result suggests that CF_2 radicals are an effective precursor to polymerization and that the higher-molecular-weight radicals lead to a porous polymer because the F atoms suppress the cross-linking of carbon atoms and these radicals have a higher sticking coefficient than the CF_2 radicals. From these results, we found that the polymer formation is very different in C_2F_4/CF_3I and C_4F_8/Ar mixture plasmas because the polymer precursor in the C_2F_4/CF_3I mixture plasma is only CF_2, whereas both CF_2 and C_xF_y cause polymer deposition in the C_4F_8/Ar mixture plasma.

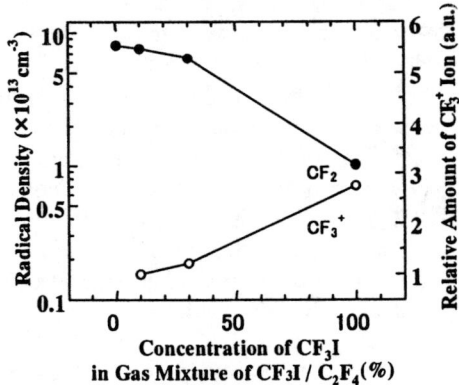

FIGURE 4. Density ratios of CF_2 radicals and CF_3^+ ions could be independently controlled by changing the gas-flow of the CF_3I and C_2F_4 mixture in UHF plasma.

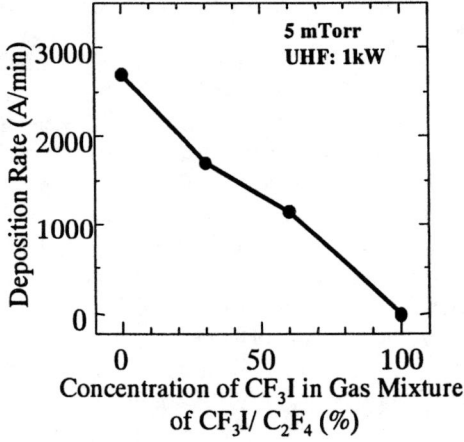

FIGURE 5. Polymer deposition rate on Si substrate as a function of gas flow ratio of the CF_3I and C_2F_4 mixture in UHF plasma.

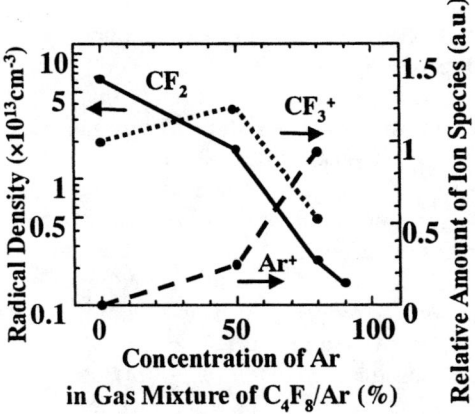

FIGURE 6. Density ratios of CF_2 radicals and CF_3^+ and Ar^+ ions could be independently controlled by changing the gas-flow of the Ar and C_4F_8 mixture in the UHF plasma.

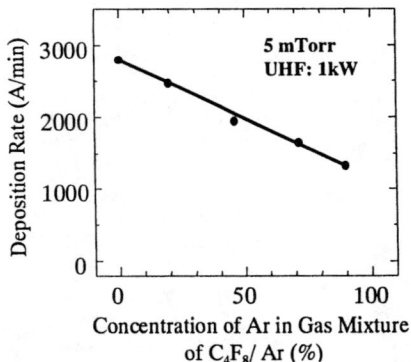

FIGURE 7. Polymer deposition rate on Si substrate as a function of gas flow ratio of the Ar and C_4F_8 mixture in UHF plasma.

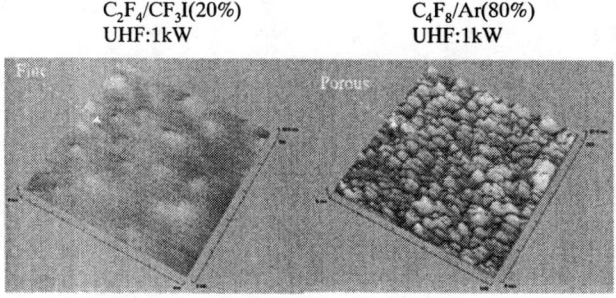

FIGURE 8. Surface roughness of deposited polymer on Si substrate in C_4F_8/Ar and CF_3I/C_2F_4 UHF plasma by using an atomic force microscope (AFM: HITACHI Wa-200). In this condition, rf bias was not supplied on substrate.

We also investigated the relationship between CF_2 radicals and polymerization in a C_2F_4/CF_3I mixture inductively coupled plasma (ICP). The density ratios of CF_2 radicals and CF_3^+ ions were independently controlled by changing the gas-flow ratio of the CF_3I and C_2F_4 mixture, even in the ICP, whereas the CF_2 radical density in the ICP became smaller than in the UHF plasma because of a higher degree of dissociation due to higher electron temperature (Fig. 9). As a result, the optical emission intensity ratio of CF_2/F in the ICP was lower than that in the UHF plasma (Fig. 10), that is, the generated F atom density in the ICP became higher than in the UHF plasma. The polymer deposition rate could thus be accurately controlled by changing the density of CF_2 radicals (Fig. 11). As polymer deposition occurred when using mainly CF_2 radicals, the higher-molecular-weight radicals can also be eliminated by using a C_2F_4/CF_3I mixture even in the ICP. As a result, a fine high-density polymer was deposited on the Si substrate in the C_2F_4/CF_3I mixture ICP (Fig. 12).

These results demonstrated that the selection of gas chemistry is very important in controlling reactive species and that lower-molecular-weight gases, such as C_2F_4, are effective in eliminating the generation of higher-molecular-weight radicals in both a UHF plasma and an ICP.

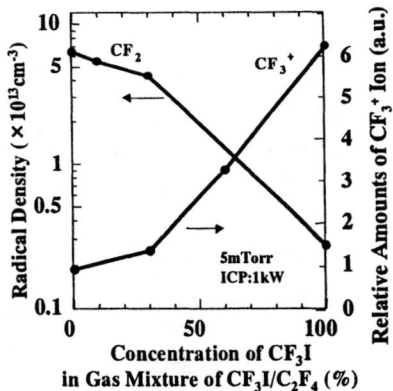

FIGURE 9. Density ratios of CF_2 radicals and CF_3^+ ions could be independently controlled by changing the gas-flow of the CF_3I and C_2F_4 mixture in ICP.

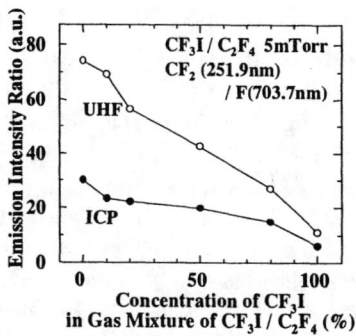

FIGURE 10. Optical emission intensity ratio of CF_2(251.9 nm)/F(703.7 nm) in UHF plasma and ICP.

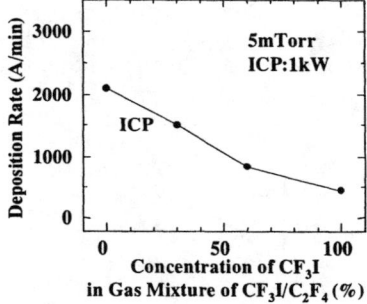

FIGURE 11. Polymer deposition rate on Si substrate as a function of gas flow ratio of the CF_3I and C_2F_4 mixture in ICP.

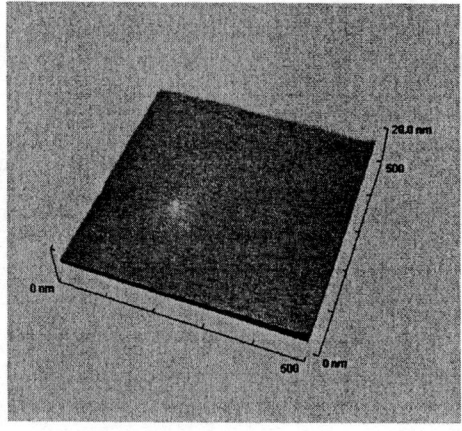

FIGURE 12. Surface roughness of deposited polymer on Si substrate in CF_3I/C_2F_4 ICP by using an atomic force microscope (AFM: HITACHI Wa-200). In this condition, rf bias was not supplied on substrate.

Relationship between Microloading Effects and Higher-Molecular-Weight Radicals

We measured the change in etching rates for hole diameters between 0.05 and 0.5 μm in a photoresist masked BPSG film using both C_2F_4/CF_3I and C_4F_8/Ar UHF plasmas (Fig. 13).[11] The etching rates were normalized for a 0.5-μm-diameter contact-hole. With our new chemistry, the pattern dependence of the SiO_2 etching rate was nearly eliminated by the optimum balance between polymerization and etching and by the suppression of higher-molecular-weight radicals. One possible mechanism for eliminating the etching-stop in the new chemistry is shown in Fig. 14.[11] The higher-molecular-weight radicals (C_xF_y) are preferentially formed in higher-molecular-weight gas plasmas. These species are more likely to form a thick film on both SiO_2 and Si, which narrows the width of the etched feature, depletes etchants from the bottoms of the features, and eventually stops the etching of the SiO_2. CF_2, on the other hand, is less likely to stick on the sides of etched features.[3,24] Selective generation of CF_2 radicals and CF_3 ions is thus advantageous for achieving etching that is free of the microloading and etching-stop problems, thus providing a high etching rate and high etching selectivity.

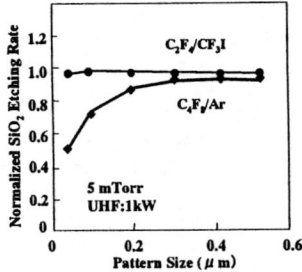

FIGURE 13. Etching rates for hole diameters between 0.05 and 0.5 μm in a photoresist masked BPSG film using C_2F_4/CF_3I and C_4F_8/Ar UHF plasmas.

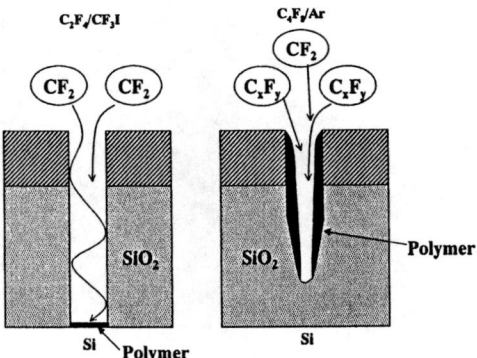

FIGURE 14. Mechanism for eliminating of microloading and etching-stop in C_2F_4/CF_3I gas mixture plasma.

To clarify the effects of higher-molecular-weight radicals on the patterning of high-aspect contact holes, we investigated the change in etching rates for hole diameters between 0.05 and 0.5 μm in a photoresist masked BPSG film using C_2F_4/CF_3I, C_3F_6/CF_3I and C_4F_8/CF_3I in the UHF plasma (Fig. 15). In this experiment, we controlled the gas mixture ratio so that the CF_2 radical and CF_3 radical densities were fixed at 6×10^{13} cm^{-3} and 2×10^{13} cm^{-3}, respectively. Furthermore, the generation of F atoms could be suppressed by addition of CF_3I in the UHF plasma, because high-energy electrons in the plasma were reduced by a low ionization threshold of CF_3I. The SiO_2 etching rate was also fixed at 4400-4800 Å/min by controlling the bias voltage on the substrate. By increasing the gas molecular weight from C_2F_4 to C_4F_8, the microloading effects were enhanced. Because a higher-molecular-weight gas is usually thought to form a larger amount of higher-molecular-weight radicals in this fluorocarbon plasma, this result indicates that higher-molecular-weight radicals cause the sidewall polymerization in high-aspect contact holes. Using C_2F_4 gas is a more effective way to eliminate higher-molecular-weight radicals than using C_3F_6 and C_4F_8. Even in C_4F_8 and C_3F_6 gas chemistries, it is possible that the higher-molecular-weight radicals can be reduced by optimizing the conditions (mainly, gas mixture). For example, the optimum addition of O_2 in the C_4F_8/Ar gas mixture plasma is effective for eliminating the microloading, whereas the etching selectivity to photo-resist and underlying-Si becomes lower. It is speculated that the higher-molecular-weight radicals in the plasma are reduced through the reaction of C_xF_y and O_2 in the gas phase.

These results show that lower-molecular-weight gases are useful for eliminating higher-molecular-weight radicals and etching-stop during SiO_2 etching processes.

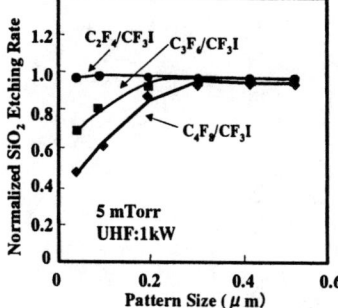

FIGURE 15. Change in etching rates for hole diameters between 0.05 and 0.5 μm in a photoresist masked BPSG film using C_2F_4/CF_3I and C_4F_8/CF_3I in UHF plasma.

CONCLUSION

The problem of microloading and etching-stop in high-aspect-ratio contact-hole patterning are caused by sidewall polymerization during SiO_2 etching processes. In conventional gas chemistries, such as C_4F_8, the CF_2 radicals and other radicals (higher-molecular-weight radicals: C_xF_y) lead to polymerization and cause microloading in contact holes with diameters of less than 0.1 μ m. However, a combination of novel feed gases (CF_3I and C_2F_4) and non-Maxwellian electron energy distributions in a UHF plasma enable the selective generation of CF_2 radicals and the elimination of microloading for 0.5-μ m contact holes. From our results, we found that higher-molecular-weight radicals (C_xF_y) play an important role in sidewall polymerization in contact holes because these radicals have a higher sticking coefficient than that of CF_2 radicals. The novel feed gases (CF_3I/C_2F_4) are promising candidates for use in future precise etching processes for fabricating ULSI devices.

ACKNOWLEDGMENTS

The authors would like to thank Prof. Toshio Goto and Associate Prof. Masaru Hori of Nagoya University for their useful advice about IR-LAS measurements. We are also grateful to Mr. Hiroshi Aoyama and Dr. Mitutugu Itano of Daikin Industries, Ltd., for their useful advice on the gas-dissociation processes and the gas preparation for our experiments.

REFERENCES

1. Kazumi, H., and Tago, K., Jpn. J. Appl. Phys. **34**, 2125 (1995).
2. Suzuki, C., Kawai, Y., Sasaki, K., and Kadota, K., Proceedings of the Third Conference on Reactive Plasmas (The Japan Society of Applied Physics, Nara, Japan, 1997).
3. Miyata, K., Hori, H., and Goto, T., Jpn. J. Appl. Phys. **36**, 1540 (1997).
4. Samukawa, S., Nakagawa, Y., Tsukada, T., Ueyama, H., and Shinohara, K., Jpn. J. Appl. Phys. **34**, 6805 (1995).
5. Tachi, S., Miyake, K., and Tokuyama, T., Jpn. J. Appl. Phys. **21**, Supplement 21-1, 141 (1981).
6. Sakai, T., Hayashi, H., Abe, J., Horioka and Okano, H., Proceedings of the 15th Dry Process Symposium (The Institute of Electrical Engineering of Japan, 1993) p. 193.
7. Itoh, H., J. Phys. D: Appl. Phys. **24**, 277 (1991).
8. Samukawa, S., and Tsuda, K., Jpn. J. Appl. Phys. **37**, L1095 (1998).
9. Samukawa, S., Mukai, T., and Tsuda, K., J. Vac. Sci. Technol. A**17**, 2551 (1999).
10. Samukawa, S., Mukai, T., and Noguchi, K., Mater. Sci. Semicond. Proc. **2**, 203 (1999).
11. Samukawa, S., and Mukai, T., J. Vac. Sci. Technol. **18B**, 166 (2000).
12. Samukawa, S., Nakagawa, Y., Tsukada, T., Ueyama, H., and Shinohara, K., Appl. Phys. Lett. **67**, 1414 (1995.
13. Samukawa, S., Nakagawa, Y., Tsukada, T., Ueyama, H., and Shinohara, K., Jpn. J. Appl. Phys. **34**, 6805 (1995).
14. Nakano, T., Ohtake, H., and Samukawa, S., Jpn. J. Appl. Phys. **35**, L338 (1996).
15. Samukawa, S., and Nakano, T., J. Vac. Sci. Technol. **14**, 1002 (1996).
16. Samukawa, S., and Tsukada, T., Appl. Phys. Lett. **69**, 1056 (1996).
17. Samukawa, S., and Tsukada, T., Jpn. J. Appl. Phys. **36**, 7646 (1997).

18. Malyshev, M.V., Donnelly, V.M., and Samukawa, S., J. Appl. Phys. **84**, 1222 (1998).
19. While these are thermodynamic bond strengths, we expect the relative trends to hold for electron impact dissociation thresholds. In addition, we expect the peak cross section energies to follow the same trends, so that those processes with the weakest bond strength will have the largest electron impact dissociation rate constants for the relatively cold bulk electron temperatures in the UHF plasma.
20. Taoyoda, H., Iio, M., and Sugai, H., Jpn. J. Appl. Phys. **36**, 3730 (1997).
21. Takahashi, K., Hori, M., Maruyama, K., Kishimoto, S., and Goto, T., Jpn. J. Appl. Phys. **32**, L694 (1993).
22. Takahashi, K., Hori, M., and Goto, T., Jpn. J. Appl. Phys. **32**, L1088 (1993).
23. Miyata, K., Takahashi, K., Kishimoto, S., Hori, M., and Goto, T., Jpn. J. Appl. Phys. **34**, L444 (1995).
24. Inayoshi, M., Ito, M., Hori, M., Goto, T., and Hiramatsu, M., J. Vac. Sci. Technol. A **16**, 233 (1998).

D. Atmospheres

Atmospheric Pollutant Removal by Non-Thermal Plasmas: Basic Data Needs for Understanding and Optimization of the Process

S. Pasquiers[1], M. Cormier[2], O. Motret[2]

[1] *Laboratoire de Physique des Gaz et des Plasmas,*
Université Paris-Sud, Bât.210, 91405 Orsay cedex, France
[2] *Groupe de Recherches sur l'Energétique des Milieus Ionisés*
Université d'Orléans, B.P. 6744 45067 Orléans cedex 2, France

Abstract. Since fifteen years, an increasing interest has been devoted to removal of atmospheric pollutant by non-thermal plasmas achieved using e-beams or pulsed discharges, for the nitrous oxides the so-called de-NO_X process, or for Volatils Organic Compounds, the so-called de-VOC process. However the physical and chemical mechanisms involved are not easy to understand: molecules or gas mixtures are quite complex, and the transient plasma created by the type of discharge often used, dielectric barrier or corona ones, is non homogeneous in space. In this paper is discussed some data needs for understanding of the NO-removal process and the destruction of some selected VOC molecules, TCE and TCA, by pulsed discharge plasmas. Some experimental studies performed to get insight into the discharge plasma kinetic involved in the pollutant removal are presented, in particular about the hydroxyl radical OH which play an important role in this kinetic.

INTRODUCTION

Researches on atmospheric pollutant removal using non-thermal plasmas are performed since beginning of the eighties [1, 2]. The electron beam technique has been first used and it has been shown that it is very efficient for destruction of most pollutants like nitrogen oxide or Volatile Organic Compounds (VOCs) [1-5]. Pulsed discharges are much well suited than electron beams for some industrial or domestic applications because there are easier to handle and of low energy cost. Thus a growing interest has developed for removal of pollutants using dielectric barrier (DBD) [6] or corona discharges [7]. Whatever the technique used in the cleaning process, the non-thermal plasma kinetic needs to be studied in detail for the understanding and the improvement of such a process. Usually this is done through comparison between experimental results on the pollutant removal and predictions obtained from the modelling of the electrically energised reactive mixture. This has been widely performed in studies about NO removal [8-16], and to a certain extend in studies about VOC removal [5, 17, 18]. It can bring new information about the plasma kinetic. However development of effective kinetic models, to be used for predictions [19-26], needs basic knowledge of rate constants for the various reactions in which the chemical species created in the plasma are involved.

Typically the discharge is running in humid air, or $N_2/O_2/H_2O$ with oxygen and water vapour concentration values less than 20% and 3% respectively, at near ambient temperature and at atmospheric pressure. This gas mixture contains one or several pollutant molecules at a low concentration value, of the order of 1000 ppm or less. In some cases, such as treatment of car engine exhaust gases in which NO has to be removed [27], the temperature of the mixture is of several hundred Kelvin before creation of the plasma. Moreover hydrocarbons are present and the H_2O concentration can be as high as 12%. The physical and chemical mechanisms involved in the pollutant removal are not easy to understand in such conditions: molecules and gas mixtures are quite complex, and the transient plasma created by the DBD or by the corona discharge is non homogeneous in space. This has motivated the use of the homogeneous photo-triggered discharge in order to achieve an efficient comparison between experimental results and predictions of a 0D self-consistent model taking into account all the necessary chemical reactions [13-16]. As for other types of pulsed discharges, the production rates of ions, molecular excited states, atoms and radicals by the discharge, as well as the rate constants for reactions of these species with each other and with molecules of the mixture, and the temperature dependence and these rates, are key parameters for understanding of the pollutant removal kinetics.

The goal of the present work is not to give an overall description of the kinetics involved in non-equilibrium plasmas used for atmospheric pollutant removal. The reader can refer to the comprehensive list of books and articles given here. Our objective is rather to put in light some open questions about such kinetics induced by pulsed discharges. The first part deals with the production of atoms, radicals, and excited states, through electron collisions on the gas mixture major components. Whatever is the pollutant to be removed in the gas stream, these species are initiators for the various chain reactions which take place during the afterglow and which induce the decay of the pollutant concentration at the exit of the plasma reactor. In the second part is discussed the NO removal kinetic in mixtures of the type $N_2/O_2/H_2O/HC/NO$, where HC is either ethene or propene. The third part is devoted to the removal kinetic of VOC, with emphasis on trichloroethylene and trichloroethane. Moreover the discussion is focussed on the neutral kinetic, i.e. reactions in which ions are involved are not considered here. Recent works have emphasized that combination of non-thermal plasmas with catalysis can be necessary to complete the cleaning process [28-34]. As a result knowledge about chemical reactions are needed for some of the secondary species which are created in the discharge afterglow and which could help in the catalytic activity. This is briefly discussed for NO removal.

PRODUCTION OF ATOMS, EXCITED STATES AND RADICALS THROUGH ELECTRON COLLISIONS

Apart from the various atomic or molecular ions and excited states, mainly N, O, $O(^1D)$, H atoms and OH radicals are produced through collisions of electrons on molecules during the discharge. Roughly speaking, production rates of these species depend on the shape of the electron energy distribution function, which is mainly governed by electron collisions on the nitrogen molecules for practical situations in non-thermal plasma reactor for pollution control. Many works have been performed on electron collision cross sections on N_2, O_2, and H_2O, see for example references [35-37], and such data are now well-known in particular for production of nitrogen and oxygen atoms, and metastable excited states of molecules. Production of OH radicals is still the subject of studies [38], and it appears that more experiments are necessary to validate predictions of models for the OH density growth during the development of the discharge. As it is discussed in the followings, this radical is very important for flue gases plasma treatment.

Collisions of electrons on VOC, in particular hydrocarbons, have been studied for determination of their ionisation cross section, and for identification of the type of ions produced [39, 40]. Electron attachment processes have been also the subject of a lot of researches [41]. However very few is known about dissociative processes giving neutral fragments. Such processes involved excited states of the molecules which can be created either through forbidden or through allowed transitions from the electronic ground state. Even if the pollutant concentration is low, it would be useful to determine the production rates of secondary neutral species through dissociative collisions because such species should initiate chain reactions leading to formation of others which could be useful for catalysis (alcohols, aldehides, ...). For example, it is noteworthy that few data exist for electron collisions on unsaturated hydrocarbons [42], which are present in car engine exhaust gases for which combination of non-thermal plasmas and catalysis appears very attractive to reduce NO emissions [28, 31].

REMOVAL OF NITROGEN OXIDE

The NO removal has been studied using DBD or corona [1, 2, 5-7, 8-11], photo-triggered [13-16] or microwave [12] discharges as far as non-equilibrium plasmas are concerned. The discharge duration is in the range 10-100 ns for the first three types, while it is higher than 1 µs for the last one. Mixtures studied are of the type $N_2/O_2/NO$ where the initial NO concentration does not exceed several thousands ppm, sometimes with addition of water vapour (up to 12% at a temperature higher than the ambient one) or hydrocarbons (at an initial concentration of the same order than that of NO).

Depending on the gas composition, with or without oxygen or water vapour, NO destruction follows in part from the well-known reduction process which involves N [15, 43], or oxidation processes which involve either O, ozone O_3, or peroxy radicals RO_2 which appear during the afterglow [19-23]. Ozone participates to oxidation at low gas temperature and hydrocarbon concentration values. Excited states of molecules are also produced in the plasma, among which nitrogen metastables play an important role in the kinetic of the reactive medium for the destruction of NO when the reduction channel dominates [14-16], i.e. in a mixture with very low percentage of O_2.

The Reduction Channel

Mainly two kinetic processes are involved in the NO destruction when there is no molecular oxygen in the mixture before the discharge is running, first of all the reduction reaction [43],

$$N + NO \rightarrow N_2 + O \tag{1}$$

and second the dissociative reaction,

$$N_2(a') + NO \rightarrow N_2 + N + O \tag{2}$$

where $N_2(a')$ denotes the singlet states $N_2(a'^1\Sigma)$, $N_2(a^1\Pi)$ and $N_2(w^1\Delta)$. Importance of reaction (2) has been emphasised in the photo-triggered discharge [15, 16]. Value of the rate constant is known [45]. However there is no study, to authors knowledge, specifically devoted to determine products of reaction (2). All what can be said is that the dissociative path is the most probable one because it is the only way to explain experimental results obtained with an homogeneous plasma. It will be now of interest to check the influence of reaction (2) on the NO removal using the electron beam technique, and also in models describing the streamer development and the NO destruction in and out of the streamer in corona and DBD discharges.

Addition of hydrocarbons or water vapour leads to decrease the NO removal owing to de-excitation of $N_2(a')$ on these molecules, i.e.

$$N_2(a') + C_2H_4 \rightarrow \text{products} \tag{3}$$

$$N_2(a') + H_2O \rightarrow \text{products} \tag{4}$$

These two reactions has been recently revealed and rate constant values are approximately identical: $k_3 = (4\pm2)\times10^{-10}$ cm^{3}s^{-1} [15] and $k_4=(3.0\pm1.5)\times10^{-10}$ cm^{3}s^{-1} [16], close to the rate for collisions of $N_2(a')$ with methane [45]. An open question is the nature of end products, which remain unknown at the present time, both for (3) and (4). Only one work has been performed to analyze the stabilized products in the N_2/C_2H_4 mixture using mass spectrometry [14], which seems to indicate that the reaction path for (3) is not a dissociative one. On the other hand data on reactions between excited states of N_2 with H_2O are somewhat scarce in the litterature [46, 47]. It is known that OH is produced when the metastable A state is involved, but there is no quantitative yield data [48]. Reaction (4) has never been studied up to now.

The Oxidation Channel

Addition of O_2 to the $N_2/H_2O/HC/NO$ mixture leads to enhance the NO removal owing to oxidation reactions [8-11]. These reactions predominate, with respect to the reduction one, as soon as the O_2 concentration increases above a value, typically between 0.1 and 1% for the photo-triggered discharge [14], which depends on parameters such as the hydrocarbon (HC) concentration, the deposited electrical energy in the mixture during the discharge, and the initial temperature of that mixture.

The oxidation of NO is, first of all, due to the production of O atoms by electron collisions on O_2, which leads to

$$O + NO + N_2 \rightarrow NO_2 + N_2 \tag{5}$$

But production of NO_2 is in part counterbalanced by a reduction process, i.e.

$$O + NO_2 \rightarrow NO + O_2 \tag{6}$$

Secondly, formation of ozone leads to

$$O_3 + NO \rightarrow NO_2 + O_2 \tag{7}$$

but the density of O_3 rapidly decreases as soon as the temperature increases owing to the decrease of the formation rate and owing to dissociative processes. This well-known kinetic [19-21] explains that, without HC in the mixture, the NO removal decreases when the temperature increases. This is completely different in presence of an unsaturated hydrocarbon like ethene or propene. In that case two mixture types can be considered, i.e. with or without moisture. In $N_2/O_2/HC/NO$, oxidation of the hydrocarbon take place, for example [49],

$$O + C_2H_4 \rightarrow H + CH_2CHO \tag{8a}$$
$$\rightarrow CH_3 + HCO \tag{8b}$$
$$\rightarrow H_2 + CH_2CO \tag{8c}$$
$$\rightarrow CH_2 + HCHO \tag{8d}$$

and thereafter a reaction chain follows which leads to production of OH and HO_2 radicals in the afterglow.

The radical HO_2 is a direct oxidising agent for NO,

$$NO + HO_2 \rightarrow NO_2 + OH \tag{9}$$

whereas OH reacts with hydrocarbon molecules and induces the formation of the peroxy radical HOC_2H_4OO via the following reaction chain,

$$OH + C_2H_4 + N_2 \rightarrow HOC_2H_4 + N_2 \tag{10}$$

$$HOC_2H_4 + O_2 \rightarrow HOC_2H_4OO \tag{11}$$

For propene the reaction path is very similar, but two peroxy radicals are produced, $CH_3CH(OH)CH_2OO$ and $CH_3CH(OO)CH_2OH$. Oxidation of NO to produce NO_2 follows,

$$RNO_2 + NO \rightarrow NO_2 + \text{products} \tag{12}$$

where RNO_2 is for the peroxy radicals. When H_2O is added to the $N_2/O_2/HC/NO$ mixture, OH is directly produced during the discharge through electron collisions on H_2O. This can have an effect on the NO removal only if the maximum density of OH produced by electron collisions is higher than the OH density produced by the neutral kinetic in the afterglow. It depends on values of the various parameters involved (initial molecule concentrations, deposited energy, excitation electric field, ...).

At the present time, the oxidation chain reaction described above is well-known and it has been demonstrated that it can explain experimental results on NO removal obtained with various discharge types [8-11, 22, 23]. With or without H_2O in the mixture, it is clear that addition reaction of OH on the hydrocarbon molecule (reaction 10 for C_2H_4) is a key step in the NO removal process. Owing to this reaction, the NO removal efficiency is greatly enhanced when the hydrocarbon is present in the mixture, in particular for a value of the initial temperature higher than the ambient one. As a typical example, Fig. 1 shows the computed NO concentration one second after the excitation current pulse (of duration 60 ns) of the photo-triggered discharge, for an initial value of 500 ppm and for two values of the initial temperature, 20°C and 300°C, plotted against the initial ethene concentration [14]; other conditions are given in the figure caption.

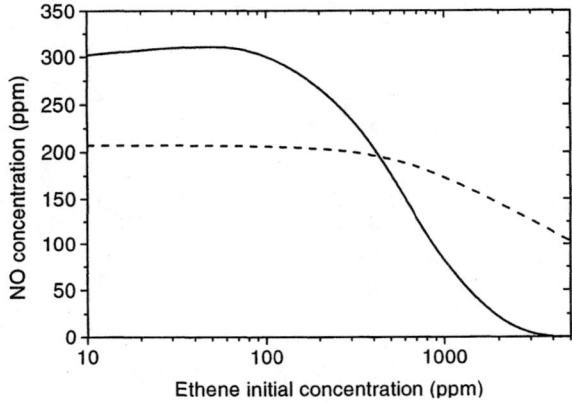

FIGURE 1. NO concentration one second after the photo-triggered discharge, as given by a self-consistent 0D kinetic model. Initial concentrations: $[O_2]$ = 10%, [NO] = 500 ppm. Mixture: $N_2/O_2/C_2H_4/NO$. Total pressure: 1 bar. Initial temperature: 20°C (dashed line) and 300°C (full line). Taken from [14].

However it is noteworthy that the temperature dependence of rate constants for the addition reaction of OH on hydrocarbons are known on a very limited range of temperature values. For ethene, the rate is known between 200 and 300 K [50], and for propene it is known between 250 and 425 K [51]. Experimental validation of the model used to obtain predictions plotted in Fig. 1 has been recently performed at the LPGP laboratory (France) for reactions of OH with C_2H_4 and C_3H_6 [52], through time resolved LIF measurement of the OH density decrease in the afterglow of the photo-triggered discharge in $N_2/H_2O/HC$ mixtures. In Fig. 2 is displayed some results obtained for ethene. As predicted by the model, the temperature of the gas mixture during LIF measurements, which are performed between 1 and 20 µs after the current pulse excitation, is roughly constant and equal to 380 K. Figure 2 emphasizes that the temperature dependence of the rate constant used [50] is also valid in conditions of this work. However it will be useful to have more information on such a rate, and on others for same type of reactions, for a temperature value as high as 600 K in order to achieve a complete validation of kinetic models used to predict NO removal in combustion effluents.

In case of mixtures with hydrocarbons, NO molecules are not only oxidise to NO_2 but also add to various radicals produced during the afterglow to form nitrosoethanal $NOCH_2CHO$ or other compounds such as CH_2OHNO, CH_3NO, ... Moreover some of the NO_2 molecules disappear to form nitro-alcane such as CH_3ONO, CH_3NO_2, CH_3ONO_2 , ... Reactivity of all these molecules with atoms and radicals are not well known at the present time, in particular at a temperature value higher than 300 K. Such data are needed to achieve a comprehensive understanding of the de-NO_X processes using non-thermal plasmas.

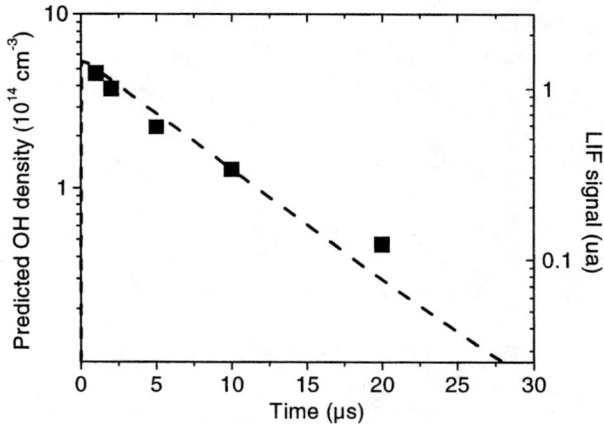

FIGURE 2. Time evolution of the OH density in the afterglow of the photo-triggered discharge. Symbols: LIF measurements. Dashed line: predictions of a self-consistent 0D kinetic model. Initial concentrations: $[H_2O] = 1.2\%$, $[C_2H_4] = 1000$ ppm. Mixture: $N_2/H_2O/C_2H_4$. Total pressure: 460 mbar. Initial temperature: 27°C. Deposited energy: 92 J/l.

Useful Species For Catalysis

Various types of chemical species are produced in the late afterglow of the pulsed discharge in the $N_2/O_2/H_2O/HC/NO$ mixture (HC: hydrocarbons), i.e. alcohols (CH_3OH, ...), aldehides (CH_2O, CH_3CHO, ...), and also molecules of the type RNO_X (nitro-alcane, ...). Some of these species should play a role on a catalytic surface placed after the plasma reactor in order to complete the deNO$_X$ process, i.e. the total reduction of NO and NO_2 to produce N_2 and O_2. In particular it has been recently shown that the plasma of a DBD, leading to formation of both NO_2 and $C_XH_YO_Z$ intermediate molecules, could substitute for the first two functions of a three function catalyst [31]. Optimisation of the catalyst formula should be achieved providing that the concentrations of the $C_XH_YO_Z$ compounds at the plasma reactor exit could be predicted with sufficient accuracy for the various parameters values involved (initial HC concentration, temperature of the gas mixture, electrical deposited energy, ...). It needs an accurate knowledge of the kinetics for these compounds, in particular at a temperature value higher than the ambient one.

REMOVAL OF VOLATILS ORGANIC COMPOUNDS

Research on the removal of VOC is still in its early stage as far as a whole understanding of the kinetic involved is concerned. A lot of experimental studies has been performed on numerous molecules [7], using corona or dielectric barrier discharges, but very few detailed kinetic models have been developed [25, 26] comparatively to studies on the NO removal. However some kinetic scheme have been proposed to explain experimental results in many publications, for example for chlorinated molecules [5, 18]. It appears that the OH radical play a very important role in the removal of this type of pollutant, as for the nitrogen oxide, owing to its very high activity in comparison with other species, for example O atoms.

OH is produced during the discharge, in great part through dissociative electron collisions on the water molecule. Hydrogen abstraction from H_2O by the oxygen excited states $O(^1D)$ can also participate to this production

$$O(^1D) + H_2O \rightarrow OH + OH \tag{13}$$

but several reactions with molecules of the mixture deplete this specie

$$O(^1D) + M \rightarrow O(^3P) + M \tag{14}$$

where M is either N_2, O_2, or H_2O. On the other hand, recent studies on OH production processes in the N_2/H_2O mixture [53] indicate that reaction (4), see the previous part, could be dissociative such as

$$N_2(a') + H_2O \rightarrow OH + H + N_2 \tag{15}$$

which emphasises that metastable states of N_2 can be also of importance in the OH production in mixtures without O_2, or with a low concentration of this molecule. This has to be more studied in the future, as well as interactions between metastable states and VOC.

As for other VOC such as aldehydes, the removal of trichloroethylene (TCE) is due to a series of chain reactions. Reactions with OH and O are the first step of this chain,

$$OH + C_2HCl_3 \rightarrow products \tag{16}$$

$$O + C_2HCl_3 \rightarrow products \tag{17}$$

Rates for these processes, at 25°C, are $k_{16} = 2.2 \times 10^{-12}$ cm^{3}s^{-1} and $k_{17} = 10^{-13}$ cm^{3}s^{-1} [25, 54]. As a result it can be conclude that TCE is mainly oxidised by OH, at condition that the O atom density created by the discharge is of the same order of magnitude than the OH density. Model predictions show that it is the case, for example, in the homogeneous phototriggered discharge [14]. However the rate for reaction (16) decreases when the temperature increases [54], whereas the temperature dependence for the rate of (17) is unknown. As a result the effect of the increase of the temperature on the relative importance of (16) and (17) on TCE removal can not be accurately predicted and experiments are necessary. At ambient temperature the influence, on TCE removal, of chemical species produced in a DBD has been for example studied in ref. [55]. It emphasises that it is easier to remove TCE in wet air than in dry air. The faster reaction of TCE with OH than with O explains the good efficiency of the removal process in presence of water. Other chlorinated species like trichloroethane (TCA) present prohibitive energy cost of removal compared to TCE [7]. Explanation should be found in different reactivities with OH. Indeed the rate for the reaction of TCA with this radical is about two orders of magnitude lower that the rate of reaction (16) at 25°C. Particular attention must be focussed on the breaking bond mechanisms by the mean of specific fundamental studies.

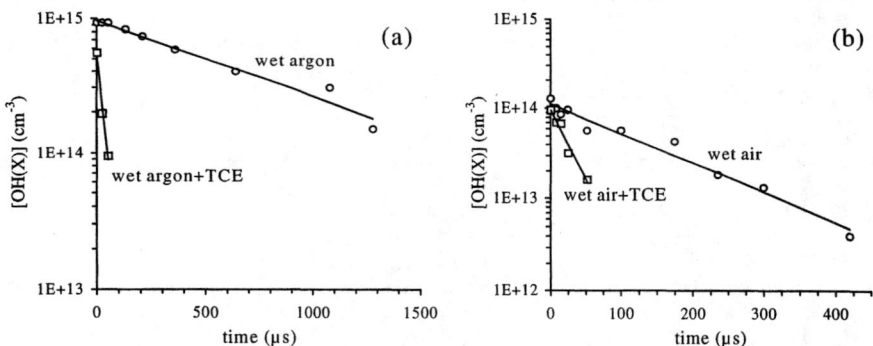

FIGURE 3. Time evolution of the average OH density in the afterglow of a Dielectric Barrier Discharge in wet argon (a) and wet air (b), with 0.27% of water vapour, and with and without 500 ppm of TCE, at atmospheric pressure [56].

Experiments specially dedicated to study elementary key processes involved in the VOC removal, such as the reaction of OH with the pollutant, are very important. As example, the rate for reaction of OH with TCE was estimated in mixtures of argon with water vapor and TCE, and next in mixtures of wet air and TCE [56]. This rate was deduced from the comparison between the OH density decay frequency with and without TCE, see Fig. 3. The values obtained were $(2.7 \pm 1.7) 10^{-12}$ cm^{3}s^{-1} and $(2.2 \pm 1.7) 10^{-12}$ cm^{3}s^{-1}, in argon and air respectively. These values are very close and in good agreement with values of literature [54]. This result shows the fundamental role played by the OH radical in the pollutant oxidation processes.

REFERENCES

1. Penetrante, B., and Schultheis, S. (eds), *Non-Thermal Plasma Techniques for Pollution Control*, NATO ASI Series Vol.G34, parts A and B, Berlin: Springer-Verlag, 1993.
2. Van Veldhuizen, E-M. (ed), *Electrical discharges for environmental purposes. Fundamentals and applications*, New York: NOVA Science Publishers, 2000.
3. Tokunaga, O., and Suzuki, N., *Radiat. Phys. Chem.* **24**, 145-165 (1984).
4. Vitale, S., Hadidi, K., Cohn, D., Bromberg, L., *Phys. Lett. A* **232**, 447-455 (1997).
5. Penetrante, B., Hsiao, M., Bardsley, J., Merritt, B., Vogtlin, G., Kuthi, A., Burkhart, C., Bayless, J., *Plasma Sources Sci. Technol.* **6**, 251-259 (1997).
6. Kogelschatz, U., Eliasson, B., and Egli, W., J. Phys. IV Colloque C4 **7**, 47-66 (1997).
7. Yan, K., van Heesch, E., Pemen, A., and Huijbrechts, P., *Plasma Chem. Plasma Proc.* **21**, 107-137 (2001).
8. Mok, Y., Ham, S., and Nam, I-S., *IEEE Trans. Plasma Sci.* **26**, 1566-1574 (1998).
9. Niessen, W., Wolf, O., Schruft, R ., and Neiger, M., *J. Phys. D: Appl. Phys.* **31**, 542-550 (1998).
10. Orlandini, I., and Riedel, U., *J. Phys. D: Appl. Phys.* **33**, 2467-2474 (2000).
11. Dorai, R., and Kushner, M., *J. Phys. D: Appl. Phys.* **34**, 574-583 (2001).
12. Baeva, M., Gier, H., Pott, A., Uhlenbusch, J., Höschele, J., and Steinwandel, J., *Plasma Chem. Plasma Proc.* **21**, 225-247 (2001).
13. Rozoy, M., Postel, C., and Puech, V., *Plasma Source Sci. Technol.* **8** 337-348 (1999).
14. Fresnet, F., PhD Thesis, Université d'Orsay (2001).
15. Fresnet, F., Baravian, G., Magne, L., Pasquiers, S., Postel, C., Puech, V., and Rousseau, A., *Appl. Phys. Lett.* **77**, 4118-4120 (2000).
16. Fresnet, F., Baravian, G., Magne, L., Pasquiers, S., Postel, C., Puech, V., and Rousseau, A., *Plasma Sources Sci. Technol.* **11**, 152-160 (2002).
17. Yamamoto, T., Chang, J-S., Berezin, A., Kohno, H., Honda, S., and Shibuya, A., *J. Adv. Oxid. Technol.* **1**, 67-78 (1996).
18. Falkenstein, Z., *J. Adv. Oxid. Technol.* **2**, 223-238 (1997).
19. Lowke, J., and Morrow, R., *IEEE Trans. Plasma Sci.* **23**, 661-671 (1995).
20. Gentile, A., and Kushner, M., *J. Appl. Phys.* **78**, 2074-2085 (1995).
21. Eichwald, O., Yousfi, M., Hennad, A., and Benabdessadok, M., *J. Appl. Phys.* **82**, 4781-4794 (1997).
22. Dorai, R., and Kushner, M., *J. Appl. Phys.* **88**, 3739-3747 (2000).
23. Filimonova, E., Amirov, R., Kim, H. and Park, I., *J. Phys. D: Appl. Phys.* **33**, 1716-1727 (2000).
24. Storch, D., and Kushner, M., *J. Appl. Phys.* **73**, 51-55 (1993).
25. Evans, D., Rosocha, L., Anderson, G., Coogan, J., and Kushner, M., *J. Appl. Phys.* **74**, 5378-5386 (1993).
26. Gentile, A., and Kushner, M., *J. Appl. Phys.* **78**, 2977-2980 (1995).
27. Hammer, T., Proceedings of the International Symposium on High Pressure Low Temperature Plasma Chemistry, Greifswald, Germany, 2000, pp.234-241.
28. Shimizu, K., and Oda, T., *IEEE Trans. Plasma Sci.* **35**, 1311-1317 (1999).
29. Bröer, S., and Hammer, T., *Appl. Catal. B: Environ.* **28**, 101-111 (2000).
30. Kim, H., Takashima, K., Katsura, S., and Mizuno, A., *J. Phys. D: Appl. Phys.* **34**, 604-613 (2001).
31. Gorce, O., Jurado, H., Thomas, C., Djega-Mariadassou, G., Khacef, A., Cormier, J-M., Pouvesle, J-M., Blanchard, G., Calvo, S., and Lendresse, Y., SAE Paper 2001-01-3508 (2001).
32. Yamamoto, T., Mizuno, K., Tamori, I., Ogata, A., Nifuku, M., Michalska, M., and Prieto, G., *IEEE Trans. Ind. Appl.* **32**, 100-105 (1996).
33. Mizumo, A., Kisanuki, Y., Noguchi, M., Katsura, S., Lee, S., Hong, Y., Shin, S., and Kang, J., *IEEE Trans. Ind. Appl.* **35**, 1284-1288 (1999).
34. Kang, M., Kim B-J., Cho, S., Chung, C-H., Kim, B-W., Han, G., and Yoon, K., *J. Mol. Catal. A: Chem.* **180**, 125-132 (2002).
35. Pitchford, L., and Phelps, A., Technical Report n°26, JILA Information Center, University of Colorado, Boulder, Colorado, USA (1996).
36. Itikawa, Y., *J. Phys. Chem. Ref. Data* **18**, 23-42 (1989).
37. Yousfi, M., Benabdessadok, M., *J. Appl. Phys.* **80**, 6619-6630 (1996).

38. Harb, T., Kedzierski, W., and McConkey, J., *J. Chem. Phys.* **115**, 5507-5512 (2001).
39. Hwang, W., Kim, Y-K., Rudd, M., *J. Chem. Phys.* **104**, 2956-2966 (1996).
40. NIST Standard Reference Database Number 69 - July 2001.
41. Christophorou, L., *Contrib. Plasma Phys.* **27**, 237-281 (1987).
42. Fresnet, F., Pasquiers, S., Postel, C., and Puech, V., *J. Phys. D: Appl. Phys.* **35**, 882-890 (2002).
43. Penetrante, B., Hsiao, M., Merritt, B., Vogtlin, G., and Wallman, P., *IEEE Trans. Ind. Appl.* **23**, 679-687 (1995).
44. Atkinson, R., Baulch, D., Cox, R., Hampson Jr., R., Kerr, J., and Troe, J., *J. Phys. Chem. Ref. Data* **21**, 1125-1147 (1992).
45. Piper, L., *J. Chem. Phys.* **87**, 1625-1629 (1987).
46. Callear, A., and Wood, P., *Trans. Faraday Soc.* **71**, 598 (1971).
47. Golde, M., Ho, G., Tao, W., and Thomas, J., *J. Phys. Chem.* **93**, 1112-1118 (1989).
48. Herron, J., *J. Phys. Chem. Ref. Data* **28**, 1453-1483 (1999).
49. Baulch D., Cobos, C., Cox, R., Frank, P., Hayman, G., Just, Th., Kerr, J., Murrells, T., Pilling, M., Troe, J., Walker, R., and Warnatz, J., *J. Phys. Chem. Ref. Data* **24**, 1609-1630 (1995).
50. Atkinson, R., Baulch, D., Cox, R., Hampson Jr., R., Kerr, J., Rossi, M., and Troe, J., *J. Phys. Chem. Ref. Data* **28**, 191-393 (1999).
51. Atkinson, R., *J. Phys. Chem. Ref. Data* **26**, 215-290 (1997).
52. Baravian, G., Fresnet, F., Magne, L., Pasquiers, S., Postel, C., Puech, V., Rousseau, A., Proceedings of the 54th Gaseous Electronics Conference, State College, PN, USA (9-12 Octobre 2001).
53. Fresnet, F., Baravian, G., Magne, L., Pasquiers, S., Postel, C., Puech, V., Rousseau, A., to be presented at XIV International Conference on Gas Discharges and Their Applications, Liverpool, UK, 1-6 Sept. 2002.
54. Atkinson, R., Baulch, D., Cox, R., Hampson Jr., R., Kerr, J., Rossi, M., and Troe, J., *J. Phys. Chem. Ref. Data* **26**, 521-1011 (1997).
55. Gaurand, I., PhD Thesis, Université d'Orléans (2000).
56. Hibert, C., Gaurand, I., Motret, O., and Pouvesle, J-M., *J. Appl. Phys.* **85**, 7070-7075 (1999).

E. Astrophysics

The Effect of High-Lying Configurations and Ionization and Recombination Processes on Analyses of Solar and Stellar Coronal Spectra

Rami Doron[*†], Ehud Behar[¶], George A. Doschek[†] and Uri Feldman[†]

Institute for Computational Sciences and Informatics, George Mason University, Fairfax, VA 22030
[†]*E.O. Hulburt Center for Space Research, Naval Research Laboratory, Washington, DC 20375*
[¶]*Columbia Astrophysics Laboratory and Department of Physics, Columbia University, New York, NY 10027*

Abstract. This work addresses two topics important for the appropriate interpretation of astrophysical spectra. The first is the effect of high-lying levels on the atomic models and the second is the importance of ionization and recombination processes in forming line emission. In the first part of the work we study the influence of high-lying configurations on the calculated intensities of UV lines, particularly of O-like ions, observed by the Solar Ultraviolet Measurements of Emitted Radiation (SUMER) spectrometer aboard the SOHO satellite. The high-lying configurations alter the line intensities through radiative cascades and configuration interaction effects. We find that cascades can significantly enhance the intensities of some lines of the considered ions by up to 65% at temperatures of the ion maximum fractional abundance. The enhancement due to cascades increases with increasing temperature and charge state. The configuration mixing effects can either enhance or reduce the line intensities. In a second study, we calculate the theoretical intensities of the soft X-ray Fe^{16+} lines arising from $2l$-$3l'$ transitions using a three-ion collisional-radiative model that includes the contribution of recombination and ionization processes to line formation. Dielectronic recombination is found to be particularly important. The newly calculated line intensities can explain the high values of the 2p-3s / 2p-3d intensity ratios, which are often obtained in astrophysical observations. Observed intensity ratios among the 2p-3s lines are also better reproduced.

INTRODUCTION

Spectroscopic analysis is a primary tool in astrophysical research. Recent missions such as the Solar and Heliospheric Observatory (SOHO), Chandra, or XMM-Newton, provide a wealth of high quality new spectroscopic data. Using spectral line intensities and profiles, one can obtain the most fundamental properties of the plasma: ionization state, electron temperature (T_e), electron density (n_e), elemental abundances, emission measure, and non-thermal motions. Information can also be obtained by comparing images recorded using filters sensitive to different wavelength bands, corresponding to different temperature regimes. The level of accuracy to which the plasma properties are determined by each method, can sometimes hold the key for the appropriate interpretation of the observation. Good examples of such cases can be found in the works by Flower & Jordan [1], Feldman et al. [2], or Pinfield et al. [3]. In the first two papers above, discrepancies between electron temperature derived by temperature sensitive line intensity ratios and temperature derived from ionization balance considerations, are suggested to be an indication of the transient nature of the plasma. In the third paper, an enhanced line arising from the decay of a relatively high-lying level in Si III is interpreted as an evidence for the presence of non-thermal hot electrons. It is therefore desirable to refine the atomic models as much as possible.

The errors in the theoretical calculations basically emerge from two sources. The first source is inaccuracies in the basic atomic quantities (i.e., energy levels, radiative rates, collisional cross-sections etc.). The second source of error is the simplicity or the invalidity of the atomic models that are being used for deriving the line intensities. In the present work, we address two aspects of the second source of errors, associated with the expansion of atomic models.

The first aspect is the effect of inclusion of high-lying configurations in the atomic models. Radiative cascades from high-lying levels are already proven to be important populating mechanism in several atomic systems. (e.g., in Fe XVII [4] and Be-like ions [5]). The effect of including high-lying configurations in the atomic model is not limited to that resulting from radiative cascades. Introducing additional configurations into the atomic model, in the stage in which the raw atomic data are generated, might also significantly alter the basic atomic quantities due to configuration interaction (mixing) effects. In section 2, both effects of radiative cascades and configuration interaction involving high-lying configurations are demonstrated through the example of UV lines of O-like ions observed by the Solar Ultraviolet Measurements of Emitted Radiation (SUMER) spectrometer [6] aboard the SOHO satellite.

The second aspect discussed here concerns the possible influence of ionization and recombination processes on line intensities. In the commonly used atomic model for deriving line intensities relevant to solar-like coronae, it is assumed that the energy level populations are mainly determined by atomic processes that occur within each ionization state. In Section 3, we present a model that takes into account ionization and recombination processes involving neighboring ionization states. The model is then applied to calculate the X-ray emission of the astrophysically important Ne-like iron ion (Fe^{16+}).

THE EFFECT OF HIGH-LYING LEVELS ON SPECTROSCOPIC ANALYSES OF THE SOLAR UV SPECTRA

The SUMER spectrometer provides a wealth of solar UV spectral features in the 500-1600 Å wavelength range. In particular, this spectral range contains lines associated with ions that have their maximum abundance at the temperature range $10^5 - 10^6$ K. The SUMER has a spectral dispersion of ~44 mA pixel^{-1} and a spatial resolution along its slit direction of 1 arcsec that corresponds to ~700 km on the sun. More information on SUMER can be found in Refs. [6, 7, 8].

UV transitions of O-like ions observed in the SUMER spectra provide a good example for the importance of considering high-lying configurations in the atomic models.

Theoretical Method

A collisional-radiative model that takes into account all collisional excitations and de-excitations and radiative decays among the levels of each ionization state is employed to determine the level populations. Assuming a steady-state plasma, the corresponding set of rate equations for the density (population) n_j^{q+} of an ion with charge $q+$ in a level j can be written as:

$$\frac{dn_j^{q+}}{dt} = n_e \sum_{k \neq j} n_k^{q+} Q_{kj}(T_e) + \sum_{k>j} n_k^{q+} A_{kj} - n_j^{q+} \left(n_e \sum_{k \neq j} Q_{jk}(T_e) + \sum_{k<j} A_{jk} \right) = 0 \quad (1)$$

The terms $Q(T_e)$ represent the rate coefficients for collisional excitations or deexcitations and the terms A denote the rate coefficients for spontaneous radiative decays. It should be noted that resonant excitations, i.e., an electron capture into an autoionizing level followed by autoionization to an excited level, are not considered here. The set of equations (1) is then solved using the normalization condition $\sum_j n_j^{q+} = 1$.

All the basic atomic quantities used in this work were generated by means of the multiconfiguration, relativistic HULLAC (Hebrew University Lawrence Livermore Atomic Code) computer package developed by Bar-Shalom, Klapisch, and Oreg [9]. Using this efficient code we are able to keep adding higher configurations to the atomic model and systematically study the accumulating effect. By implementing atomic data generated with configuration mixing with high-lying configurations in a collisional radiative model that includes only the low-lying levels, we are able to distinguish between the cascades and mixing effects.

UV Lines of O-like Ions

Numerous UV lines from several ions belonging to the O I sequence (ground configuration $2s^22p^4$) are clearly observed in the SUMER spectra. These line arise mainly from three types of transitions, $2s^22p^33l$ - $2s^22p^33l'$ (l=s, p; l'=p, d), resonance transitions $2s^22p^4$ - $2s2p^5$, and forbidden transitions within the ground configuration. As a basic atomic model for O-like ions we consider an 86-level model comprised of the following configurations: $2s^22p^4$, $2s2p^5$, $2p^6$, $2s^22p^33l$ (l=s, p, d). Calculations are first performed for O-like S (S^{8+}) in a temperature of kT_e=90 eV, which corresponds to the temperature of maximum ionization abundance (T_e^{max}) under coronal conditions [10], and a typical coronal electron density of n_e=10^9 cm^{-3}.

By including the $2s2p^43l$ configurations (additional 140 levels) we allow cascades towards the $2s^22p^33l$ levels through the relatively strong 2s-2p transitions, which subsequently increase the intensity of some of the $3l$-$3l'$ transitions by up to 30%. Adding the $2s2p^44l$ and the $2s^22p^34l$ configurations enhances these line intensities by up to an additional 5% and 15%, respectively. Configurations from the n=5 shell could contribute up to 5% to the $3l$-$3l'$ line transitions. The contribution from configurations with $n=6$ or above, are very small. It is important to note that the resonance transitions $2s^22p^4 - 2s2p^5$, as well as the transitions within the ground configuration, are hardly affected by cascades.

We have also studied the behavior of the cascades effect as a function of temperature and Z. In Fig. 1 we show the cascades effect in a model that includes configurations up to $n=5$ on one of the strongest $3l$-$3l'$ transitions; the $2s^22p^33s\ ^3D_2$ - $2s^22p^33p\ ^3F_3$ (at 882.68 Å in S^{8+}), as a function of the electron temperature, normalized to the ionization energy. The results are given for O-like Si, S, and Ca ions. The small vertical segments indicate the relevant temperature of maximum abundance for each ion, assuming equilibrium in coronal conditions. As expected, for each of the ions the effect increases with increasing temperature. It can further be seen that at T_e^{max} for each ion, the effect increases with Z. The increase in intensity rises from ~48% in Si^{6+}, to ~64% in S^{8+}, and to ~87% in Ca^{12+}. This isoelectronic trend is also found in other isoelectronic sequences from LiI to NaI. In some cases it was found that for the same *normalized* temperature the cascades effect decreases with Z; however at the most relevant temperature T_e^{max} of each ion, the effect still increases with increasing Z. It would therefore be more accurate to say that the *effective* cascades effect increases with increasing Z.

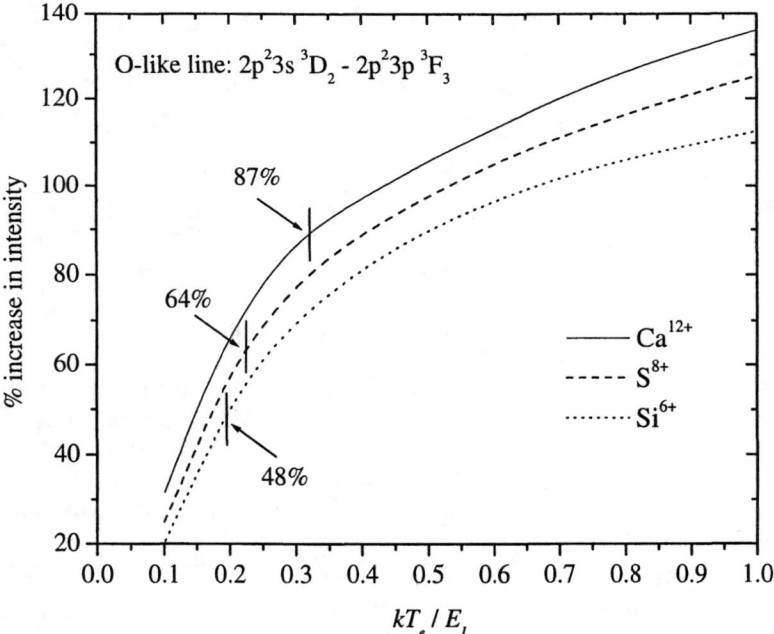

FIGURE 1. Increase in calculated intensity of the $2s^22p^33s\ ^3D_2$ - $2s^22p^33p\ ^3F_3$ O-like line due to cascades in a model that includes configurations up to $n=5$ as a function of the normalized temperature. The vertical segments indicate the temperature of maximum abundance in ionization equilibrium. The percent increases are relative to the $n=2$ basic model.

Interestingly, when the configuration interaction between the basic model configurations and the 2s2p^{4}3l configurations is included, the intensity of some transitions diminishes. Indeed, the corresponding lines of these transitions are not identified in the observations. The mixing effect increases the intensities of other transitions by 20-50%. We find that the mixing with the 2s2p^{4}3l configurations also significantly improves the calculated wavelengths. As in the case of the cascades effect, both resonance 2s-2p transitions and those arising from forbidden transitions within the ground configuration are only marginally affected (by 8% at the most for S^{8+}) by mixing effects.

Kink et al. [11] used the intensity ratios between some of the 3l-3l' transitions and the forbidden transition within the ground configuration (2s^{2}2p^4 ^{3}P$_1$ - 2s^{2}2p^4 ^{1}S$_0$) in O-like Si and S in order to estimate the temperature in different regions of the solar corona. In their analysis they used HULLAC calculations that include the 9 configurations comprising the *basic* model used here. Their results indicate a much higher temperature than expected from coronal equilibrium conditions. Specifically, for O-like S the obtained temperature was 3-4 times T_e^{max}(S^{8+}). The present calculations that include higher configurations result in an intensity increase of many of the 3l-3l' transitions by more than 30% (while leaving the forbidden transition unchanged) and suggest much lower temperatures, closer to T_e^{max}(S^{8+}).

THE ROLE OF IONIZATION AND RECOMBINATION PROCESSES IN FORMING THE Fe^{16+} LINE EMISSION

Soft X-ray lines from Fe^{16+} ions are prominent in many spectra of hot astrophysical objects. This line emission mostly arises from 2s^{2}2p^6-2s^{2}2p^{5}3s and 2s^{2}2p^6-2s^{2}2p^{5}3d transitions. Even though Fe^{16+} is one of the most extensively investigated spectroscopic systems in astrophysics, the interpretation of the relative observed intensities of the above transitions is still challenging. One of the longstanding problems associated with this system is the relative intensity of the 2p-3s lines that often appear to be enhanced relative to theoretical predictions, when compared to the intensities of the 2p-3d lines. In an effort to improve the existing atomic models, which usually neglect atomic processes that involve neighboring ions, in this work we consider a 3-ion collisional-radiative model that includes levels of the F-like (Fe^{17+}), Ne-like (Fe^{16+}), and Na-like Fe (Fe^{15+}) ions. Such a model allows one to consider collisional excitations as well as the effect of recombination and ionization simultaneously.

Theoretical Method

The set of rate equations include all electron impact excitations and de-excitations, as well as radiative transitions, among levels belonging to each of the 3 ionization states. In addition, we consider the following ionization and recombination processes that couple the different ionization states: Collisional ionization from Fe^{15+} and Fe^{16+}, autoionization from Fe^{16+}, radiative recombination from Fe^{16+} and Fe^{17+}, and finally dielectronic recombination from Fe^{17+}. For Fe^{16+}, we also consider resonance excitations (de-excitations) from (to) the ground state to (from) the $2p^53l$ (l=s, p) levels. The rates for the resonance excitations are incorporated from Chen & Reed [12].

In a straightforward, all-inclusive 3-ion model, in which rate equations for the three ionization states are solved simultaneously, the fractional ion abundances are a self-consistent result. This would clearly be erroneous since in order to obtain correct results for the ionization balance, one must include more than three ionization states, as well as numerous levels for each charge state, which makes this approach impractical. In order to circumvent this problem, we solve a set of rate equations by imposing the following two constraints:

$$\sum_{j'} n_{j'}^{(q-1)+} / \sum_j n_j^{q+} = f_{q-1}/f_q, \quad \sum_j n_j^{q+} / \sum_{j''} n_{j''}^{(q+1)+} = f_q/f_{q+1} \quad (2)$$

where f_q, f_{q-1}, and f_{q+1} represent the relevant ionization fractions, and are calculated independently. In other words, we impose the ionization balance and only seek for the consequential population distribution among the individual levels. We are primarily interested in levels j of the ion with charge $q+$. In the coronal approximation collisional transitions among excited levels can be neglected. For Fe^{16+}, this approximation holds for $n_e \leq 10^{12}$ cm^{-3}. Additional information on the 3-ion model can be found in Ref. [13].

For Fe^{16+} we include in the model the configurations $2s^22p^6$, $2s^22p^5nl$, $2s2p^6nl$ (n=3 to 7), $2s^22p^43l3l'$, $2s2p^53l3l'$, $2s^22p^43l4d$, $2s^22p^43d4l$. For Fe^{17+}, the configurations $2s^22p^5$ and $2s^22p^43l$ are taken into account. The ground configuration $2s^22p^63s$ of Fe^{15+} is also included. All of these configurations add up to 2111 levels. The doubly excited configurations included for Fe^{16+} are expected to produce the majority of the DR effect [14].

Line Intensity Ratios in Fe^{16+}

Dielectronic recombination (DR) and resonant excitation (RE) are found to be very important in the relevant temperature range 300-600 eV. They contribute ~40% to the 2p-3s line emission. Collisional ionization and radiative recombination contribute only at a level of several percent. At relatively lower temperatures the RE effect dominates. However, as the temperature rises and the Fe^{17+} abundance increases, DR takes over. The large impact of the additional atomic processes considered in the 3-ion model directly affects the line ratios in this system. In Figure 2 we present the calculated intensity ratio of the sum of the three 2p-3s lines at 16.78, 17.05, and 17.1 Å, to the strongest 2p-3d line at 15.01 Å as a function of T_e, using two different models. The solid curve represents the results of the current 3-ion model, whereas the dashed curve represents the results of a traditional single-ion model that includes direct excitations and subsequent radiative decays only. It can be seen in the figure that the additional atomic processes, considered in the 3-ion model, significantly enhance the relative overall intensity of the 2p-3s lines. These new higher ratios are much closer to those observed in many astrophysical sources.

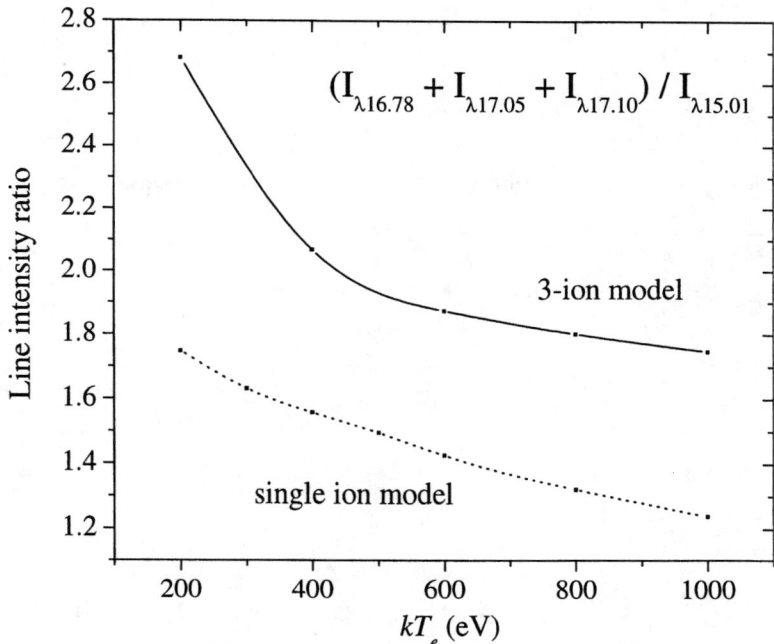

FIGURE 2. Intensity ratios of the three 2p - 3s lines at 16.78, 17.05 and, 17.10 Å to the 2p - 3d resonant line at 15.01 Å. The solid and dotted curves represent the results of the 3-ion and the single-ion model, respectively.

Among the 2p-3s lines, we find that the calculated intensity of the line at 17.10 Å is the most affected by the 3-ion model. Therefore, it is also interesting to compare theoretical ratios among the 2p-3s lines with observations. In Figure 3 we show the ratio $I_{\lambda 17.05}/I_{\lambda 17.10}$ as a function of temperature calculated by the 3-ion model and by a single-ion model. The figure shows that the ratio in each model has totally different temperature sensitivity, due to the fact that DR favors the 17.10 Å line. Obviously, it is impossible to have a similar plot for observational data since the exact temperature is unknown. However, it is possible to plot the observed $I_{\lambda 17.05}/I_{\lambda 17.10}$ ratio as a function of another temperature-sensitive ratio that would serve as a rough temperature indicator. Indeed, Rugge and McKenzie [15] plotted the ratio $I_{\lambda 17.05}/I_{\lambda 17.10}$ obtained from a series of observations on solar flares and active regions as a function of the ratio $I_{\lambda 14.2}/I_{\lambda 15.01}$; the line at 14.2 Å being the 2p-3d transition of Fe^{17+}. This latter ratio is an indication of the temperature and, even more directly, an indication of the ionic abundance ratio Fe^{17+}/Fe^{16+}. The observed ratios of $I_{\lambda 17.05}/I_{\lambda 17.10}$ presented in Fig. 5 of Ref. [15] are in much better agreement with the theoretical curve obtained by the 3-ion model than with that of the single-ion model. We suggest interpreting this result as evidence for the importance of the dielectronic recombination processes onto Fe^{16+} that correlates with the increasing abundance of Fe^{17+}.

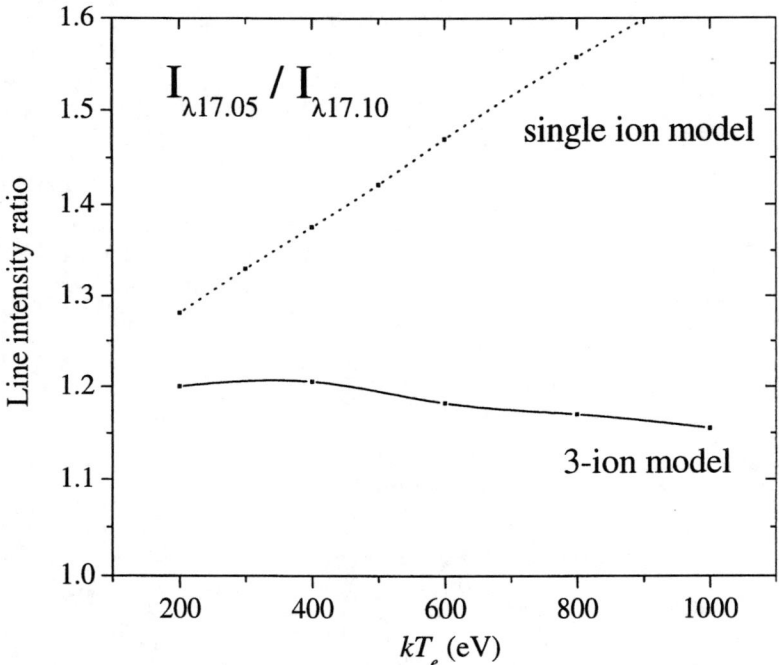

FIGURE 3. Intensity ratio of the 2p $^1S_0$ - 3s $^3P_2$ line at 17.10 Å to the 2p $^1S_0$ - 3s $^1P_1$ line at 17.05 Å. The solid and dotted curves represent the results of the 3-ion and the single-ion model, respectively. The solid curve is in much better agreement with solar observations [15].

CONCLUSIONS

We investigate the effect of including high-lying configurations in atomic models used to derive line intensities in the UV range observed in the solar SUMER spectra. The study mainly focuses on O-like ions with ground configuration $2s^2 2p^4$. The UV transitions in this sequence can be divided into three groups: Resonance transitions 2s-2p, forbidden transitions within the ground configuration, and 3l-3l' transitions. We find that cascades could enhance some of the 3l-3l' transitions in S^{8+} at T_e^{max} by as much as ~60%. The importance of cascades generally increases with increasing charge state. The mixing effects with the high-lying configurations can either enhance or reduce the calculated line intensities. Resonance transitions, as well as the forbidden transitions within the ground configurations are not significantly affected. Similar behavior is found also in other isoelectronic sequences from Li I to Na I.

A 3-ion model that takes into account also ionization and recombination processes is constructed. Simulation of the soft X-ray line emission of Fe^{16+} using this model enables to reproduce the systematically high values of the 2p-3s / 2p-3d ratios, observed in many astrophysical sources. The intensity ratios among the 2p-3s lines are also better reproduced. Recombination processes are found to be much more important than ionization in forming the Fe^{16+} line emission. Recombination processes are most likely to play even a more important role in higher charge states.

ACKNOWLEDGMENTS

The NRL portion of the work was supported by a NASA Solar Physics Guest Investigator Grant S137816.

REFERENCES

1. Flower, D.R., and Jordan, C., *A&A* **14**, 473 (1971).
2. Feldman, U., Laming, J.M., Mandelbaum. P., Goldstein, W.H., and Osterheld, A.L., *ApJ*, **398**, 692 (1992).
3. Pinfield, D. J., Keenan, F. P., Mathioudakis, M., Phillips, K. J. H., Curdt, W., and Wilhelm, H., *ApJ*, **527**, 1000 (1999).
4. Liedahl, D.A., "The completeness Criterion in Atomic Modeling" in *Atomic Data Needs for X-ray Astronomy*, Edited by M.A. Bautista, T.R. Kallman, and A.K. Pradhan, NASA/CP-2000-209968, P.151 (2000).
5. Landi, E., Doron, R., Feldman, U., & Doschek, G.A., *ApJ*, **556**, 912 (2001).
6. Wilhelm, K., et al., *Sol. Phys.*, **162**, 189 (1995).
7. Wilhelm, K., et al., *Sol. Phys.*, **170**, 75 (1997).
8. Lemaire, P., et al., *Sol. Phys.*, **170**, 105 (1997).
9. Bar-Shalom, A., Klapisch, M., and Oreg, J., *J. Quant. Spectros. Radiat. Transf.*, **71/2-6**, 169 (2001).
10. Mazzotta, P., Mazzitelli, G., Colafrancesco, S., and Vittorio, N., *A&AS*, **133**, 403 (1998).
11. Kink, I., Jupén, C., Engström, L., Feldman, U., Laming, J.M., and Schühle, U., *ApJ*, **487**, 956 (1997).
12. Chen, M.H., and Reed, K.J., *Phys. Rev. A*, **40**, 2292 (1989).
13. Doron, R., and Behar, E., *ApJ* **574**, in press (July 2002).
14. Chen, M.H., *Phys. Rev. A*, **38**, 2332 (1988)
15. Rugge, H.R., and McKenzie, D.L., *ApJ*, **297**, 338 (1985)

Atomic and Molecular Data Needs for Astrophysics

Robert L. Kurucz

Harvard-Smithsonian Center for Astrophysics, Cambridge, MA 02138, U.S.A.

Abstract. We need a list of all the energy levels of all atoms and molecules that matter (qualifiers below). Except for the simplest species, it is impossible to generate accurate energy levels or wavelengths theoretically. They must be measured in the laboratory. From the list of energy levels can be generated all the lines. Given the low accuracy required, 1 - 10%, all the other data we need can eventually be computed or measured. With the energy levels and line positions known, one can measure gf values, lifetimes, damping, or one can determine a theoretical or semiempirical Hamiltonian whose eigenvalues and eigenvectors produce a good match to the observed data, and that can then be used to generate additional radiative and collisional data for atoms or molecules.

For atoms and ions, we need all levels, including hyperfine and isotopic splittings, for $n \leq 9$ below the lowest ionization limit and as much as practicable above. Lifetimes and damping constants depend on sums over the levels. Inside stars there are thermal and density cutoffs that limit the number of levels, but in circumstellar, interstellar, and intergalactic space, photoionization and recombination can populate high levels, even for high ions. We need all stages of ionization for elements at least up through Zn.

In the sun there are unidentified asymmetric triangular features that are unresolved multiplets of light elements with $n \leq 20$. Simple spectra should be analyzed up to $n = 20$. Levels that connect to the ground or to low levels should be measured to high n, say $n = 80$. The high levels are necessary to match line series merging into continua.

All the magnetic dipole, electric quadrupole, and maybe higher-pole, forbidden lines are required as well. Most of the universe is low density plasma or gas. If the Hamiltonian is well determined, forbidden lines should be reliably computable.

For molecules, we need all levels below the first dissociation limit and as much as is practicable above, especially levels of all states that connect to the ground state. Stars populate levels to high V and to high J. In the sun there are many broad bumpy features that are molecular bands that are not in the line lists. For the cooler stars we need all the diatomics among all the abundant elements, and, essentially, the hydrides and oxides for all elements (especially ScO, TiO, VO, YO, ZrO, LaO). For M stars triatomics also become important. Much more laboratory and computational work is needed for H_2O. In the brown dwarfs and "planets" methane is important and it needs more laboratory and computational work.

We can produce more science by investing in laboratory spectroscopy rather than by building giant telescopes that collect masses of data that cannot be correctly interpreted.

INTRODUCTION

Astrophysicists work on "Important", "Big" problems and they think that the basic physics that they require to solve their problems has already been done, or, if it has not been done, it is easy and can be readily produced, as opposed to the hard problems they are working on. They have it backward. Getting the basic data is the hard part. When all the basic physics is known, pushing the "state-of-the-art" becomes straightforward.

Half the lines in the solar spectrum are not identified. All the features are blended. Most features have unidentified components that make it difficult to treat any of the identified components in the blend. And even the known lines have hyperfine and isotopic splittings that have not yet been measured. Is an asymmetry produced by a splitting. or by a velocity field, or both? It is very difficult to determine abundances, or any property, reliably when you do not know what you are working with.

For planetary and telluric atmosphere projects the solar irradiance spectrum is required as the input at the top of the atmosphere. It has never been observed. People ask me to compute it. I can compute it theoretically using both known and predicted lines and get agreement averaged over a nanometer but there is no way to predict the resolved spectrum when only half the lines are known. In other stars the situation is worse because the signal-to-noise and resolution of the observations are worse. Logically one has to know a priori what is in the spectrum in order to interpret it; there is not enough information in the observed spectrum itself.

Basically we need a list of all the energy levels of all atoms and molecules that matter (qualifiers below). From that list can be generated all the lines. With the energy levels and line positions known, one can measure gf values, lifetimes, damping, or one can determine a theoretical or semiempirical Hamiltonian whose eigenvalues and eigenvectors produce a good match to the observed data, and that can then be used to generate additional radiative and collisional data for atoms or molecules.

For atoms and ions, we need all levels, including hyperfine and isotopic splittings, for $n \leq 9$ below the lowest ionization limit and as much as practicable above. This is the only element that does not have splitting. Lifetimes and damping constants depend on sums over the levels. Inside stars there are thermal and density cutoffs that limit the number of levels, but in circumstellar, interstellar, and intergalactic space, photoionization and recombination can populate high levels, even for high ions.

One very important problem is diffusion of heavy elements inside stars because it changes the density and reaction rates. The radiative acceleration is computed by integrating over the line spectrum. At the surface some elements can be enhanced by a factor of 10^4. If the diffusion is deep inside the star, spectra for high stages of ionization are required.

In the sun I see unidentified asymmetric triangular features that are unresolved multiplets of light elements with $n \leq 20$. Simple spectra should be analyzed up to n = 20. Levels that connect to the ground or to low levels should be measured to high n, say n = 80. The high levels are necessary to match line series merging into continua.

All the magnetic dipole, electric quadrupole, and maybe higher-pole, forbidden lines are required as well. Most of the universe is low density plasma or gas. If the Hamiltonian is well determined, forbidden lines should be reliably computable.

For molecules, we need all levels below the first dissociation limit and as much as is practicable above, especially levels of all states that connect to the ground state. Except for H_2(BX,CX), far ultraviolet bands have been ignored unless they appear as interstellar lines. We see H_2 lines in stars as hot as 8000K when the stars have low metal abundances so that the lines are not masked.

In the sun we see, and have linelists for, C_2(AX,ba,da,ea), CN(AX,BX), CO(AX,XX), H_2(BX,CX), CH(AX,BX,CX), NH(AX,ca), OH(AX,XX), MgH(AX,BX), SiH(AX), SIO(AX,EX,XX). The isotopomers are included. Some stellar spectroscopists have more recent linelists than I do. Mine are based on old laboratory data and were computed with rotationless RKR potentials. They all have to be brought up to date, or even further improved, and expanded to higher V and J levels. In many cases there are new analyses based on FTS spectra. Ions and a few minor molecules have to be added to the linelist as well. In the sun there are many broad bumpy unidentified features that are molecular bands that are not in the line lists. Most of them are probably just high-V transitions. It is important that the laboratory analyses include all the isotopomers. They are needed to interpret the stellar spectra. When they are not measured in the laboratory we have to make up our own predicted linelists for them.

For the cooler stars we need all the diatomics among all the abundant elements, and, essentially, the hydrides and oxides for all elements (such as ScO. TiO, VO, YO, ZrO, LaO, etc.). Ca appears as CaOH and CaH, not CaO. I use the TiO linelist from Schwenke (1998) [1] with 38 million lines.

Stars that are evolved and have high C abundances from nuclear burning can bind all the O into CO so that there are no other oxides, just C-bearing molecules. CN and C_2 bands are everywhere.

For M stars cooler than 3500K triatomics also become important. Much more laboratory and computational work is needed for H_2O. I currently use the linelist from Partridge and Schwenke (1997) [2] with 66 million lines.

In the brown dwarfs and "planets" methane is important and it needs more laboratory and computational work. This is too cool and too hard for me.

58, 154, 500 MILLION LINES

Here is the background starting with my calculations at the San Diego Supercomputer Center in the 1980s. I have computed line data for 42 million lines of the iron group elements [3] plus I have all the data from the literature for all elements. I have computed line data for 16 million diatomic molecular lines (Some as much as 20 years ago.) I have tabulated opacities for more than 30 abundances, for temperatures from 2000K to 200000K using all 58 million lines at 3.5 million wavelengths from 10 nm to 10000 nm. I have computed more than 9000 models for a wide range of abundances for 3500K to 50000K effective temperature. I have computed a solar model that matches the observed energy distribution. I have computed fluxes and colors for the models. I have distributed all of this as it was produced, to supernova modelers, to galaxy modelers, to interior modelers, to stellar atmosphere modelers, to photometrists, etc. My line data are used as input to modelling codes for atmospheres, novas, and supernovas that are completely independent of my codes. They are basic data. Only 1 per cent (i.e., 600,000) of my computed lines have accurate wavelengths between known levels because the laboratory analyses have not yet found the levels and need improvement. When published theory or laboratory f values or broadening data seem better than mine, I use the better data. This "good" line list is the input for spectrum synthesis programs. I put the programs and data on CD-ROMs and I have distributed 26 titles so far. They are now on my web site.

I have added the TiO and H_2O line lists from Schwenke and I have thrown out my old TiO linelist. That leaves me with 154 million lines with which I can compute reasonable models for M stars down to 3500K.

To compute the iron group line lists I made Slater-expansion model Hamiltonians that included as many configurations as I could fit into the Cray. I used Hartree-Fock Slater integrals (scaled) for starting guesses and for higher configurations that had no laboratory energy levels. All configuration interactions were included. I then determined the Slater integrals for the observed configurations by least squares fitting the eigenvalues computed from the Hamiltonian matrix to the observed energies. The complication was that the eigenvalues and the observed energies had to be correlated by hand each iteration and more than a hundred interations were often required for convergence. My computer programs for these procedures have evolved from Cowan's (1968) programs [4]. Transition integrals were computed with scaled-Thomas-Fermi-Dirac wavefunctions and the whole transition array was produced for each ion. Radiative, Stark, and van der Waals damping constants and Landé g values were automatically produced for each line. The first nine ions of Ca through Ni produced 42 million lines. Eigenvalues were replaced by measured energies so that lines connecting measured levels have correct wavelengths. Most of the lines have uncertain wavelengths because they connect predicted rather than measured levels.

I am now computing or recomputing all the atoms and diatomic molecules. My old Cray programs from the 1980s were limited to 1100 x 1100 arrays in the Hamiltonian for each J. With my Alpha workstation I can easily run cases with 3000 x 3000 arrays so that I can include many more configurations and many more configuration interactions. The larger arrays produce about 3 times as many lines. At present I am limited to 61 even and 61 odd configurations and I try to include everything up through $n = 9$. I decided to test the new program on Fe I and Fe II to see whether there was any great difference in the low configurations compared to those from the Cray program. The major result was that the electric quadrupole transitions were 10 times stronger than before because the transition integrals are weighted by r^2 —they become very large for high n, and because there are numerous configuration interactions that mix the low and high configurations. As a check I was able to reproduce Garstang's (1962) lower results [5] by running his three configurations with my program. Since my model atom is still only a subset of a real Fe II ion, the true quadrupole A values are probably larger than mine. The magnetic dipole lines are affected by the mixing but the overall scale does not change.

EXAMPLES

Here I show sample statistics from my new semiempirical calculations for Fe II, Ni I, and Co I to illustrate how important it is to do the basic physics well and how much data there are to deal with. Ni, Co, and Fe are prominent in supernovas, including both radioactive and stable isotopes. There is not space here for the lifetime and gf comparisons. Generally, low configurations that have been well studied in the laboratory produce good lifetimes and gf values while higher configurations that are poorly observed and are strongly mixed are not well constrained in the least squares fit and necessarily produce poorer results and large scatter. My hope is that the predicted energy levels can help the laboratory spectroscopists to identify more levels and further constrain the least squares fits. From my side, I check the computed gf values in spectrum calculations by comparing to observed spectra. I adjust the gf values so that the spectra match. Then I search for patterns in the adjustments that suggest corrections in the least squares fits.

As the new calculations accumulate I will put on my web site the output files of the least-squares fits to the energy levels, energy level tables, with E, J, identification, strongest eigenvector components, lifetime, A sum, C_4, C_6, Landé g. The sums are complete up to the first (n = 10) energy level not included. There will be electric dipole, magnetic dipole, and electric quadrupole line lists. Radiative, Stark, and van der Waals damping constants and Landé g values are automatically produced for each line. Hyperfine and isotopic splitting are included when the data exist but not automatically. Eigenvalues are replaced by measured energies so that lines connecting measured levels have correct wavelengths. Most of the lines have uncertain wavelengths because they connect predicted rather than measured levels. Laboratory measurements of gf values and lifetimes will be included.

When computations with the necessary information are available from other workers, I am happy to use those data instead of repeating the work.

Fe II

Based on Johansson (1978) [6] and on more recent published and unpublished data. Johansson has data for more than 100 energy levels that I do not yet have.

d^7

$d^6 4s$	$d^5 4s^2$	$d^6 4d$	$d^5 4s4d$			$d^4 4s^2 4d$	$d^5 4p^2$
$d^6 5s$	$d^5 4s5s$	$d^6 5d$	$d^5 4s5d$	$d^6 5g$	$d^5 4s5g$	$d^4 4s^2 5s$	
$d^6 6s$	$d^5 4s6s$	$d^6 6d$	$d^5 4s6d$	$d^6 6g$	$d^5 4s6g$		
$d^6 7s$	$d^5 4s7s$	$d^6 7d$	$d^5 4s7d$	$d^6 7g$	$d^5 4s7g$	$d^6 7i$	$d^5 4s7i$
$d^6 8s$	$d^5 4s8s$	$d^6 8d$	$d^5 4s8d$	$d^6 8g$	$d^5 4s8g$	$d^6 8i$	$d^5 4s8i$	$d^5 4s9l$
$d^6 9s$	$d^5 4s9s$	$d^6 9d$	$d^5 4s9d$	$d^6 9g$	$d^5 4s9g$	$d^6 9i$	$d^5 4s9i$	$d^6 9l$

$d^6 4p$	$d^5 4s4p$	$d^6 4f$	$d^5 4s4f$			$d^4 4s^2 4p$	$d^4 4s^2 4f$
$d^6 5p$	$d^5 4s5p$	$d^6 5f$	$d^5 4s5f$			$d^4 4s^2 5p$	
$d^6 6p$	$d^5 4s6p$	$d^6 6f$	$d^5 4s6f$	$d^6 6h$	$d^5 4s6h$		
$d^6 7p$	$d^5 4s7p$	$d^6 7f$	$d^5 4s7f$	$d^6 7h$	$d^5 4s7h$		
$d^6 8p$	$d^5 4s8p$	$d^6 8f$	$d^5 4s8f$	$d^6 8h$	$d^5 4s8h$	$d^6 8k$	$d^5 4s8k$
$d^6 9p$	$d^5 4s9p$	$d^6 9f$	$d^5 4s9f$	$d^6 9h$	$d^5 4s9h$	$d^6 9k$	$d^5 4s9k$

configurations	46 even	39 odd
levels	19771 even	19652 odd
largest J matrix	2965 even	3007 odd
known levels	403 even	492 odd
metastable levels	72 even	1 odd

[The odd metastable level $(^2 I)4sp(^3 P)\ ^4 K_{8.5}$ is predicted at 103122 ± 150 cm^{-1}.]

Hamiltonian parameters	2645 even	2996 odd
free LS parameters	58 even	51 odd
standard deviation	56 cm^{-1} even	75 cm^{-1} odd
total E1 lines saved	7719063	old K88 [3] 1264969
between known levels	81225	old K88 45815
total M1 lines saved	1852641 even	2468074 odd
between known levels	28102 even	41374 odd
between metastable	1180 even	0 odd
total E2 lines saved	10347332 even	13179033 odd
between known levels	49019 even	71225 odd
between metastable	1704 even	0 odd

[My intuition tells me to keep all the forbidden lines, not just the ones connecting metastable levels. I do not have time to think about it now, but since the quadrupole A values get larger as n gets larger and since there are more than 10 million lines, they must somehow make our lives more complicated.]

isotopic components	^{54}Fe	^{55}Fe	^{56}Fe	^{57}Fe	^{58}Fe	^{59}Fe	^{60}Fe
fractional abundances	.059	.0	.9172	.021	.0028	.0	.0

There are 4 stable isotopes. ^{57}Fe has not yet been measured because it has hyperfine splitting. Rosberg, Litzén, and Johansson (1993) [7] have measured ^{56}Fe–^{54}Fe in 9 lines and ^{58}Fe–^{56}Fe in one line. I split the computed lines by hand.

Ni I
Ni I mostly based on Litzén, Brault, and Thorne (1993) [8] with **isotopic splitting**.

configurations	46 even	48 odd
levels	3203 even	4800 odd
largest J matrix	517 even	840 odd
known levels	130 even	153 odd
metastable levels	13 even	1 odd

[The odd metastable level is $(^3F)4sp(^3P)\ ^5G_6$ at 27260.894 cm^{-1}.]

Hamiltonian parameters	2446 even	2996 odd
free LS parameters	33 even	33 odd
standard deviation	60 cm^{-1} even	88 cm^{-1} odd
total E1 lines saved	529632	old K88 149926
between known levels	9637	
total M1 lines saved	67880 even	159049 odd
between known levels	2227 even	5272 odd
betweem metastable	41 even	0 odd
total E2 lines saved	453222 even	929692 odd
between known levels	3776 even	7539 odd
between metastable	24 even	0 odd

isotope	^{56}Ni	^{57}Ni	^{58}Ni	^{59}Ni	^{60}Ni	^{61}Ni	^{62}Ni	^{63}Ni	^{64}Ni
fraction	.0	.0	.6827	.0	.2790	.0113	.0359	.0	.0091

There are 5 stable isotopes. There are measured splittings for 326 lines from which I determined 131 energy levels relative to the ground. These levels are connected by **11670 isotopic lines**. Hyperfine splitting was included for 61Ni but only 6 levels have been measured which produce 4 lines with 38 components. A pure isotope laboratory analysis is needed.

Ni I lines are asymmetric from the splitting. When the isotopic calculation was first checked against the solar spectrum it did not look right. Subsequently, I found a program error and recomputed the splittings. Now the profiles match the observed. Observed stellar spectra are generally not high enough quality to show that there are such errors.

Co I
Co I based on Pickering and Thorne (1996) [9] and on Pickering (1996) with **hyperfine splitting** [10]. This calculation was made before my programs were

expanded. I will rerun this with twice as many configurations.

configurations	32 even	32 odd
levels	3546 even	5870 odd
largest J matrix	748 even	1130 odd
known levels	139 even	223 odd
metastable levels	31 even	1 odd

[The odd metastable level is $(^4F)4sp(^3P)\ z^6G_{6.5}$ at 25138.806 cm^{-1}.]

Hamiltonian parameters	1446 even	1762 odd
free LS parameters	27 even	26 odd
standard deviation	129 cm^{-1} even	126 cm^{-1} odd
total E1 lines saved	1729299	old K88 546130
between known levels	15481	
total M1 lines saved	396174 even	602458 odd
between known levels	3497 even	11993 odd
betweem metastable	286 even	0 odd
total E2 lines saved	1218019 even	2468646 odd
between known levels	5094 even	15943 odd
between metastable	410 even	0 odd
isotopic components	^{56}Co ^{57}Co ^{58}Co ^{59}Co	
fractional abundances	.0 .0 .0 1.00	

^{59}Co is the only stable isotope. Hyperfine constants have been measured in 297 levels which produce **244264 component E1 lines**. I have not yet computed the M1 or E2 components. The new calculation greatly improves the appearance of the Co I lines in the solar spectrum.

U I

In this volume, Wyart and Hubbard describe their web site for actinides. The U I directory is a good example of what can be accomplished with hard work. There are 1426 even levels and 536 odd levels, many with Landé g and isotopic splitting U^{238}-U^{235}.

Cr IV, Mn IV-V, Fe IV-VI, Co IV-VII, Ni IV-VIII

Sugar and Corliss (1985) [11] found no laboratory energy levels for n = 5, 6, 7, 8, 9 for these ions. The laboratory sources used were not able to populate the high upper levels to produce emission lines. However in hot stars lines to the excited levels appear in absorption shortward of Lyman α and through the Lyman continuum. For absorption it is necessary to populate only the lower level of a transition. The observed lines cannot be identified or analyzed. When I compute all these excited levels the uncertainty in the energies is too great. This is a problem that has to be solved by building new laboratory sources and by measuring the spectrum from the infrared to the extreme ultraviolet.

TiO

Schwenke calculated energy levels for TiO including in the Hamiltonian the 20 lowest vibration states of the 13 lowest electronic states of TiO (singlets a, b, c, d, f, g, h and triplets X, A, B, C, D, E) and their interactions. He determined parameters by fitting the observed energies or by computing theoretical values. Using Langhoff's transition moments [12] Schwenke generated a linelist for J = 0 to 300 for the

isotopomers	$^{46}Ti^{16}O$	$^{47}Ti^{16}O$	$^{48}Ti^{16}O$	$^{49}Ti^{16}O$	$^{50}Ti^{16}$
fractional abundances	.080	.073	.738	.055	.054

My version has 37744499 lines.

Good laboratory analyses and a similar semiempirical treatment are needed for CaOH, ScO, VO, YO, ZrO, LaO, etc. Better laboratory data could be used to further improve TiO.

I_2

I_2 is not an astronomical molecule but it is dear to the hearts of people who search for planets [13]. I_2 absorption cells are the standard against which radial velocities are measured. The I_2 transmission spectrum [14] is imposed on the stellar spectrum by passing the light from the star through an absorption cell maintained at a constant temperatue above 300K. By various reduction techniques the motion of the stellar spectrum relative to the I_2 spectrum is determined as a function of time. Since thousands of lines are compared, weak signals can be found. However a problem that people ignore is that the resolved spectrum of I_2 has never been observed. $^{127}I_2$ has 1/3 the doppler width of $^{16}O_2$. The resolving power required is in the millions but FTS spectra are in the range 300000 to 500000. Observers have been using FTS I_2 templates to reduce their data but the templates are so underresolved that an FTS line that has a depth of 1/3 is black in reality. A number of techniques have produced high resolution spectra of I_2 [15] that show the hyperfine structure in small wavelength intervals. We need a list of all the hyperfine energy levels. It would be absolutely fabulous if someone then would write a computer program that can generate the resolved transmission spectrum of an I_2 absorption cell for any temperature.

REFERENCES

1. Schwenke, D.W., *Faraday Discussions* **109**, 321-334 (1998).
2. Partridge, H., and Schwenke, D.W., *J. Chem. Phys.* **106**, 4618-4639 (1997).
3. Kurucz, R.L., in *Trans. Internat. Astron. Union* **XXB**, edited by M.McNally, Kluwer, Dordrecht, 1988, pp. 168-172.
4. Cowan, R.D., *J. Opt. Soc. Am.* **58**, 808-818 (1968).
5. Garstang, R.H., *Mon. Not. R. Astron. Soc.* **124**, 321-341 (1962); *Comm. Univ. of London Obs.* **57** (1962).
6. Johansson, S., *Physica Scripta* **18**, 217-265 (1978).
7. Rosberg, M., Litzén, U., and Johansson, S., *Mon. Not. R. Astron. Soc.* **262**, L1-L5 (1993).
8. Litzén, U., Brault, J.W., and Thorne, A.P., *Physica Scripta* **47**, 628-673 (1993).
9. Pickering, J.C., *Astrophys. J. Suppl.* **107**, 811-822 (1996).
10. Pickering, J.C., and Thorne, A.P., *Astrophys. J. Suppl.* **107**, 761-809 (1996).
11. Sugar, J., and Corliss, C., *J. Phys. Chem. Ref. Data* **14, Supp. 2** (1985).
12. Langhoff, S.R., *Astrophys. J.* **481**, 1007-1015 (1997).
13. Marcy, G.W., and Butler, R.P., *Pub. Astron. Soc. Pacific* **104**, 270-277 (1992).
14. Gerstenkorn, S., and Luc, P., *Atlas du Spectre d'Absorption de la Molecule de l'Iode (14800-20000 cm^{-1})*, C.N.R.S., Paris, 1978.
15. Sansonetti, C.J., *J. Opt. Soc. Am. B* **14**, 1913-1920 (1997).

Charge Transfer Data Needs for Cometary X-ray Emission Modeling

P. C. Stancil*, J. G. Wang*, M. J. Raković[†], D. R. Schultz[†] and R. Ali**

*Department of Physics and Astronomy and Center for Simulational Physics, The University of Georgia, Athens, GA 30602-2451
[†]Physics Division, Oak Ridge National Laboratory, P.O. Box 2008, Oak Ridge, TN 37831-6372
**Department of Physics, University of Nevada, Reno, NV 89557-0058

Abstract. The emission of x-rays has been observed from nearly twenty comets and evidence exists that it is the result of radiative decays from highly-excited, highly-charged solar wind ions following charge exchange with neutral species, mostly water, in the cometary atmosphere. We review the progress to date in constructing models of the x-ray emission. However, the construction of accurate models is impeded by the near lack of reliable state-selective charge exchange cross section data for the relevant neutral species (H_2O, CO, CO_2, etc.). The progress, and difficulties, of theoretical studies for the relevant collision processes is discussed with a particular focus on the breakdown of approximations made in the emission models.

INTRODUCTION

In 1996 the Röntgen x-ray satellite *(ROSAT)* made the unexpected observation of strong x-ray emission from comet Hyakutake [1]. Subsequent searches of the *ROSAT* all-sky survey archival data found that four other comets had also been observed emitting x-rays [2]. Additionally, x-ray emission has been detected from C/1996 Q1 [3], 6P/d'Arrest [4], Hale-Bopp [5], 2P/Encke [6], C/1999 S4 [7], C/1999 T1 [8], C/2000 WM1 [9], and C/2002 C1 [10]. The total to date is about 18 and the number can only be expected to grow!

A number of models have been put forth to explain cometary x-ray emission, as summarized by Krasnopolsky [11], but the predicted luminosities of most of the models are too small by two to three orders of magnitude. Only two mechanisms appear to reproduce the observed magnitude of the x-ray luminosity: scattering of solar x-ray radiation by small attogram dust particles in the coma [12] and electron capture of heavy solar wind ions to highly excited states following collisions with cometary neutrals [13]. However, the abundance of attogram dust is unknown and variations of the cometary x-ray intensity is found not to be correlated with cometary dust production rates or the solar x-ray intensity. On the other hand, the cometary x-ray emission has been found to correlate with the solar wind flux and cometary gas production rates [6, 14]. Further, observations of comet Hyakutake [15] with the *Extreme Ultraviolet Explorer (EUVE)* have revealed emission lines from multiply charged ions (O^{4-6+}, $C^{4,5+}$, and Ne^{7+}) and a broad line in *Chandra X-ray Observatory (CXO)* spectra of Linear S4 [7] appears to be attributable to O^{5+}. *CXO* observations of comet McNaught-Hartley (C/1999 T1) revealed nine narrow emission features between 195 and 940 eV due to highly-charged

C, O, Ne, Mg, and Si [8]. These three observations provide very convincing proof of the solar wind charge exchange (SWCX) mechanism for x-ray emission from comets.

After Cravens [13] put forth the SWCX mechanism to explain x-ray emission from comets, the atomic collision community became active in trying to meet the data demands for modeling this interesting solar system phenomenon. However, the complexity of the SW environment and cometary atmosphere and the relevant collision energies makes construction of a complete and accurate database a daunting challenge. Further, the required data not only include total single electron capture (SEC) cross sections, but various levels of final-state-selectivity. Multiple electron capture processes add yet another level of complexity, but for brevity will not be addressed here (See for example [16, 17]). In this article, we discuss the construction of cometary x-ray emission models, the approximations inherit in them, and point-out the failings of many of these approximations. We review the hierarchy of theoretical methods for charge transfer calculations and their success in comparison to recent experiments. Finally, we summarize with a list of needed atomic collision data, but first we briefly describe the solar wind and cometary atmosphere environments. A recent review of the topic has been given by Cravens [18].

THE ENVIRONMENT OF THE COMET-SOLAR WIND INTERACTION

The SW is a highly ionized, low density (~ 7 cm^{-3} at 1 AU) plasma which flows radially outward from the sun. Its element and ionization distribution reflects that in the 10^6 K solar corona. While it consists predominately of protons, alpha particles, and He$^+$, the heavy elements compose about 0.1% by volume [19]. Due to the high temperature of the solar corona, the heavy elements are highly charged often with only one or two bound electrons (e.g., C^{5+} or O^{6+}) or fully-stripped (e.g., N^{7+}), but multielectron ions are also common (e.g., Mg^{10+} or Fe^{9+}) [20].

The SW is neither uniform in time or isotropic in heliospheric latitude, but can be roughly partitioned into two types: the *slow* and *fast* SW. The slow SW, which has speeds averaging near 300 km/s but ranging from 200 to 400 km/s (~ 0.2-0.8 keV/u), originates from low heliospheric latitudes (i.e., near the plane of the solar system), while the fast SW, with speeds up to 800 km/s (~ 3 keV/u), is emitted from corona holes and near the solar poles. Further, the compositions are somewhat different with the slow SW being more highly ionized and enriched with elements of low first ionization potentials (e.g., Mg, Si, Fe).

On the other hand, the region near the comet known as the coma is similar to a planetary atmosphere where the gas density is high and primarily neutral. The coma is composed mostly of H$_2$O ($\sim 75\%$), its dissociation products OH, O, and H, and CO$_2$, CO, hydrocarbons, other trace species, and dust. The coma density falls off as $1/r^2$, r being the distance from the comet center, and extends to 10^5-10^6 km [21].

THE SOLAR WIND CHARGE EXCHANGE MECHANISM

The currently favored mechanism for the cometary x-ray emission suggests that it originates from highly-excited, high-charge-state, heavy SW ions whose levels are populated through charge changing collisions with ambient cometary neutral molecules and atoms. The primary process is the transfer of a single electron from the neutral cometary species (or target) to the SW ion (or projectile), which can be represented by the following reaction equation for the collision of a fully-stripped nitrogen ion with a water molecule, for example,

$$N^{7+} + H_2O \rightarrow N^{6+*}(nl\ ^{2S+1}L_J) + H_2O^+. \qquad (1)$$

The electron can be captured into a range of electron configurations, nl, and possibly into different terms of the multiplet $^{2S+1}L_J$ (in this example $2S+1=2$). However, charge exchange is a fairly state-selective process and the important values of n, the principal quantum number of the product SW projectile ion, are somewhat limited. In fact the n for the dominant capture channel can be estimated with the relation obtained from the classical over-the-barrier (COB) model

$$n_{\max} \leq q\left[2I_t\left(1 + \frac{q-1}{2\sqrt{q}+1}\right)\right]^{-1/2} \qquad (2)$$

[22, 23] which is similar to

$$n_{\max} \approx \frac{q^{3/4}}{\sqrt{2I_t}} \qquad (3)$$

found from unitarized distorted wave (UDW) [24], classic trajectory Monte Carlo (CTMC) [25], and multichannel Landau-Zener (MCLZ) [26] calculations for fully-stripped ion collisions with H. Here q is the initial charge of the incident SW ion and I_t the ionization potential of the target (in atomic units). These two relations give $n_{\max} = 5$ and 4, respectively. The former results in an internal excitation energy of 640 eV, all of which is available for radiative decay possibly contributing to the x-ray emission.

Now, within a given n-manifold there is a range of possible orbital angular momenta, l, from $l = 0$ to $n-1$, and the resulting emission depends on the population in each l, or l-distribution. For intermediate-energy collisions ($E > 10$ keV/u), it is well known that the l-distribution is statistical and the distribution function can be written as

$$W_{nl}^{st} = (2l+1)/n^2. \qquad (4)$$

Therefore, for $n = 5$, the dominant capture configuration would be $5g$. However, the $5g$ cannot decay directly to the ground state, but only by a series of $\Delta n = \Delta l = -1$ Yrast cascade transitions, i.e. $5g \rightarrow 4f \rightarrow 3d \rightarrow 2p \rightarrow 1s$. Only the last step will result in a x-ray, but of only 500 eV, while the preceding transitions result in visible or IR photons. However, captures to other l can result in complicated cascade schemes. For example, if the dominant capture is to the $5p$, then this can either decay directly to the ground state emitting a 640 eV x-ray, cascade through the $4d$ or $4s$ before arriving at the $3p$ which can decay with a 592 eV x-ray to the ground state, or other cascade paths. In fact, since all of the l-levels will receive some population, the resulting x-ray spectrum would be expected to be intermediate between these two extremes.

X-RAY EMISSION MODELS

To simulate the cometary x-ray emission spectra, a number of models have been constructed, but at various levels of sophistication [13, 27, 28, 20, 29, 30]. However, each was forced to make a number of assumptions due to the lack of the relevant charge exchange cross section data. In particular, data are primarily available for H, He, and H_2 targets colliding with multiply charged ions with the majority for total cross sections, few results being available for n-selective, fewer for l-selective cross sections. Little data of any kind are available for the neutral cometary species. The major approximation was in the estimation of the l-distribution with three possibilities being adopted: (i) an equal population in all l-levels [28, 20], (ii) the statistical population of equation (4) [27], or (iii) use of state-selective data from experimental and theoretical studies of H, He, and H_2 targets [29, 30]. They also assumed that the cross sections are independent of the collision energy and considered only $n = n_{max}$ where n_{max} was obtained by equations (2) or (3). The earlier models [27] considered a limited number of SW ions (C, N, and Ne) though larger ion sets where included in later work. Due to a lack of transition probability data for most SW ions, all of the calculations adopted hydrogenic cascade models. The most realistic atomic physics was included in the work of Kharchenko and Dalgarno [29, 30] who also considered forbidden transitions from He-like product SW ions. However, the ratio of triplet to singlet states following SEC was assumed to be statistical (i.e., 3:1).

The theoretical spectra and predicted total luminosities generated with these approximate SWCX models are found to be in good agreement with the pre-*CXO* low-resolution observations. Comparison with the high-resolution *CXO* results have yet to be undertaken, but preliminary indications of discrepancies, possibly due to approximations in the models, have been found in the *EUVE* spectra of Hyakutake [15]. Krasnopolsky and Mumma [15] found strong correlations in line positions, but less so for the line intensities when comparisons were made to the most advanced models of Kharchenko and Dalgarno [29, 30].

BREAKDOWN OF APPROXIMATIONS

The approximations that are made in the cometary x-ray models are necessary because of the lack of charge transfer data for the relevant neutral cometary species. We will concentrate here on examining the issues of (a) the l-distributions and (b) the triplet-singlet ratio.

The l-distribution

The approximation (i) of assuming that all l-levels are equally populated following SEC, as discussed in the previous section, can be discounted as unphysical. Such a behavior is not observed in measurements or calculations in any energy regime. However, the approximation (ii) of adopting the statistical distribution of equation (4) is valid at intermediate energies ($E > 10$ keV/u) giving $l = n - 1$ as the maximum populated level. CTMC calculations find that the l-distribution is only statistical for $n < n_{max}$, while for $n > n_{max}$, a peak is observed at $l \approx q - 1$.

Unfortunately, the SW velocity is in the low-energy regime ($E < 10$ keV/u) and a number of recent studies have demonstrated that the l-distribution is not statistical, but peaked toward smaller l [31, 16, 32]. The departure from a statistical l-distribution becomes more pronounced with decreasing collision energy and therefore would be expected to be a poor approximation for slow SW conditions.

The Triplet-Singlet Ratio

There is a significant population of SW ions which initially have one or more electrons. For the case of an H-like SW ion, it becomes He-like after SEC. The spins of the two electrons can couple to give either singlet ($S = 0$) or triplet ($S = 1$) states. It is assumed purely by statistical arguments that the relative populations of the triplet to singlet states following SEC is 3:1. However, triplet states are spin-forbidden from decaying to the ground-state so that it is expected that the emission spectra from He-like ions should be very different from H-like. Typically, forbidden transitions have been neglected in models, but Kharchenko and Dalgarno [30] have pointed out that they can be some of the most intense lines.

However, studies for charge exchange of N^{4+} with H [33, 34] and O^{3+} + H [35, 36] suggest that for slow SW conditions there can be significant departures from a statistical triplet-singlet ratio. For N^{4+} with H, the total triplet-singlet ratio is found by quantum molecular-orbital close-coupling (QMOCC) calculations [33] to be ~ 2 at 1 keV/u, but to approach 3 for fast SW conditions (3 keV/u) in agreement with the measurements of Bliek et al. [34]. For O^{3+} with H measurements are only available for capture to $O^{2+}(3s)$ [35] which agree with the QMOCC calculations of Wang et al. [36] predicting the triplet-singlet ratio to be ~ 2 for slow SE conditions as shown in Figure 1. On the other hand, the triplet-singlet ratio for capture to $O^{2+}(3p)$ is find by the QMOCC calculations to approach 10.

No studies have done on cometary neutrals, but measurements on H_2 for O^{3+} collisions show behaviour similar to H [35], but the departure from the statistical triplet-singlet ratio is less pronounced for N^{4+} collisions [34]. More work is needed, particularly for H_2O, to gauge whether this effect is important.

CHARGE EXCHANGE CALCULATIONS

Clearly, to produce accurate cometary x-ray emission models it is desirable to have absolute cross sections at the n-, l- and S-selective level for all the relevant SW ions and cometary neutrals at the required collision energies. Because of the vast array of relevant collision systems, one would hope that simple analytical relations, or at least theoretical methods with minimal computational effort, would be available to predict the cross sections. However, it has been shown that at the final-n-state-resolved level,

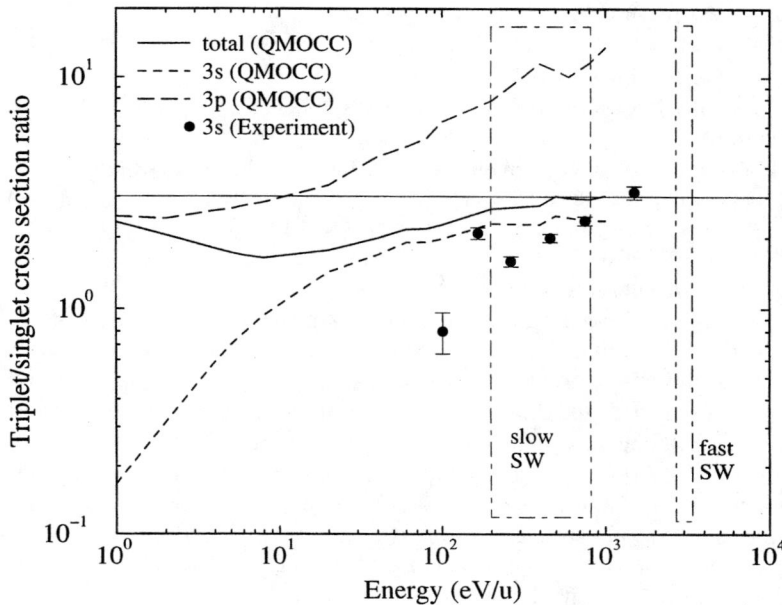

FIGURE 1. Triplet-singlet ratios for $O^{3+} + H \rightarrow O^{2+} + H^+$. See the text for cross section references.

many such techniques (e.g., COB, MCLZ, and CTMC) are not in agreement with recent accurate measurements [37].

Simple theoretical relations either fail or are unable to reproduce the experimental l-distributions. For the COB model, no theory has been developed to predict the l-distribution. However, Burgdörfer et al. [38] have extended the COB model to obtain an average l-value, $<l>$, which has been found to be in reasonable agreement with $<l>$ deduced from x-ray spectra following SEC in collisions of O^{8+} and Ne^{10+} with He, H_2, H_2O and CO_2 [16]. Though, the x-ray spectra might be contaminated by multiple capture processes which would tend to enhance $<l>$.

For fully-stripped ions, the l-distribution cannot be obtained directly from an MCLZ calculation, but by multiplying the resulting n-resolved cross section by a distribution function W_{nl}. Janev et al. [26] have developed such a technique where they used the statistical distribution function (4) for intermediate energies and a distribution function derived by Abramov et al. [39] at low energies. The low-energy distribution function is given by

$$W_{nl}^{le} = (2l+1)\frac{[(n-1)!]^2}{(n+1)!(n-1-l)!} \quad (5)$$

which has a maximum at $l = 1$ for $n < 8$, but shifts to larger l for $n > 8$. One notices that for either distribution function there is no explicit dependence on the collision energy (or the target or projectile). Therefore, there is some ambiguity as to the energy for which each of the distribution functions are applicable. Janev [40] has discussed this issue and suggests that for $E > 10$ keV/u $W_{n,l}^{st}$ should be used, while $W_{n,l}^{le}$ should be reliable for

$E < 1$ keV/u. As these distribution functions are very different, there is some region of transition which appears to occur between ~ 1 and ~ 10 keV/u. Unfortunately, again this is just the energy regime relevant for the SW.

The CTMC method, however, is very convenient in that n- and l-distributions are naturally produced using so-called binning schemes of the Becker-MacKellar type if the final-states are hydrogenic [41] or other variants for partially stripped ions [42, 43]. Recent CTMC calculations by Perez et al. [44] for a number of highly charged ion collisions with H, show that CTMC can reproduce the general trend of transitioning from an statistical l-distribution at intermediate energies, to a distribution similar to equation (4) at low-energies.

Recently, triple-coincidence measurements of x-rays, scattered projectiles, and recoil ions following SEC collisions of Ne^{10+} with various neutral species have been made [45]. The measurements provide x-ray spectra which are n-level-resolved and through a combination of cascade modeling can be used to infer the l-distributions. These spectra therefore provide the most stringent test of theoretical l-distributions as it is not possible to directly measure l-resolved cross sections, at least for fully-stripped incident ions. An example is shown in Figure 2 for a 4.54 keV/u Ne^{10+} collision with He. Hydrogenic cascade models were constructed with n, l-resolved cross sections computed with the MCLZ method, using the two l-distribution functions of equations (4) and (5), and with the CTMC method. The theoretical spectra were normalized to the total integrated experimental spectra and convolved with a Gaussian FWHM of 133 eV, corresponding to the x-ray detector's resolution. The primary capture channel is $n = 5$ followed by $n = 4$. Figure 4b shows the spectra following capture to $n = 5$. The primary peak near 1 keV is due to the $2p \rightarrow 1s$ transition following a cascade network to the $2p$. The broader peak centered near 1250 eV is due to a number of $np \rightarrow 1s$ transitions with $n = 3 - 5$. Clearly, neither the $n = 5$ nor the $n = 4$ spectra result from statistical l-distributions. The $n = 5$ spectra is intermediate between the distributions of equations (4) and (5) as a consequence of the collision energy being between the regions of applicability of the distribution functions, in agreement with the discussion of the previous section. Further, for $n = 5$, none of the presented calculations agree with the experiment. The agreement for $n = 4$ is much better appearing to indicate the low-energy distribution function (5) is preferred which is also in agreement with the CTMC result.

DISCUSSION

The problem with the theoretical methods presented here is that (i) the relevant collision energy range (~ 0.2-3 keV/u) is below the applicablity of classically-based methods (CTMC or COB) or (ii) the l-distribution functions do not have explicit energy dependence. There is some indication of a departure of the statistical l-distribution with the CTMC method, but not for the dominant n-channel. Alternatively, the $Ne^{10+} + H$ CTMC results of Perez et al. [44] suggests the statistical l-distribution departure occurs, but that it is delayed to lower collision energies. For example, they find that the distribution peaks at $l = 1$ for $E < 100$ eV/u. Apparently, the methods which might provide accurate results for total, n-, and l-resolved cross sections are of the close-coupling variety,

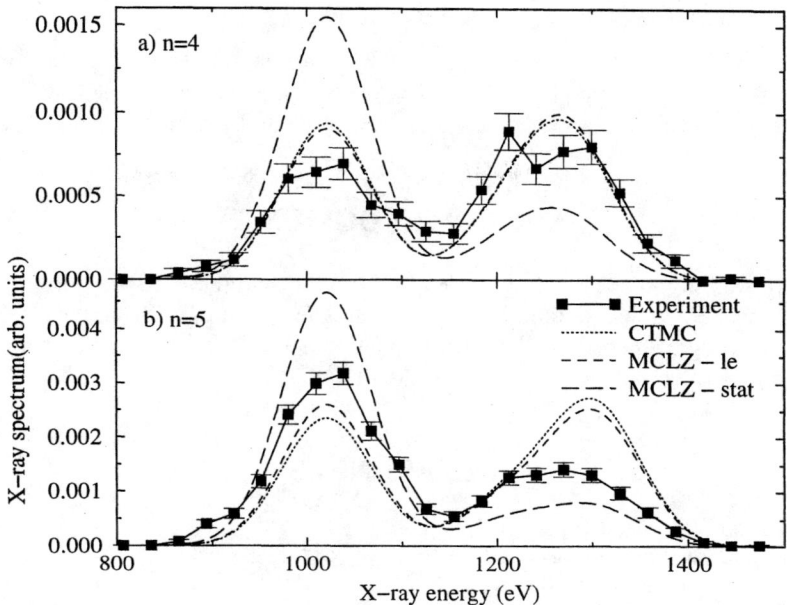

FIGURE 2. X-ray emission spectra following Ne^{10+} + He SEC. a) $n = 4$; b) $n = 5$. See the text for references.

i.e., QMOCC, semiclassical MOCC, or atomic-orbital close-coupling (AOCC), which are also the most labor-intensive methods. Hasan et al. [37] found that the n-resolved MOCC calculations of Harel and Jouin [46] gave the best agreement with their N^{7+} + He measurements while results from COB, MCLZ, and CTMC were less accurate. Close-coupling calculations can in general be readily performed for ion-atom collisions resulting in accurate n, l-distributions for the required energy range, but the number of such studies for ion-molecule collisions is extremely limited and primarily to diatomic targets. We are unaware of any close-coupling calculations for a multiply-charged ion with water. Therefore, it is a field of theoretical research which needs further work and is now being stimulated by the application to the cometary x-ray emission problem.

SUMMARY OF NEEDED DATA

To conclude, we list the range of parameters for charge transfer studies necessary for the construction of accurate x-ray emission models.

1. Energy range: $\sim$0.2-3 keV/u;
2. Neutral targets: H_2O, CO, CO_2, OH, O, and H;
3. Solar wind elements: O, N, C, He, Ne, Mg, Si, Fe, S, etc.;
4. Solar wind ion charge: $q = 4 - 13+$; and

5. Multielectron SW ions: H-like, He-like, Li-like, etc.

This results in a total of ~240 relevant collision systems. Of these, it might be possible to select the 25 most important for modeling cometary x-ray emission spectra which might be detected with future high-resolution observations from telescopes like the *CXO*.

ACKNOWLEDGMENTS

This work was supported by NASA grant NAG5-9088 and the NASA Applied Information Systems Research Program. DRS also acknowledges support from the DOE Office of Fusion Energy Sciences to Oak Ridge National Laboratory which is managed by UT-Battelle, LLC under contract No. DE-AC05-00OR22725.

REFERENCES

1. Lisse, C. M. *et al.*, *Science*, **274**, 205 (1996).
2. Dennerl, K., Englhauser, J., and Trümper, J., *Science*, **277**, 1625 (1997).
3. Dennerl, K., Englhauser, J., Trümper, J., and Lisse, C., *IAU Circular*, **6495** (1996).
4. Mumma, M. J., Krasnopolsky, V. A., and Abbott, M. J., *Astrophys. J.*, **491**, L125 (1997).
5. Owens, A. *et al.*, *Astrophys. J.*, **493**, L47 (1998).
6. Lisse, C. M. *et al.*, *Icarus*, **141**, 316 (1999).
7. Lisse, C. M. *et al.*, *Science*, **292**, 1343 (2001).
8. Krasnopolsky, V. *et al.*, *AGU Spring Meeting Abstracts* (2002), p52A-07.
9. Lisse, C. M. *et al.* (2002), in preparation.
10. Dennerl, K., and Lisse, C. M. (2002), in preparation.
11. Krasnopolsky, V., *Icarus*, **128**, 268 (1997).
12. Wickramasinghe, N. C., and Hoyle, F., *Astron. and Astrophys.*, **239**, 121 (1996).
13. Cravens, T. E., *Geophys. Res. Lett.*, **25**, 105 (1997).
14. Neugebauer, *J. Geophys. Res.*, **105**, 20,949 (2000).
15. Krasnopolsky, V., and Mumma, M. J., *Astrophys. J.*, **549**, 629 (2001).
16. Greenwood, J. B., Williams, I. D., Smith, S. J., and Chutjian, A., *Phys. Rev. A*, **63**, 062707 (2001).
17. Neil, P. A., Ali, R., Harris, C. L., Beiersdorfer, P., Schultz, D. R., Raković, M. J., Wang, J. G., and Stancil, P. C., *Bull. Am. Phys. Soc.*, **47**, JG.083 (2002).
18. Cravens, T. E., *Science*, **296**, 1042 (2002).
19. Bochsler, P., *Physica Scripta*, **T18**, 55 (1987).
20. Schwadron, N. A., and Cravens, T. E., *Astrophys. J.*, **544**, 558 (2000).
21. Mumma, M. J. *et al.*, *Science*, **272**, 1310 (1996).
22. Ryufuku, H., Sasaki, K., and Watanabe, T., *Phys. Rev. A*, **21**, 745 (1980).
23. Mann, R., Folkmann, F., and Beyer, H. F., *J. Phys. B*, **14**, 1161 (1981).
24. Ryufuku, H., and Watanabe, T., *Phys. Rev. A*, **20**, 1828 (1979).
25. Olson, R. E., *Phys. Rev. A*, **24**, 1726 (1981).
26. Janev, R. K., Belić, D. S., and Bransden, B. H., *Phys. Rev. A*, **28**, 1293 (1983).
27. Häberli, *et al.*, R. M., *Science*, **276**, 939 (1997).
28. Wegmann, R. *et al.*, *Planet. Space Sci.*, **46**, 603 (1998).
29. Kharchenko, V., and Dalgarno, A., *J. Geophys. Res.*, **105**, 1854 (2000).
30. Kharchenko, V., and Dalgarno, A., *Astrophys. J.*, **554**, L99 (2001).
31. Beiersdorfer, P. *et al.*, *Phys. Rev. Lett.*, **85**, 5090 (2000).
32. Beiersdorfer, P. *et al.*, *Astrophys. J.*, **549**, L147 (2001).
33. Stancil, P. C., Zygelman, B., Clarke, N. J., and Cooper, D. L., *J. Phys. B*, **30**, 1013 (1997).
34. Bliek, F. W., Woestenenk, G. R., Hoekstra, R., and Morgenstern, R., *Phys. Rev. A*, **57**, 221 (1998).
35. Beijers, J. P. M., Hoekstra, R., and Morgenstern, R., *J. Phys. B*, **29**, 1397 (1996).

36. Wang, J. G., Stancil, P. C., Turner, A. R., and Cooper, D. L., *Phys. Rev. A* (2002), in preparation.
37. Hasan, A. A., Eissa, F., Ali, R., Schultz, D. R., and Stancil, P. C., *Astrophy. J.*, **560**, L201 (2001).
38. Burgdörfer, J., Morgenstern, R., and Niehaus, A., *J. Phys. B*, **19**, L507 (1986).
39. Abramov, V. A., Baryshnikov, F. F., and Lisitsa, V. S., *JETP Lett.*, **27**, 464 (1978).
40. Janev, R. K., *Comments At. Mol. Phys.*, **12**, 277 (1983).
41. Beckar, R. L., and McKellar, A. D., *J. Phys. B*, **17**, 3923 (1984).
42. Raković, M. J., Schultz, D. R., Stancil, P. C., and Janev, R. K., *J. Phys. A*, **34**, 4753 (2001).
43. Schultz, D. R., Stancil, P. C., and Raković, M. J., *J. Phys. B*, **34**, 2739 (2001).
44. Perez, J. A., Olson, R. E., and Beiersdorfer, P., *J. Phys. B*, **34**, 3063 (2001).
45. Ali, R., Harris, C. L., Neill, P. A., Beiersdorfer, P., Schultz, D. R., Raković, M. J., Wang, J. G., and Stancil, P. C., *Bull. Am. Phys. Soc.*, **47**, H5.007 (2002).
46. Harel, C., and Jouin, H., *J. Phys. B*, **25**, 221 (1992).

Interstellar Molecules: The New Frontiers for Molecular Data

Lucy M. Ziurys and Aldo J. Apponi

*Dept. of Astronomy, Dept of Chemistry, and Steward Observatory,
University of Arizona, Tucson, AZ 85721*

Abstract. Although over 110 chemical species have been securely identified in the interstellar and circumstellar gas, there are still many data needs for molecular astrophysics. Among these needs are new high-resolution spectroscopic measurements of potential interstellar molecules, in particular organic radicals, metal-bearing molecules, and molecular ions. Ab initio studies that support these experimental investigations are necessary as well. Reaction rate measurements of many ion-molecule, neutral-neutral, and radiative association processes are also essential.

INTRODUCTION

Over the past 30 years, more than 110 different chemical species have been identified in the interstellar space, startling evidence that a rich gas-phase chemistry exists outside our solar system. These molecules have been found to be primarily present in giant gas clouds located throughout our galaxy and in external galaxies, as well as in the envelope remnants of old giant stars. These types of astronomical objects are characterized by cold (T ~ 10 – 50 K), dense (n ~ 10^4 – 10^6 cm^{-3}) gas such that usually only rotational energy levels of the molecules are populated, which occurs through collisions. Therefore, most molecular identifications are made on the basis of pure rotational spectra, including small splittings caused by fine and hyperfine interactions, all which constitute a "fingerprint" type pattern. These rotational transitions generally occur at millimeter and sub-millimeter wavelengths. Consequently radio and millimeter-wave astronomy has played a critical role in the identification and study of interstellar molecules.

The success of molecular astrophysics has been critically linked to experimental work in the laboratory, including spectroscopy and dynamics, as well as theoretical calculations. These studies have not only provided the critical "rest" frequencies for molecules, but also dipole moments, collisional cross-sections, intensity predictions, and other molecular properties. These latter data are extremely important in the interpretation of interstellar molecular line emission, which is used to analyze the physical properties of astronomical objects. The use of molecular lines as tracers of the dense interstellar medium has revolutionized our knowledge of star and solar system formation, galactic structure, stellar nucleosynthesis, and even cosmology. Hence, laboratory and theoretical molecular data has aided both in the chemical and physical knowledge of the universe.

THE CONTINUED NEED FOR LABORATORY AND THEORETICAL DATA

Spectroscopy

Despite such progress, there are still many laboratory and theoretical needs with applications in astrophysics. One obvious direction is the measurement of spectra for new possible interstellar species, particularly in the sub-millimeter and in the infrared. Although many chemical compounds have been successfully identified, hundreds of unidentified features, so-called "U-lines" exist, as shown in Figure 1. This figure shows a spectrum observed near 101 GHz towards the well-known molecular cloud, Orion-KL, with fairly good sensitivity, using the Steward Observatory Kitt Peak 12m telescope. Virtually every line that appears in these data cannot presently be identified.

These lines could arise from any number of chemical compounds for which there is limited spectra data. Such possibilities include metal analogs of known interstellar ring and chain silicon-carbon molecules, such as SiC_3 or SiC_2 (see ref. [1]). Substitution of the silicon with a magnesium, sodium, iron, or aluminum atom in any of the silicon structures could produce additional interstellar molecules. An example is the recent discovery of AlNC, as shown in Figure 2 [2]. Although theoretical calculations exist for some of these species (e.g. ref. [3]), laboratory data remains limited. The Ziurys group, however, has begun to make progress in this area (e.g. ref. [4,5]). Other possibilities are molecular ions, in particular open-shell species such as HCN^+, CS^+ or ions with refractory elements (SiO^+, SiH^+, MgH^+, etc.). Additional candidates are organic radicals and rings, including those with one heteroatom, and isomers of known species such as C_5H. The Thaddeus group has done much work on carbon radicals (e.g. ref. [6]), but there are still numerous species yet to be investigated.

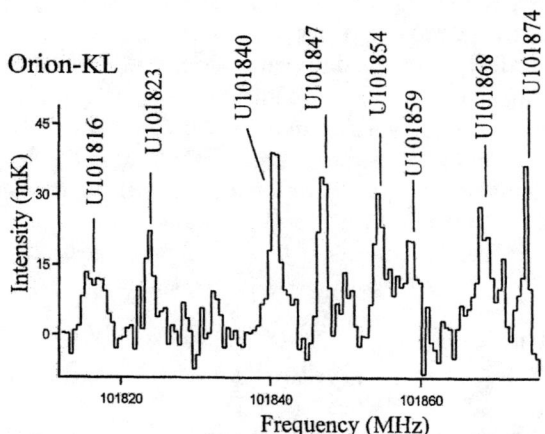

FIGURE 1. Unidentified lines observed towards Orion-KL with the Steward Observatory 12m Telescope.

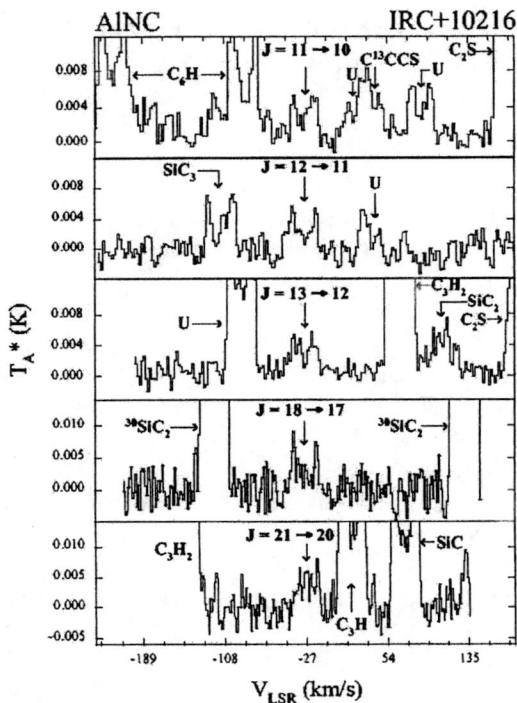

FIGURE 2. Spectra of the five rotational transitions detected for A1NC towards IRC +10216, using the IRAM 30m telescope, identifying this new circumstellar molecule [2].

Ro-vibrational measurements in the infrared are also needed. One target is gas-phase studies of polyatomic hydrocarbons (PAHs). The Saykally group, using single-photon infrared emission spectroscopy, has made significant progress in this area (e.g ref. [7]), but data for additional PAH compounds are lacking. Because long acetylenic chains are common in interstellar and circumstellar gas, it would be also useful to know the bending vibrations of species like HC_7N, C_5H, etc. More ro-vibrational studies of gas phase symmetric species like C_4, CNC and C_6 are also necessary. Finally, any data leading to the identification of the diffuse interstellar absorption bands would be of utmost relevance. Recent laboratory work suggested that negative ions such as C_7^+ might be the carriers [8], but this hypothesis has not borne fruit [9,10].

Dynamics

In addition to spectroscopy, there is a dearth of data for rates of simple ion-molecule and neutral-neutral reactions. Although complex chemical modeling is done to predict abundances of interstellar molecules, many reaction rates are only approximately known, at best, especially at low temperatures (e.g. ref. [11]). Addition of even one newly determined reaction rate can alter predicted abundances [12]. Even rates for such fundamentally important ion-molecule reactions as $H_2 + HOC^+ \rightarrow HCO^+ + H_2$ remain a debated issue. Measurements of collisional cross-sections with common interstellar molecules such as H_2 are additionally needed. These data are extremely important in establishing molecular abundances.

Theory

Along with experimental investigations, theory plays a critical role as well. There are still very simple, possible interstellar molecules whose electronic ground state is uncertain. The compounds include diatomic carbides and nitrides (MnC, CrC, MnN, for example), and triatomics with the general formula MNC (MCN), MC_2, or MNH, where M is a metal atom. There are various carbon radicals whose structure is uncertain, such as HC_6N. Dipole moments, vibrational frequencies, and IR intensities are other critical concerns.

CONCLUSIONS

Although many interstellar and circumstellar molecules have been unambiguously discovered, it is obvious from the many unidentified features observed in the ISM that additional species remain to be detected. In this sense, astronomical instrumentation has advanced far ahead of laboratory studies. It is critical for further developments in molecular astrophysics that spectroscopy, dynamics and theory regain the lead.

ACKNOWLEDGMENTS

This work has been supported by the National Science Foundation and the NASA Laboratory Astrophysics Program.

REFERENCES

1. Apponi, A. J., McCarthy, M. C., Gottlieb, C. A., and Thaddeus, P., *Ap. J. Lett.* **516**, LI 03 (1999).
2. Ziurys, L. M., Savage, C., Highberger, J. L., Apponi, A. J., Guélin, M., and Cernicharo, J., *Ap. J. Lett.* **564**, 45 (2002).
3. Redondo, P., Barrientos, C., and Largo, A., *Chem. Phys. Lett.* **335**, 64 (2001).
4. Brewster, M. A., and Ziurys, L. M., *Ap. J.* **559**, LI 63 (2001).
5. Halfen, D. T., Apponi, A. J., and Ziurys, L. M., *Ap. J.*, submitted.
6. McCarthy, M. C., Travers, M. J., Kovacs, A., Gottlieb, C. A., and Thaddeus, P., *APJS* **113**, 105 (1997).
7. Cook, D. J., Schlemmer, S., Balucani, N., Wagner, J., Harrison, J. A., Steiner, B., and Saykally, R. J., *J. Phys. Chem. A* **102**, 1465 (1998).
8. Tulej, M., Kirkwood, D. A., Pachkov, M., and Maier, J. P., *Ap. J.* **506**, L69 (1998).
9. McCall, B. J., York, D. G., and Oka, T., *Ap. J.* **531**, 329 (2000).
10. Galazutdinov, G., Musaev, F., Nirski, J., and Krelowski, J., *A&A* **377**, 1063 (2001).
11. Le Tueff, Y. H., Millar, T. J., and Markwick, A. J., *A&AS* **146**, 157 (2000).
12. Terzieva, R., and Herbst, E., *Ap. J.* **501**, 207 (1998).

F. Fusion Energy

Atomic and Molecular Processes for Heat and Particle Control in Tokamaks

Hirotaka Kubo

*Naka Fusion Research Establishment, Japan Atomic Energy Research Institute
801-01 Mukoyama, Naka-machi, Naka-gun, Ibaraki-ken, 311-0193 Japan*

Abstract. Heat and particle control is an essential issue in tokamak fusion research. Heat control using edge radiation losses and cold divertor plasmas is needed for mitigation of the severe problem of the concentrated power loading on the divertor plates. Understanding particle (hydrogen isotopes, helium and impurity) behavior in cold divertor plasmas is needed for control of fuel density, reduction of fuel dilution and control of radiation losses. Atomic and molecular processes play important roles for the edge radiation losses and particle behavior in the cold divertor plasmas. Atomic and molecular data are applied to studying the processes.

INTRODUCTION

In tokamak fusion reactors, heat and particle (hydrogen isotopes, helium and impurity) control is essential for obtaining high fusion performance and preventing damage of the plasma facing components. The poloidal divertor is the most promising method for heat and particle control. A poloidal cross-section of a tokamak with a poloidal divertor is shown in Fig. 1. With the divertor coil, the high-temperature main plasma, where the nuclear fusion takes place, is kept away from the first walls. Heat and particles from the main plasma flow along magnetic field lines to the divertor plates. Mitigation of the heat load to the divertor plates is required for reduction of divertor-plate erosion and impurity production. Edge radiation losses and a cold divertor plasma can reduce the heat load. In the divertor, some of the particles arriving at the divertor plates are pumped out, and some of them return to the main plasma. Impurities produced at the divertor plates can also be transported to the main plasma. Particle control is necessary for control of fuel density, exhaust of helium ash and reduction of impurity contamination. Understanding particle behavior in the cold divertor plasma is needed for particle control. Atomic and molecular processes play important roles for the edge radiation losses and particle behavior in the cold divertor plasma.

In this paper, some topics relating to the atomic and molecular processes for heat and particle control are presented, mainly based on recent studies in JT-60U [1,2]. The significance of the topics for ITER (International Thermonuclear Experimental Reactor) [3] is also discussed. Atomic and molecular data applications and needs for study of the heat and particle control are described. First, enhancement of the edge radiation losses by impurity injection for heat control is described. Application of

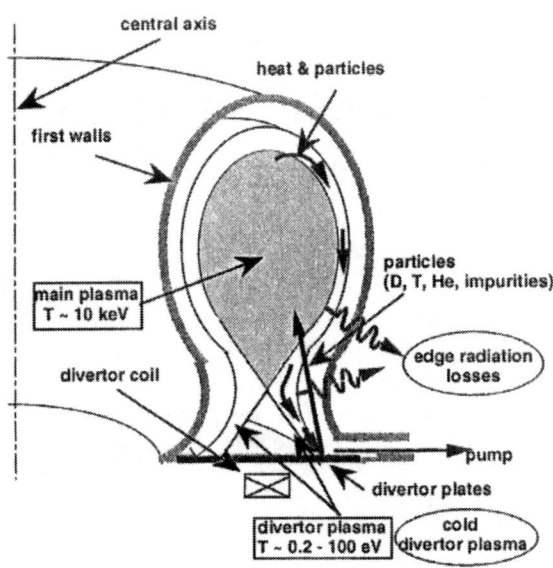

FIGURE 1. Poloidal cross-section of a tokamak with a poloidal divertor.

atomic data for investigation of heavy impurity transport and radiation in the main plasmas is presented. Second, study of particle behavior in cold divertor plasmas is described. Molecular data application and needs for investigation of hydrogen- and hydrocarbon-molecule behavior are presented. Effects of elastic collisions on helium exhaust are also described.

ENHANCEMENT OF EDGE RADIATION LOSSES

Enhancement of the edge radiation losses can reduce the heat load on the divertor plates. In ITER, a radiation-loss-power fraction relative to the heating power of 80% is desired [3]. Although intense puffing of hydrogen-isotope gas can enhance the radiation losses, the radiation losses should be controlled separately from the fuel control. Furthermore, confinement degradation is usually observed with intense hydrogen-isotope gas puffing [4]. Controlled injection of impurity gases is a promising technique for enhancement of radiation loss power. In ITER, Ar injection is being considered to enhance the radiation loss power.

High confinement plasmas with a high radiation-loss-power fraction have recently been obtained by impurity injection in some tokamaks [5]. Figure 2 shows control of edge radiation loss power with Ar injection in a JT-60U ELMy H-mode plasma [4]. (The ELMy H-mode is the basic operational regime of ITER [3].) The radiation loss power from the edge plasma was controlled with a feedback technique using the Ar gas puffing and divertor pumping. As seen in the figure, the edge radiation loss power was well controlled near the reference value by changing the Ar puffing rate. With a

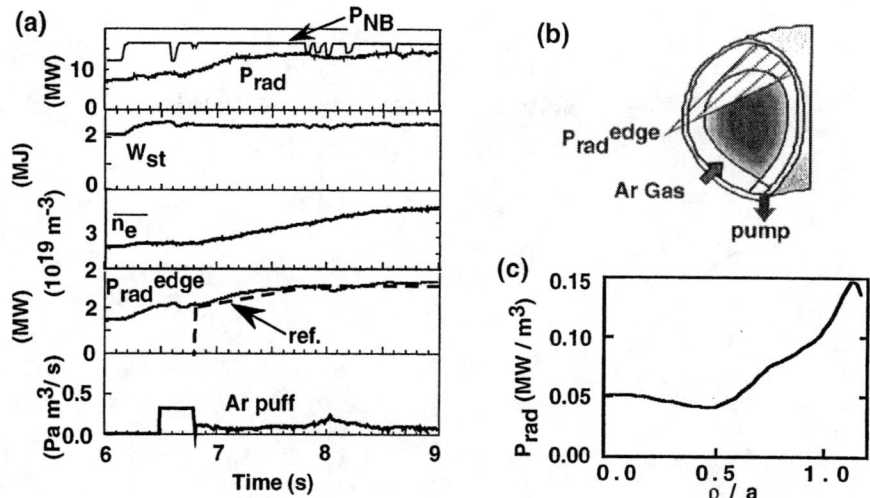

FIGURE 2. Control of the edge radiation loss power with Ar injection in JT-60U [4]. (a) Waveforms of an ELMy H-mode plasma with Ar injection. P_{NB}: neutral beam injection power, P_{rad}: total radiation loss power, W_{st}: stored energy, n_e: line-averaged electron density, P_{rad}^{edge}: radiation loss power from the edge plasma, ref: reference for the feedback control of the P_{rad}^{edge}, and Ar puff: puffing rate of Ar gas. The ratio of the Ar density to the electron density was ~0.5%. (b) Positions of the gas puff and divertor pump on the poloidal cross-section. The area observed with bolometers in order to control the P_{rad}^{edge} is also shown. (c) Profile of the radiation loss power density in the main plasma, where ρ/a is the normalized minor radius.

radiation-loss-power fraction equal to 80% of the heating power, the stored energy was maintained at a high level, and the improved energy confinement characteristics of ELMy H-mode plasmas was maintained. The impurity transport and radiation losses are investigated using impurity transport models [6]. The density (n_z) of the impurity ion with the charge of z can be calculated from the continuity equations as

$$\frac{\partial}{\partial t} n_z = -\frac{1}{r}\frac{\partial}{\partial r} r\Gamma_z + S_{z-1} n_e n_{z-1} - (S_z + \alpha_z) n_e n_z + \alpha_{z+1} n_e n_{z+1}, \tag{1}$$

where r is the minor radius, S_z and α_z are the effective ionization and recombination rate coefficients, respectively, and n_e is the electron density. Γ_z is the impurity flux density, and it is usually expressed as

$$\Gamma_z = -D_z \frac{\partial}{\partial r} n_z + v_z n_z, \tag{2}$$

where D_z and v_z are the diffusion coefficient and inward velocity, respectively. The diffusion coefficients and inward velocities are derived from experimental measurements. An example of the analysis is shown in Fig. 3, where the transport coefficients were estimated to reproduce the measured soft x-ray emission profile. In

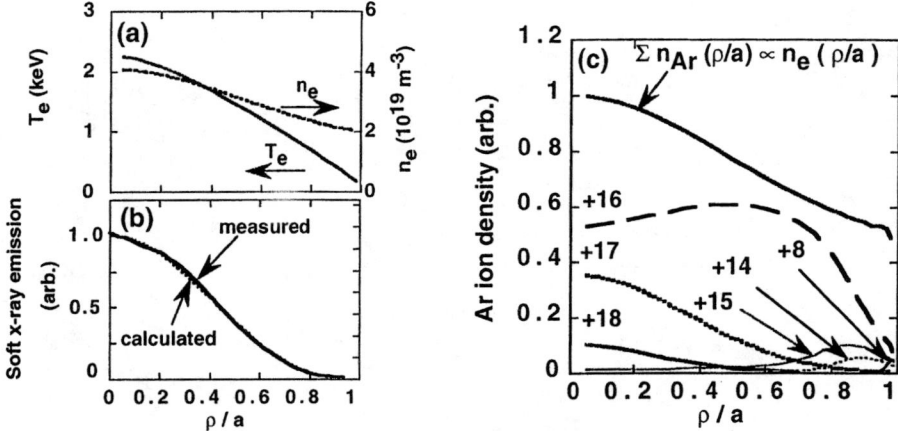

FIGURE 3. Profiles of (a) electron temperature and density, (b) soft x-ray (the energy range is 2.7 - 20 keV) emission, and Ar ion density calculated with an impurity transport code in a JT-60U ELMy H-mode plasma. The measured and calculated soft x-ray emission profiles are shown. For the calculation, it was assumed that $D = 1$ m^2/s and $v = D\, \partial(\log n_e)/\partial r$. Then, $\Sigma\, n_z\, (\rho/a) \propto n_e\, (\rho/a)$.

this case, the profile of the total Ar density was similar to the profile of the electron density, and impurity accumulation was not observed. Thus, for investigation of impurity transport and radiation losses, in addition to effective ionization and recombination rate coefficients, including charge exchange recombination, rate coefficients for soft x-ray emission, total radiation, and diagnostic line emission are required. In the future, Kr is considered to be injected for radiation loss power enhancement, and W and other heavy metals are considered to be used for the divertor plates. Therefore, heavy ion data are necessary for study of impurity transport and radiation losses.

STUDY OF PARTICLE BEHAVIOR IN COLD DIVERTOR PLASMAS

Hydrogen molecules

Most of hydrogen ions arriving at the divertor plates are eventually desorbed in the form of H$_2$ [7]. In cold divertor plasmas, the hydrogen molecules can play some role as a source and sink of hydrogen ions. In detached divertor plasmas [8], which are attractive for reduction of the heat load to the divertor plates, the ion flux to the divertor plates reduces. The ion flux reduction might be attributed to molecular assisted recombination (MAR: H$_2$ (v) + H$^+$ → H$_2^+$ + H and H$_2^+$ + e → H + H; H$_2$ (v) + e → H$^-$ + H and H$^-$ + H$^+$ → H + H) [9]. Figure 4 shows the recombination and ionization rate coefficients for hydrogen particles. The rate coefficients were obtained

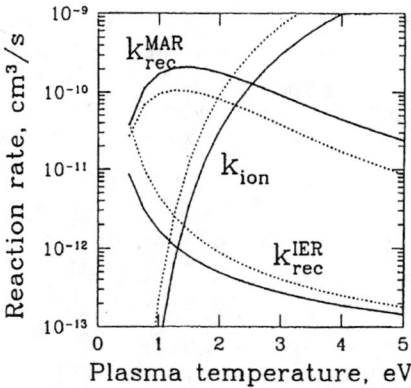

FIGURE 4. Temperature dependence of the recombination (k_{rec}^{IER}: rate coefficient for H^+-e recombination, k_{rec}^{MAR}: rate coefficient for MAR) and ionization rate coefficients (k_{ion}: rate coefficient for H ionization) for hydrogen particles [9]. Effective rates are given for a plasma density of 10^{20} m^{-3} by the solid curves and 10^{21} m^{-3} by the broken curves.

on the basis of a collision-radiative model, taking account of vibrationally excited molecules. The MAR rate is higher than the ion-electron recombination rate. The rate coefficients depend strongly on the vibrational excitation of the ground state $X\ ^1\Sigma_g^+$ [10]. Therefore, measurement of the H_2 density and vibrational population is necessary for understanding the roles of H_2.

For measurement of the H_2 density and vibrational populations, Fulcher lines (d $^3\Pi_u$

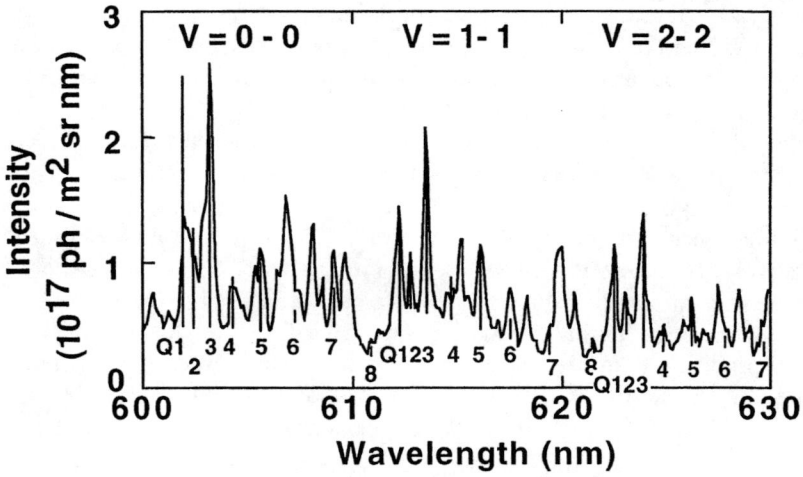

FIGURE 5. Fulcher lines observed near the divertor plates in JT-60U. The vertical lines indicate intensities of Q branches of v=0-0, 1-1, 2-2 transitions, calculated on the assumption that $T_{rot}(0) = 0.04$ eV, $T_{rot}(1) = 0.057$ eV, $T_{rot}(2) = 0.046$ eV, n (1) / n (0) = 0.76, and n (2) / n (0) = 0.60. $T_{rot}(v)$ and n (v) are the rotational temperature and population of the vibrational level v of d $^3\Pi_u$, respectively.

→ a $^3\Sigma_g^+$) have recently been observed in divertor plasmas [11]. Figure 5 shows Fulcher lines observed near the divertor plates in JT-60U. The Q branches with v = 0,1, and 2 were used for the analysis. The rotational temperature was assumed to be ~ 0.05 eV and the vibrational population ratios n (1) / n (0) and n (2) / n (0) were assumed to be 0.76 and 0.60, respectively. It is expected that the population of the d $^3\Pi_u$ level is predominantly produced by excitation from the ground state and the vibrational population of the ground state can be estimated from the population of the d $^3\Pi_u$ level. However, it is also expected that collisional transitions between the d $^3\Pi_u$ level and other levels, especially n=3 levels, affect the population of the d $^3\Pi_u$ level. For a quantitative estimation of the H_2 density and the vibrational population of the ground state, collisional-radiative models are being developed [11]. The models are also needed for analysis of H I line emission, since some of the emission can be attributed to dissociation of H_2 molecules [12,13]. However, lack of the molecular data causes difficulty in developing the models. The excitation cross sections between the excited levels are required especially.

Hydrocarbon molecules

For high-heat-load divertor plates, carbon materials are used in present and next generation tokamaks (such as ITER) because of their high thermal-shock resistance, high thermal conductivity and low atomic number. In cold divertor plasmas, although the physical sputtering can be suppressed, the chemical sputtering due to production of hydrocarbon molecules (C_mH_n/C_mD_n) cannot be suppressed. Thus, chemical sputtering can be the dominant impurity production and divertor-plate erosion process. In ITER, the chemical sputtering is one of the most important processes that determine the lifetime of the carbon divertor plates [14]. In addition, tritium-co-deposition due to production of hydrocarbon particles is an issue of great importance, since the amount of tritium in the vacuum vessel should be controlled for safety [15]. In ITER, the sputtering yield is expected to decrease as the particle flux increases. Therefore, it is necessary to measure the chemical sputtering yield at real divertor plates under conditions of high particle flux. In most tokamak experiments, only production of CH_4/CD_4 has been considered, and production of C_2H_n/C_2D_n has not been investigated quantitatively. However, it is expected that production of C_2H_n/C_2D_n (where n = 2, 4 and 6) becomes important in low-temperature (T_e < 15 eV) divertor plasmas [16].

The chemical sputtering yields due to C_2H_n/C_2D_n production in addition to CH_4/CD_4 production have recently been measured at the carbon divertor plates in JT-60U [17]. Emission of the CH/CD band (A $^2\Delta \rightarrow$ X $^2\Pi$, v = 0 - 0) and the C_2 band (A $^3\Pi_g \rightarrow$ X $^3\Pi_u$, v = 0 - 0) from the divertor plasmas were observed simultaneously. Figure 6 shows the CD and C_2 band spectra. Assuming that all of the C_mH_n produced at the divertor plates are dissociated near the divertor plates, the measured band emission intensity (I_{CH} and I_{C2}) can be expressed using the C_mH_n flux (Γ_{CmHn}) as follows,

$$I_{CH} = \Gamma_{CH_4} / (D/BX)_{CH_4,CH} + \sum \Gamma_{C_2H_n} / (D/BX)_{C_2H_n,CH} , \quad (3)$$

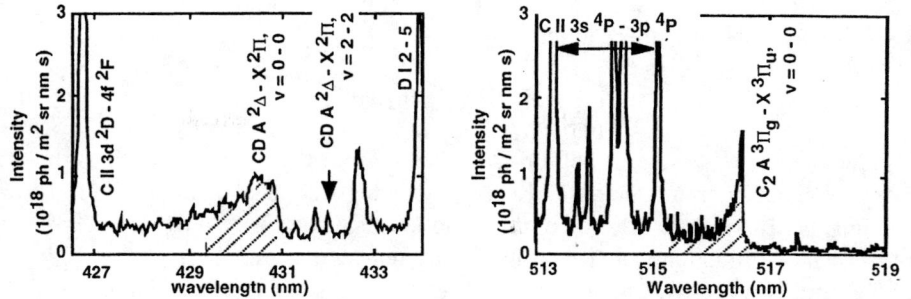

FIGURE 6. Spectra of (a) CD A $^2\Delta \rightarrow$ X $^2\Pi$ and (b) C_2 A $^3\Pi_g \rightarrow$ X $^3\Pi_u$ observed in the JT-60U divertor plasmas [17]. The shaded areas indicate the band emission intensities defined in the paper.

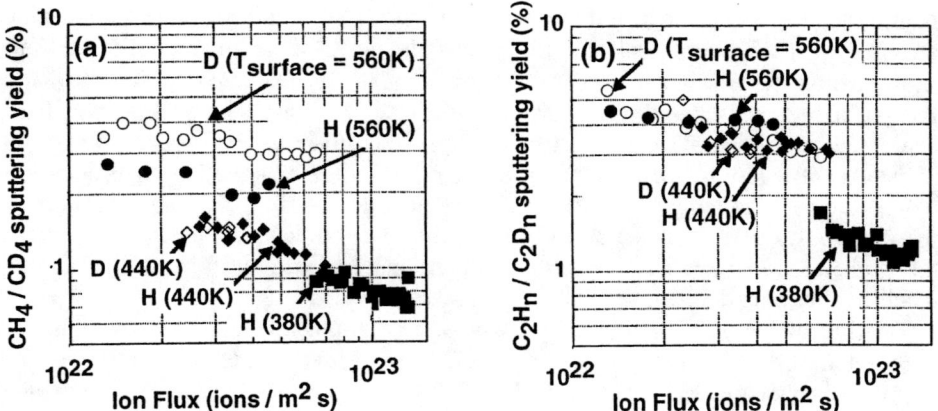

FIGURE 7. Sputtering yield of (a) CH_4/CD_4 and (b) C_2H_n/C_2D_n (n = 2,4,6) against the ion flux to the divertor plates [17]. Closed circles, diamonds and squares indicate the sputtering yields by hydrogen ions at the surface temperature of the carbon divertor plates of 380K, 440K and 560K, respectively. Open diamonds and squares indicate the sputtering yields by deuterium ions at the surface temperatures of 440K and 560K, respectively.

$$I_{C_2} = \sum \Gamma_{C_2H_n} / (D/BX)_{C_2H_n, C_2}, \qquad (4)$$

where $(D/BX)_{C_mH_n,CH}$ and $(D/BX)_{C_mH_n,C_2}$ indicate CH and C_2 band emission per C_mH_n dissociation. Here, D is the dissociation rate, B is the branching ratio, and X is the excitation rate. The intensities were defined as the shaded areas in Fig. 6. The (D/BX) coefficients presented as functions of electron temperature in Ref. 18 were used for the analysis. The obtained sputtering yields of CH_4/CD_4 and C_2H_n/C_2D_n, which are defined as the ratios of the C_mH_n/C_mD_n fluxes to the ion flux to the divertor plates, are plotted in Fig. 7. Dependences of the sputtering yields on the ion flux and the surface temperature of the divertor plates were obtained. As shown in the figure, production of C_2H_n/C_2D_n, in addition to CH_4/CD_4, was significant. In the analysis, the D/BX coefficients for CD_4 and C_2D_n were assumed to be the same as those for CH_4 and C_2H_n, since the coefficients were not available. Thus, reliable data for all relevant hydrocarbon molecules are required.

Helium atoms

Fundamental processes of helium atom behavior in the divertor plasma should be understood in order to design a divertor with high exhaust capability for fusion-produced helium particles. Some of the important rate coefficients for helium particles are shown as functions of electron or ion temperature in Fig. 8 (a). In a low temperature (< 15 eV) range, the rate coefficient of elastic collisions is larger than that of ionization. The rate coefficients of charge exchange with helium ions are rather large. However, the charge exchange processes are not dominant, since the helium ion density is typically ~1/10 of the electron density in the divertor plasma. In Fig. 8 (b), helium particle behavior in a divertor is illustrated. While a part of the helium ions arriving at the divertor plates are reflected, most of them are absorbed and subsequently desorbed as low-energy helium atoms with a kinetic energy near that of the surface temperature of the divertor plates. Helium atoms with such a low kinetic energy hardly penetrate the divertor plasmas, since their ionization length is short. However, at a low temperature, they are expected to be scattered by hydrogen ions before their ionization. The elastic collisions heat up the helium atoms, and increase the probability of their penetration through the divertor plasmas. On the other hand, the collisions also scatter the helium atoms, and increase the probability of their reaching pumping ducts.

Heating of helium atoms by elastic collisions has recently been observed using a high-resolution spectrometer in JT-60U [19]. A spectrum of the He I line observed in a JT-60U divertor plasma is shown in Fig. 9. In the figure, the spectrum is compared with spectra simulated with and without consideration of elastic collisions. As shown

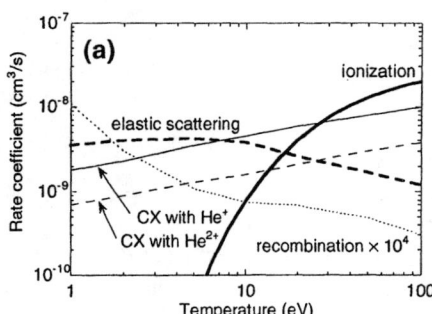 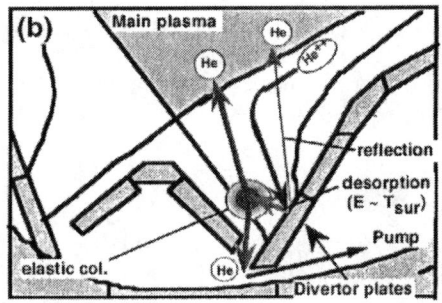

FIGURE 8. (a) Rate coefficients for helium particles against electron or ion temperature [19]. Thick solid line: ionization of helium atoms by electron collisions; thick broken line: elastic scattering by proton collisions; thin solid line: charge exchange with He^+; thin broken line: double charge exchange with He^{2+}; thin dotted line: collisional-radiative recombination of He^+ with electrons at the electron density of 10^{20} m^{-3}. The lines for elastic scattering and charge exchange indicate the rate coefficients calculated for helium atoms with no initial kinetic energy. (b) He particle behavior in a divertor.

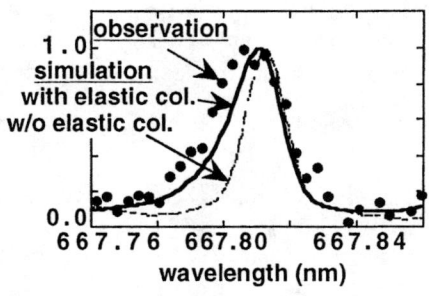

FIGURE 9. Spectra of the π-component of the He I line (667.82 nm) in a JT-60U divertor plasma [19]. The points indicate the observation. The solid line is the simulation, and the broken line is the simulation neglecting elastic collisions. In the discharge, He gas was injected for study of He particle behavior.

in the figure, the broadening of the observed spectral profile can be attributed to the elastic collisions. The kinetic energy derived from the line broadening was 0.9 eV, which was much higher than the surface temperature of the divertor plates. Recently, helium transport in ITER has been simulated with consideration of the elastic collisions. A significant (factor 3-5) reduction in the helium density at the main plasma separatrix was obtained by considering the elastic collisions [20]. In cold and dense divertor plasmas, atomic data not only for inelastic processes but also for elastic processes are needed.

SUMMARY

Edge radiation losses and cold diverter plasmas are effective for heat and particle control, which is essential in tokamak fusion research. Atomic and molecular processes play important roles for the edge radiation losses and particle behavior in the cold divertor plasmas. Enhancement of the edge radiation losses by impurity injection for heat control with maintaining improved confinement has recently been demonstrated. In the future, Kr is considered to be injected for radiation loss power enhancement, and W and other heavy metals are considered to be used for the divertor plates. For investigation of the radiation losses and impurity transport, data for heavy ions are necessary. In the cold divertor plasmas, hydrogen molecules can play some roles as a source and sink of hydrogen ions. Spectra of hydrogen molecules have recently been observed in divertor plasmas, and collisional-radiative models are being developed for quantitative analysis. Lack of the molecular data causes difficulty in developing the models. Chemical erosion is a concern for carbon divertor plates. Recently, sputtering yields due to C_2H_n/C_2D_n production in addition to CH_4/CD_4 production have systematically been obtained. In the analysis, the reliable molecular data are necessary. Thus, requirements for molecular data are increasing to facilitate model calculations and experimental data analyses. Data for molecules including deuterium and tritium are required especially for fusion research. It has been found that elastic collisions with the plasma ions can affect helium exhaust in the cold divertor plasmas. In the cold divertor plasmas, atomic data not only for inelastic processes but also for elastic processes are needed. For the above reasons, there is a continuing need for improved atomic and molecular data, especially for new regimes such as the cold divertor plasmas.

ACKNOWLEDGMENTS

The author would like to express his sincere thanks to Drs. S. Higashijima, T. Nakano, H. Takenaga, T. Sugie, and Y. Miura of Japan Atomic Energy Research Institute, Prof. K. Sawada of Shinshu University, Dr. A. S. Kukushkin of ITER International Team, and Dr. K. W. Hill of Princeton Plasma Physics Laboratory for useful discussions and suggestions. He is also grateful to the JT-60 team.

REFERENCES

1. Kamada, Y., and the JT-60 team, *Nucl. Fusion* **21**, 1311-1325 (2001).
2. Kubo, H., and the JT-60 team, *Phys. Plasmas*, to appear (2002).
3. *ITER-FEAT outline design report*, ITER EDA documentation series No.18, Vienna: International Atomic Energy Agency, 2001.
4. Kubo, H., et al., *Nucl. Fusion* **41**, 227-233 (2002).
5. Ongena, J. et al., *Phys. Plasmas* **8**, 2188-2198 (2001).
6. Hulse, R. A., "Modeling of Impurity Transport in the Core Plasma," in *Atomic and Plasma-Material Interaction Processes in Controlled Thermonuclear Fusion*, edited by R. K. Janev and H. W. Drawin, Amsterdam: Elsevier, 1993, pp. 165-190.
7. Heifetz, D. B., "Neutral Particle Transport," in *Physics of Plasma-Wall Interactions in Controlled Fusion*, edited by D. E. Post and R. Nehrisch, New York: Plenum Press, 1984, pp. 695-772.
8. Matthews, G. F., *J. Nucl. Mater.* **220-222**, 104-116 (1995).
9. Pigarov, A. Yu., and Krasheninnikov S. I., "Application of the collisional-radiative, atomic-molecular model to the recombining divertor plasma ," *Phys. Lett.* **A 222**, 251-257 (1996). Fig. 4 was reprinted with permission from Elsevier Science.
10. Sawada, K., and Fujimoto, T., *Contrib. Plasma Phys.*, to be published.
11. Fantz, U., et al., *J. Nucl. Mater.* **290-293**, 367-373 (2001).
12. Sawada, K., and Fujimoto, T., *J. Appl. Phys.* **78**, 2913-2924 (1995).
13. Kubo, H., et al., *Plasma Phys. Control. Fusion* **40**, 1115-1126 (1998).
14. Pacher, H. D., et al., *J. Nucl. Mater.* **241-243**, 255-259 (1997).
15. Janeschitz, G., et al., *J. Nucl. Mater.* **290-293**, 1-11 (2001).
16. Roth, J., et al., *Nucl. Fusion Suppl.* **1**, 63-78 (1991).
17. Nakano, T., et al., *Nucl. Fusion*, to appear (2002).
18. Pospieszczyk, A., et al, *University of California at Los Angeles Report* **No. UCLA-PPG-1251**, (1989).
19. Kubo, H., et al., *Plasma Phys. Control. Fusion*, **41**, 747-757 IOP Publishing (1999).
20. Kukushkin, A. S., and Pacher, H. D., *Plasma Phys. Control. Fusion*, to be published.

Atomic Physics Processes Important to the Understanding of the Scrape-Off Layer of Tokamaks

W.P. West, B. Goldsmith*, T.E. Evans, R.E. Olson[†]

General Atomics, P.O. Box 85608, San Diego, California 92186-5608
*Student, University of California – San Diego, La Jolla, California 92093-0319
[†]University of Missouri, Dept. of Physics, Rolla, Missouri

Abstract. The region between the well-confined plasma and the vessel walls of a magnetic confinement fusion research device, the scrape-off layer (SOL), is typically rich in atomic and molecular physics processes. The most advanced magnetic confinement device, the magnetically diverted tokamak, uses a magnetic separatrix to isolate the confinement zone (closed flux surfaces) from the edge plasma (open field lines). Over most of their length the open field lines run parallel to the separatrix, forming a thin magnetic barrier with the nearby vessel walls. In a poloidally-localized region, the open field lines are directed away from the separatrix and into the divertor, a region spatially separated from the separatrix where intense plasma wall interaction can occur relatively safely. Recent data from several tokamaks indicate that particle transport across the field lines of the SOL can be somewhat faster than previously thought. In these cases, the rate at which particles reach the vessel wall is comparable to the rate to the divertor from parallel transport. The SOL can be thin enough that the recycling neutrals and sputtered impurities from the wall may refuel or contaminate the confinement zone more efficiently than divertor plasma wall interaction. Just inside the SOL is a confinement barrier that produces a sharp pedestal in plasma density and temperature. Understanding neutral transport through the SOL and into the pedestal is key to understanding particle balance and particle and impurity exhaust. The SOL plasma is sufficiently hot and dense to excite and ionize neutrals. Ion and neutral temperatures are high enough that charge exchange between the neutrals and fuel and impurity ions is fast. Excitation of neutrals can be fast enough to lead to nonlinear behavior in charge exchange and ionization processes. In this paper the detailed atomic physics important to the understanding of the neutral transport through the SOL will be discussed.

INTRODUCTION

Atomic and molecular databases have played an important role in the development of magnetic confinement fusion since the inception of research in this area in the 1950's. The importance of basic atomic processes was appreciated early on. Over the decades, magnetic confinement fusion research has progressed to the point where fusion grade plasma conditions are achieved in magnetic confinement devices for short durations. Now the focus of research is on three general areas: 1) further improvement of plasma performance to increase efficiency and economic viability, 2) the achievement of a stationary operating point with fusion grade plasma conditions, and 3) the extension of the scientific underpinnings of scaling laws and modeling codes to increase confidence in the design of future, burning plasma devices. Atomic physics continues to play a critical role in the diagnostics systems used to measure and understand today's experiments, and is a crucial element of many modeling codes, especially codes used to model the edge plasma. The need for detailed atomic data continues to grow.

At the last ICAMDATA meeting, Dr. Kurt Behringer [1] presented a complete overview of the broad range of atomic and molecular physics that is important for magnetic confinement fusion research. Because fusion grade magnetic confinement plasmas range in temperature from <1 eV up to 25 keV, data is required for low energy processes such as molecular dissociation and high energy processes such as the radiation from highly ionized high Z elements. Because these plasmas range in density from $<10^{18}$ to 10^{21} m^{-3}, nonlinear and multi-step processes must be considered. The range of atomic species is also very broad. The atomic and molecular physics of deuterium, tritium, and helium, the fuel and ash of the fusion process, is obviously critical. Plasma facing wall materials include low Z atoms, such as Li and Be. Graphite, the most commonly used material in today's most advanced research devices, brings with it the need to understand both carbon and hydrocarbon processes. Both molybdenum and tungsten have some advantages over low Z wall materials and their use is presently being actively being pursued. However contamination of the plasma by these high Z elements represents a potential loss of energy confinement via strong radiation. Mid Z elements, e.g. argon, when intentionally injected into fusion grade plasmas have been shown to simultaneously improve total energy confinement while increasing radiative dissipation of energy outflux to the vessel walls. The ionization and radiation physics of these mid and high Z elements is also needed. All of these atomic data issues have been discussed in Dr. Behringer's paper [1].

In this paper we will emphasize the deep level of detail of atomic data that is needed to accurately model fusion plasmas and interpret spectroscopic diagnostic data. In this brief paper it is not possible to go into the detail important to all the diverse processes listed above, so we will use processes important to the scrape-off layer and edge plasma of a divertor tokamak as an example. This region of the tokamak has received increased attention over the last few years due to two recent discoveries. First, both experimental results and detailed theoretical models of tokamak energy confinement have shown that the achievement of a sharp pedestal in plasma pressure at the edge of the confinement zone significantly reduces thermal transport and improves plasma performance [2]. The physics of this sharp pedestal is not well understood, and is an active area of research. Because of its' proximity to the vessel walls, atomic physics is thought to be an important element to understanding. Second, in the region between the confinement zone and the vessel walls, called the scrape-off layer, enhanced convective transport across the magnetic field has been observed in some operational regimes [3,4]. The enhancement can be sufficiently high that the total particle fluence to the main chamber can be comparable to the total particle fluence along the magnetic field to the divertor. The resulting enhanced recycling and impurity sputtering from the wall, combined with the importance of the pedestal in plasma pressure very near the plasma edge, has been a strong motivator for renewed interest in this region.

THE TOKAMAK SCRAPE-OFF LAYER AND PEDESTAL REGIONS

The scrape-off layer is a thin layer of plasma separating the hot dense well-confined plasmas from the vessel walls. It must serve two functions, to insulate the vessel walls from the high temperature plasma and to isolate the confinement region from any erosion of material from the walls. All magnetic confinement concepts have a SOL, but in this paper we will focus on the tokamak. The tokamak configuration is the leading magnetic confinement configuration for fusion energy production. It is a toroidal device, with a large, externally imposed toroidal magnetic field and a relatively weaker poloidal magnetic field created mostly from a toroidal current inductively driven in the plasma. The resulting helically shaped magnetic field lines form closed flux surfaces from the center of the current channel outward to near the vacuum vessel walls. A cross section of the vacuum vessel and flux surfaces of a high performance plasma in the DIII-D tokamak [5] is shown in Fig. 1. Toroidally directed coils are placed around the periphery of the vessel to provide vertical stability. In most modern tokamaks these coils are also used to shape the plasma, and to create a null point in the poloidal field. The null point defines a separatrix in the flux surfaces. Inside the separatrix the flux surfaces are closed, and outside of the separatrix they intersect the vessel walls. The plasma inside the separatrix is well confined and is heated to high temperature (~10 keV) using a combination of ohmic heating due to the plasma current and auxiliary heating from injected energetic neutral beams or rf power. The plasma confinement is not perfect, and heat and particles leak from the confined region, across the separatrix, and ultimately to the vessel walls. The region of plasma near the point where the open field lines intersect the wall is known as the divertor, while the region of plasma lying on the field lines just outside of confinement zone is known as the scrape-off layer (SOL). In the conventional picture of the SOL and divertor, plasma flows along the field lines much faster than across the field lines. The heat and particles lost across the separatrix are directed to the divertor plates where they recycle back into the plasma as neutrals and sputter divertor plate material into the plasma. One of the reasons behind the divertor design is the large spatial separation between the separatrix and the divertor strike plates that can be achieved. The divertor is obviously a place rich in atomic and molecular physics and has been the focus of much research for many years [6].

The SOL is a region of lower plasma density than either the core plasma or the divertor plasma, being around 10^{18} to 10^{19} m^{-3}, with electron and ion temperatures not well equilibrated and lying typically between 10 and 200 eV. Profiles of plasma electron density and temperature and ion temperature in the region of the edge pedestal and SOL are shown in Fig. 2 for a typical high-confinement mode (H-mode) plasma in DIII-D. In an H-mode plasma, a region of very low transport is established just inside the separatrix. This narrow region of low transport appears spontaneously and is related to the appearance of a strong radial electric field. It results in the development of a sharp pedestal in plasma density and pressure. Near the separatrix the plasma density and temperature are seen to decrease exponentially away from the core plasma, with decay lengths between 0.5 cm to a few cm. Since the distance to the nearest vessel wall is usually several cm, wall fluxes were thought to be low. In

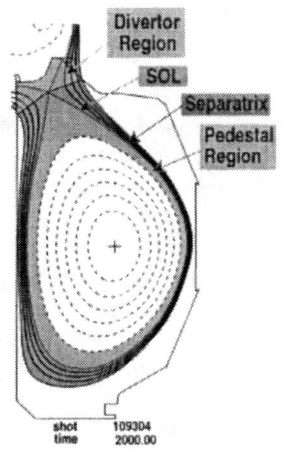
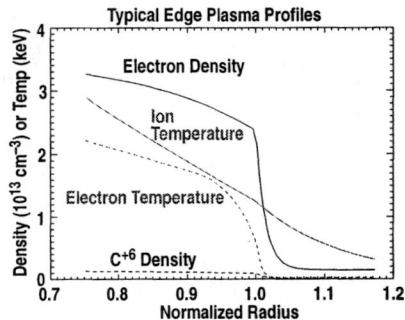

FIGURE 1. A cross section of the DIII-D vessel is shown along with reconstructed flux surfaces of an upper single null diverted discharge. The separatrix, divertor, SOL, and pedestal regions are identified.

FIGURE 2. The radial profiles of plasma density, electron and ion temperatures are shown as a function of normalized radius. The value $\rho = 1$ is at the separatrix.

addition, ionization path lengths of slow neutrals (1 eV or less) are short and so the core plasma is protected from wall sources. Recent data on SOL phenomena from the Alcator C-Mod [3] tokamak at higher than usual plasma density clearly showed that beyond the separatrix, the density and temperature scale lengths could become very long. In such cases the plasma flux across the SOL to the vessel wall was very high, completely dominating the flux to the divertor. Studies on DIII-D indicate that cross-field transport in the far regions of the SOL can result in wall flux being competitive with divertor flux at more moderate plasma density [4]. The existing data is sufficient to show that the main chamber can be a significant source of refueling and contamination, even in a diverted tokamak. Work continues to define both the regimes in which this fast cross-field transport is important, and the mechanisms responsible.

The separatrix lies at the base of a sharp edge pedestal. Most apparent in H-mode plasmas, the pedestal occupies only a few centimeters of distance, but within this narrow zone the electron density and temperature rise sharply from typical separatrix values of 1 to 2×10^{19} m^{-3} and 50-100 eV up to several 10^{19} m^{-3} and over 1 keV at the top of the pedestal. This pedestal has been seen to play important roles in enhancing energy confinement and improving stability at high plasma pressure. The pedestal arises due to a zone of reduced transport near the edge [7], however the physics that determines the pedestal height and width is not well understood. Because neutral deuterium penetrates into the pedestal, it is hypothesized that atomic physics may play a role in determining the pedestal parameters.

ATOMIC PHYSICS IMPORTANT TO NEUTRAL RECYCLING FROM THE MAIN CHAMBER WALL

The sources of atomic and molecular neutrals important for the SOL include recycling of the primary ion species at the main chamber wall, chemical and physical sputtering of the plasma facing surfaces, gas puffing of the primary species, intentionally puffed impurity species, and leakage of neutrals from the divertor region. In the plasma, the neutrals undergo many collisional processes with plasma electron, ions, and impurity ions. One of the more interesting transport problems is that of hydrogen, the primary fueling species. (We will not distinguish between hydrogen isotopes in this discussion.) The neutral hydrogen recycles from the plasma-facing wall as molecular H_2 and as atomic H. The recycling molecules result from surface recombination of adhered H and leave the wall at thermal energies. The recycling H results mostly from reflection of incident ions and will have higher initial kinetic energy. At thermal kinetic energy, the molecules have a short mean free path for electron impact dissociation, resulting in atomic H at Franck-Condon energies determined by the dissociation process. It is the transport of the atomic H that dominantly determines the fueling (re-ionization) profile and the flux of energetic charge exchange neutrals back to the wall.

At first glance, the collisional processes of atomic H with the plasma seem fairly simple. As the atoms drift away from the wall, they can be ionized by electron and proton impact, they can be ionized by impact and charge exchange with impurity ions, and they exchange velocity with the plasma ions through the symmetric charge exchange process. At the collision energies typical for the SOL, ion impact ionization is negligible, so only the electron and impurity charge exchange collisions contribute to the loss of atomic H and the net refueling of the plasma. Since the plasma ion temperature typically exceeds the energy of the neutrals, the symmetric charge exchange process serves to change the velocity of the neutrals and increase their average energy. This process increases the depth of penetration of neutrals and results in a flux of energetic neutrals back to the main chamber wall.

The atomic physics is significantly enriched due to the fast electron excitation rates for atomic H near the separatrix and in the pedestal region [8]. In the SOL, the excitation process is sufficiently fast to produce an average excitation fraction of about 10^{-3} in the first several principal quantum levels of the neutral H [9]. Because ionization rates of atomic H are a very strong function of excitation level, even this rather small excitation fraction will lead to a significant increase in the total ionization rate. The fractional populations of the n=2 and n=4 levels of atomic H as a function of electron density at electron temperatures of 10, 100 eV, and 1000 eV as obtained from the ADAS data base are shown in Fig. 3. The charge exchange cross section and electron impact ionization cross sections for n=2-5 are shown as a function of electron temperature in Fig. 4. These data are obtained from scaling relations found in the data series entitled "Atomic and Plasma-Material Interaction Data for Fusion", Volume 4, published by the IAEA. The cross sections increase dramatically with n. The charge exchange cross sections increase very dramatically, scaling as n^4 at collision energies relevant to the SOL and pedestal region.

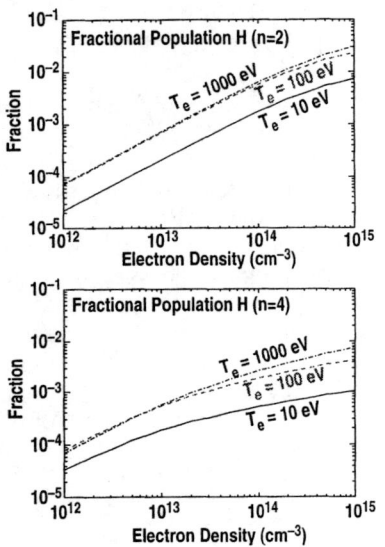

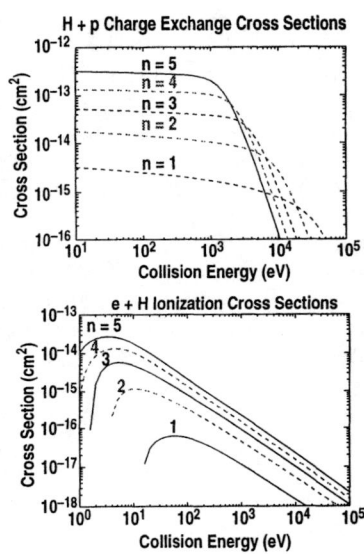

FIGURE 3. Excitation fraction of the n=2 and n=4 levels of atomic H as a function of electron density at an electron temperature of 100 eV.

FIGURE 4. The electron impact ionization and proton charge exchange cross sections for collisions with excited atomic H (n=1-5) is shown as a function of collision energy.

The total electron ionization and charge exchange (proton and carbon impurity) rates including the excited state contributions are shown in Fig. 5 as a function of plasma density. In the rather modest densities typical of the tokamak pedestal region, the inclusion of the excited state contribution is seen to double the total charge exchange rate. In this region, the proton charge exchange rate becomes comparable to, and in some cases greater than, the electron impact ionization rate. The ionization rate

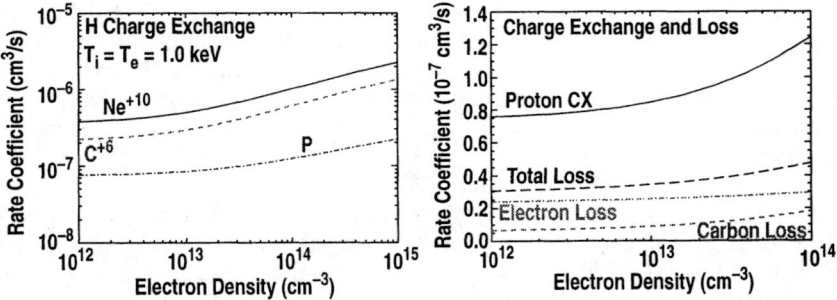

FIGURE 5. The rate coefficients for electron impact ionization and total charge exchange for protons and C^{+6} with atomic H including the effect of excited states up to n=5 are shown as a function of plasma density at an electron temperature of 1.0 keV. The scaling relations found in the IAEA data series are used, along with excitation fractions from the ADAS database

due to charge exchange with carbon impurity ions is also large in this region. Even though the carbon concentration is typically around 3%, it can contribute over 50% to the total loss rate of atomic H.

The accuracy of the data should also be of some concern. As an example of the typical level of accuracy, the charge exchange cross section of atomic H with fully stripped neon is shown as a function of energy in Fig. 6. Neon was first used in magnetic fusion to enhance the radiation in the core plasma and reduce the heat flux across the separatrix and to the plasma facing surfaces. During these enhanced radiation experiments, it was noticed that impurity injection also improved the energy confinement and overall plasma performance [10]. Cross sections obtained from the scaling relation are compared to classical trajectory Monte Carlo calculations [11]. The scaling relations are derived from a critical evaluation of available theory and experimental data. The charge exchange cross sections for the ground state obtained using the scaling relations are found to be somewhat higher than those from the CTMC calculations in the energy range of interest to the SOL.

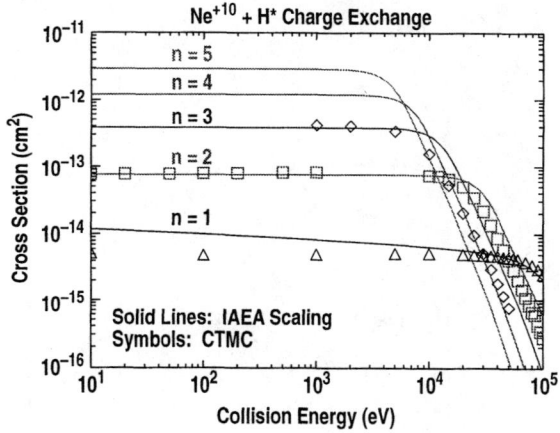

FIGURE 6. The charge exchange cross section for collisions of atomic H (n=1-5) with Ne^{+10} are shown as a function of collision energy from the scaling relations in the IAEA data series and from CTMC theoretical calculations.

DISCUSSION

Using the relatively simple example of the transport of atomic neutrals through the SOL and edge plasma, we find that much more atomic data than just the ionization and charge exchange cross sections for the ground state is needed. Because the plasma density and temperature are sufficient to provide excited state population fractions $\sim 10^{-3}$, and because charge exchange cross sections scale as n^4, we need these cross sections and their energy dependence for excited states of the neutral up to at least n=5.

Interpretation of spectroscopic data from impurities in the edge and core plasma requires another level of detail. Impurity line emission in the visible region of the spectrum is frequently used to monitor impurity ion species in the edge plasma.

Usually these lines result from transitions from highly excited states. Charge exchange between impurity ions and excited atomic H may be a significant contributor to the production of such line emission, so state resolved charge exchange cross sections are needed to accurately model the spectroscopic data. Visible emission of helium-like carbon ions in the divertor of DIII-D resulting from charge exchange with excited states of atomic H has been used to measure the parallel flow velocity of carbon between the divertor and SOL [12]. Also, excited states of atomic H in energetic neutral beams used for charge exchange spectroscopy have been shown to make a significant contribution to the photon production cross section in collisions of Ne^{+10} [13]. An example of product state resolved cross sections for H (n=2) collisions with Ne^{+10} calculated using the CTMC model is shown in Fig. 7.

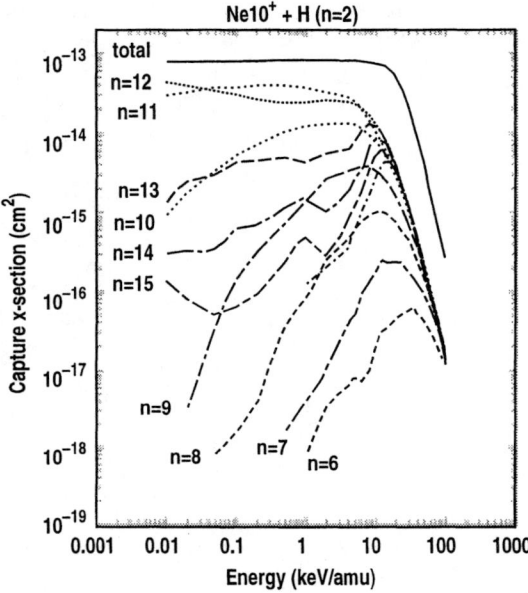

FIGURE 7. The product state resolved charge exchange cross sections for H (n=2) + Ne^{+10} from CTMC calculations are shown as a function of collision energy.

The need for such detailed charge exchange data to accurately model fusion plasmas and to interpret spectroscopic diagnostics has been understood for some time. Because magnetic confinement fusion plasmas cover a broad range of plasma density and temperature, such detail is needed in all the processes occurring in the plasma. The atomic physics community has been very responsive to the needs of plasma modelers for many years, and the databases, such as the ADAS database and the IAEA data center, continue to grow in response to the needs of the fusion community and to the resources provided by the atomic physicists.

In a similar fashion, a deep level of detail is needed to understand molecular processes in the edge plasma. Such processes are important for recycling, recombination, chemical sputtering, co-deposition and tritium entrapment.

ACKNOWLEDGMENTS

This is a report of work supported by the U.S. Department of Energy under Contract DE-AC03-99ER54463.

REFERENCES

1. Behringer, K., "Atomic Physics Requirements in Fusion Research," Atomic and Molecular Data and Their Applications, edited by K.A. Berrington and K.L. Bell, CP543, American Institute of Physics 1-56396-971-8/00, Melville, New York, 2002, p. 129.
2. Groebner, R. J., Baker, D. R., Burrell, K. H., Carlstrom, T. N., Ferron, J. R., Gohil, P., Lao, L. L., Osborne, T. H., Thomas, D. M., West, W. P., Boedo, J. A., Moyer, R. A., McKee, G. R., Deranian, R. D., Doyle, E. J., Rettig, C. L., Rhodes, T. L., and Rost, J. C., *Nucl. Fusion* **41**, 1789 (2001).
3. LaBombard, B., Umanski, M. V., Boivin, R. L., et al., *Nucl. Fusion* **40**, 2041 (2000).
4. Boedo, J. A., Rudakov, D. L., Moyer, R. A., Krasheninnikov, S., Whyte, D. G., McKee, G. R., et al., *Phys. Plasmas* **8**, 4826 (2001).
5. Luxon, J. L., and Davis, L. G., *Fusion Technol.* **8**, Part 2A, 441 (1985).
6. Stangeby, P. C., *The Plasma Boundary of Magnetic Fusion Devices*, edited by P. Stott and H. Wilhelmsson, Institute of Physics Publishing, Bristol, United Kingdom and Philadelphia, Pennsylvania, 2000.
7. Wagner, F., et al., *Phys. Rev. Lett.* **49**, 1408 (1982).
8. Johnson, L. C., *Astrophys. J.* 174 (1972).
9. Summers, H. P., *ADAS, Atomic Data and Analysis Structure*, Originally developed by the Jet Joint Undertaking, 1st Edition User Manual, 1994.
10. Ongena, J., Messiaen, A. M., Unterberg, B., et al., *Plasma Phys. and Control. Fusion* **41**, A397 (1999).
11. Olson, R. E., *J. Phys. B* **13**, 483 (1980).
12. Zaniol, B., Isler, R. C., Brooks, N. H., West, W. P., "Comparison of Experimental Measurement of Carbon Flow Velocities and Temperatures to UEDGE Predictions in the DIII-D Divertor," *Bull. Am. Phys. Soc.* **46**, 148 (2001).
13. Whyte, D. G., Wade, M. R., Finkenthal, D. K., et al., *Nucl. Fusion* **38**, 387 (1998).

II. ATOMIC AND MOLECULAR PHYSICS AND DATA

A. Atomic and Molecular Physics

Atomic Processes in Plasmas – An Overview

Hans R. Griem

Institute for Research in Electronics and Applied Physics
University of Maryland
College Park, Maryland 20742-3511

Abstract. Natural and man-made plasmas almost always not only contain free electrons and completely-stripped ions, but also bound-state atoms and positive, or even negative, ions, not to mention molecules, clusters, dust particles, etc. All of these constituents interact more or less strongly with electro-magnetic fields, and with each other. The often rather complex atomic processes or interactions to be discussed here are essential for our understanding of basic plasma properties (e.g., of the equation of state) and for many plasma applications, including materials processing. Special emphasis is placed on diagnostics and modeling under extreme plasma conditions and on the interpretation of unusual measurements, e.g., polarization spectroscopy, spectral line broadening, charge-exchange and beam-emission spectroscopy.

I. INTRODUCTION

Atoms and ions containing one or more bound electrons, immersed in plasmas, continue to play important roles in astrophysics and laboratory plasma physics. By means of spectroscopy, atomic physics enables us to study the composition and densities, temperatures and even electric and magnetic fields of and in these plasmas. Perhaps less appreciated, but equally important for such and other diagnostic applications, are the requirements for reasonably complete and accurate atomic data in the calculations of plasma equations of state, radiative opacities and kinetic models for non-equilibrium (non-LTE) situations. Much progress has been made in all these areas by the development of new experimental approaches and advanced computing, and in at least one case, also by analytic theory.

II. ELECTRON-ATOM (OR ION) COLLISIONS

Assuming that all atomic structure and radiative transition probabilities are available, the next common requirement is usually for the cross sections of electron collisions causing excitation and ionization, because cross sections and rates for the inverse processes can usually be inferred from them using the principle of detailed balancing. As shown, e.g., for Ar I [1] and Kr I [2] groundstate excitation, agreement between (electron loss or fluorescence) measurements and calculated cross sections for neutral atoms is usually within 25% for Ar I, but less satisfactory for Kr I, an even more complicated system. Especially serious are the disagreements between various Kr I excitation cross sections at high electron energies.

The energy resolution in excitation cross section measurements for ions has been much improved by the MEIBEL (merged electron-ion beam electron loss) method, e.g., for the 2s-2p excitation of C IV [3] and 3s-3p of Si III [4]. In both cases, agreement with close-coupling R-matrix calculations is excellent (after convolution with 0.17 or 0.24 eV FWHM Gaussians representing the experimental resolutions, and notwithstanding the importance of dielectronic resonances in case of Si III). Very similar agreement [5] was obtained between a recent [6] measurement of the Al III groundstate ionization cross section and R-matrix (with pseudo states) and CCC (converging close coupling) calculations. While reliable inelastic cross sections thus have become rather accessible, not much can be said about elastic cross sections, except that they are needed for the interpretation of spectral line broadening and polarization spectroscopy experiments (see Sec. IV and VI, respectively).

III. DIELECTRONIC AND THREE-BODY RECOMBINATION

Since most plasmas containing multiply-ionized atoms are far from LTE conditions, thus limiting the utility of the principle of detailed balancing, direct measurements and calculations of recombination rates continue to be of great interest. Examples for such situations are photo-ionized cosmic plasmas surrounding accretion-powered x-ray binaries or active galactic nuclei. In such plasmas the electron temperature can be as much as a factor ten lower for a given charge-state distribution compared with electron-ionized plasmas, necessitating measurements and calculations of dielectronic recombination, e.g., of Fe XX at $kT \gtrsim 1$ eV electron energies [7]. Such measurements can and have been performed using the heavy-ion test storage ring at Heidelberg, yielding energy-resolved dielectronic and radiative recombination rates. They clearly show the important role of $\Delta n = 0$ core excitations and allow detailed comparisons with various codes. The corresponding Maxwell-averaged rate coefficients are significantly larger than previous predictions for $kT_e < 100$ eV.

Another mechanism for accelerated recombination at high electron densities but also relatively low electron temperatures has been demonstrated recently by incorporating doubly-excited states and their resonant collisional deexcitation into time-dependent CR (collisional radiative) model calculations for carbon ions [8]. Considering, e.g., a beam of fully ionized carbon ions injected into a $N_e = 3 \times 10^{19}$ cm^{-3}, $kT_e = 3$ eV, background plasma, the CR code is used to predict the recombination under conditions where dielectronic recombination would normally be assumed not to be important. However, it turns out that the inclusion of C V and C IV doubly-excited states and their resonant deexcitation does have a significant effect on the calculated time histories, suggesting again that dielectronic rate coefficients may have been underestimated at low temperatures.

Also of great current interest is three-body recombination at extremely low temperatures and densities [9]. Although collisional capture into principal quantum number $n \approx 200$ levels is then very rapid, the captured electrons almost entirely go into the highest angular momentum ℓ states. Their radiative cascade decay is much too slow to be effective. However, ion-(Rydberg) atom collisions have large cross sections for elastic transfer to lower ℓ states, which then complete the recombination by radiative decay. Because of the invariance of total angular momentum **L** and of the Runge-Lentz vector **A**, elegant expressions can be derived for the $n\ell \rightarrow n\ell'$ transfer probabilities [10,11] for applications in future kinetic models. They should also be useful in spectral line broadening calculations for hydrogen and Rydberg atoms in general at low densities, at which the impact approximation [12] is valid even for ions as perturbers.

IV. COLLISIONAL BROADENING OF SPECTRAL LINES

In most sufficiently dense plasmas for Stark broadening to be important, only the electron effects can be described in the impact approximation just mentioned. However, besides inelastic (and super elastic) cross sections, one now also needs elastic cross sections. If both upper and lower levels of the line of interest are significantly perturbed, the various elastic cross sections enter the expressions for such line widths in a rather delicate way [13], making it difficult to estimate corresponding theoretical errors. For most (isolated) lines, the resulting line shapes are slightly shifted Lorentzians, the widths and shifts increasing linearly with electron density, until lines from different levels begin to overlap. Many measurements and (mostly semiclassical) calculations are listed in the NIST Physics Laboratory Physical Reference Data and have been critically reviewed since over about 25 years.

In contrast to semiclassical calculations, which tend to agree with measurements on lines from neutral atoms and low charge state ions, often to within 20%, most fully quantum-mechanical calculations are rather recent, especially those based on the CCC (convergent close-coupling) method [14]. A very recent example are the calculations [15] for 3s-3p transitions in Li-like ions. Two trends seem evident in the comparisons with measurements and semiclassical calculations: for low charge states there is reasonable agreement with measurements and some previous calculations. Two of the semiclassical calculations, however, yield overestimates by as much as 40%. For high charge state ions, on the other hand, both measurements and semiclassical calculations give significantly larger widths than the CCC calculations, by as much as a factor of 3. This may be due to motional Doppler effects in the experimental light source for these lines, which are relatively insensitive to Stark broadening.

V. HEAVY PARTICLE COLLISIONS; CHARGE EXCHANGE

Besides the elastic ℓ-changing ion-atom (or ion) collisions just mentioned, electron transfer from donor atoms (or ions) to receptor ions may also have large cross sections. These are, however, very state-specific, whereas the former cross sections tend to increase rapidly with principal quantum number [10,11]. Interest in ion-ion charge-transfer into specific receptor states was generated recently by observations of anomalous intensities of normally weak He- and Li-like satellite lines to the Ly_α and He_α resonance lines of Al in a laser-produced plasma [16]. These observations were inconsistent with kinetic models allowing only for electron collisions; but charge transfer into the particular upper states involved here was soon recognized as a likely cause for the discrepancies [17,18]. This possibility was supported by spatial resolution in these measurements, indicating that these satellites were especially pronounced near the target surface where highly ionized ions might interact with lower charge states and even neutrals and where K_α lines were observed. Furthermore, charge exchange cross sections between ions can perhaps surprisingly be rather large, e.g., for the C^{3+} and He^{2+} system as measured in the ion-ion cross-beam experiments [19] in Giessen. Two-electron transfer cross sections may be assumed of comparable magnitude, e.g., for fully-ionized Si ions interacting with L-shell ions below the target surface, leading to doubly-excited He-like Si, etc. Including such reactions, opacities, and nonthermal electrons in the kinetic model then yields much improved agreement with the observed satellite intensities at reasonable ion beam intensities [17,18].

VI. POLARIZATION SPECTROSCOPY

Most plasma spectroscopy measurements are still made without much attention to polarization, i.e., assuming implicitly that the radiation is unpolarized. However, it has long been known that, e.g., macroscopic magnetic fields can lead to polarization. As in astronomy, one can therefore infer the strength of magnetic fields along the line of sight by measuring more or less subtle differences between (mostly Doppler broadened) Zeeman patterns for parallel (π) and perpendicular (σ) polarizations. For laser-produced plasmas, early measurements on the C V $2s^3S-2p^3P$ transitions of this kind were especially successful for ring-shaped laser focal spots [20], in part because the stronger radial gradients would drive larger currents but also because the cooler plasma core would absorb the line emission from the receding portion of the radially- and axially-expanding plasmas.

To interpret the observed line profiles, it was necessary to calculate the Zeeman shifts for the magnetic sublevels J',M' of 2^3P_3 for arbitrary fields and also the relative intensities of the π and σ components (J',M',M), M being the magnetic quantum number of the 2^3S_1 states. The observed differences in widths and intensities of the Doppler-smeared π and σ emissions indicated an azimuthal magnetic fieldstrength of about 20 T.

Besides such polarization due to anisotropy caused by macroscopic fields (termed polarization of the first kind [21]), there may also be polarization due to alignment of atoms or ions in particular M-levels due to beam-like, rather than isotropic, electron velocity distributions. Such beams are especially strong in pico-second (high intensity) laser-produced plasmas. This polarization of the second kind was the interpretation [22] for a significant dependence of the Al XII resonance line intensity on polarization, while the intercombination line was not polarized, presumably because its upper level is mostly populated by recombination with low-energy, isotropic thermal electrons.

Returning to polarization-dependent line shapes, Al XIII Ly-β and Ly-γ lines emitted from a high current z-pinch where assumed [23] to be mainly Stark-broadened by electric fields in electron-plasma waves generated by currents in the axial direction. Polarization was reported only for the γ line, for which Doppler and opacity effects are less important than for the β line.

VII. CHARGE-EXCHANGE RECOMBINATION SPECTROSCOPY (CERS)

Two practical difficulties in diagnosing high temperature laboratory plasmas spectroscopically can be overcome by injecting diagnostic neutral beams, usually hydrogen or deuterium atoms. These firstly provide electrons to completely ionized atoms into selected states, which then radiate experimentally convenient lines through strong $\Delta n = 1$ transitions. Secondly, the emission volume is reasonably well defined by the intersection of the beam with the sight line of the collection optics, in contrast to the usual integration along the line of sight in standard emission spectroscopy. Any shift of the CERS emission lines should thus indicate the flow velocity component along the sight line at the point of intersection.

However, since the radiating ions execute gyro-motion in a confining magnetic field in a tokamak device, for example, and since the transfer cross sections are dependent on the relative beam-atom, ion velocities, shifts associated with this effect must be carefully evaluated and subtracted [24], accounting for the delay between electron capture and line emission. These considerations not only require realistic atom-ion charge transfer cross sections, but also a rather detailed CR model, with results again being sensitive to ℓ mixing, elastic electron-ion (C IV, $n=8$) cross sections. Because the poloidal velocities v_θ are much smaller than the toroidal velocities, v_ϕ, such corrections are especially important for v_θ and had therefore to be tested by comparing shifts obtained from upward and downward poloidal views in the torus.

VIII. BEAM EMISSION SPECTROSCOPY (BES)

Line emission from excited neutral beam atoms is also of considerable interest for the measurements of magnetic fields via the motional Stark effect in the $\underline{v}_B \times \underline{B}$ $\underline{E}$-field in the atoms' frame of reference. The ensuing spectra [25] resemble classical Stark spectra, except that the individual components overlap because of Doppler effects due to the beam energy spread. At the relatively low fields in this reversed-field pinch experiment, $B \approx 0.5$ T, polarization effects are not very important, and fine structure of the Balmer-α line observed here may not be entirely negligible. Statistical distribution of populations in the sublevels was assumed, but may of course not be entirely achieved at the relatively low electron densities.

IX. CONCLUSIONS

In the over 50 years of research, the author has enjoyed fruitful interactions with many colleagues from his own and other physics and astronomy subfields and many students. Also, in the present case, several colleagues were very helpful with their comments and contributions. Clearly, much improved measurements and computations for many processes have been achieved, facilitating astrophysical and laboratory plasma applications over very large parameter ranges in electron density from 0.1-10^{25}cm^{-3}. Still, there is a lot more work to be done on electron-neutral atom collisions, kinetic models for recombining plasmas and on elastic collisions for line broadening and polarization spectroscopy. Last but not least, cross sections for charge transfer and double charge transfer between ions will hopefully be measured and calculated in the near future.

ACKNOWLEDGEMENTS

Partial support by the National Science Foundation and the U.S. Department of Energy is gratefully acknowledged.

REFERENCES

1. Dasgupta, A., Blaha, M., and Giuliani, J. L., *Phys. Rev. A* **61**, 012703 (2000).
2. Dasgupta, A., Bartschat, K., Vaid, D., Grune-Grzhimaito, A.N., Madison, D. H., Blaha, M., and Guiliani, J. L., *Phys. Rev. A* **64**, 052710 (2001).
3. Bannister, M. E., Durić, Woitke, O, Dunn, G. H., Chung, Y.-S., Smith, A. C. H., and Wallbank, B, 11th APS Topical Conference on Atomic Processes in Plasmas, edited by E. Oks and M. S. Pindzola, AIP Conference Proceedings 443, New York, 1998, p. 149. See also Bannister, M. E, et al., *Phys. Rev. A* **57** 278 (1998) and Janzen, P. H. et al., *Phys. Rev. A* **59**, 4821 (1999).
4. Wallbank, B., N. Durić, N., Woitke, O., Zhou, S., Dunn, G. H., Smith, A.C.H., and Bannister, M. E., *Phys. Rev. A* **56**, 3714 (1997). See also Reisenfeld, D. B. et al., *Phys. Rev. A* **60**, 1153 (1999).
5. Bartschat, K., AIP Conference Proceedings 443, 11th APS Topical Conference on Atomic Processes in Plasmas, edited by E. Oks and M. S. Pindzola, New York, 1998, p. 121.
6. Thomason, J.W.G., and Peart, B., *J. Phys. B* **31**, L201 (1998).
7. Savin, D.W. et al., *Astrophys. J. Suppl.* **138**, 377 (2002).
8. Ralchenko, Yu. V., and Maron, Y., *J. Quant. Spectrosc. Radiat. Transfer* **71**, 601 (2001).
9. Flannery, M. R., and Vrinzeanu, D., 11th APS Topical Conference on Atomic Processes in Plasmas, edited by E. Oks and M. S. Pindzola, AIP Conference Proceedings 443, New York, 1998, p. 317.
10. Vrinzeanu, D., and Flannery, M. R., *Phys. Rev. A* **63**, 03270 (2001).
11. Vrinzeanu, D., and Flannery, M. R., *J. Phys. B* **34**, L1 (2001).
12. Griem, H. R., *Principles of Plasma Spectroscopy*, Cambridge University Press, 1997.
13. Baranger, M., *Phys. Rev.* **112**, 855 (1958).
14. Bray, I., *Phys. Rev. A* **49**, 1066 (1994).
15. Ralchenko, Yu. V., Griem, H. R., and Bray, I., to be published.
16. Elton, R. C., Cobble, J. A., Griem, H. R. Montgomery D. S., Mancini, R. C., Jacobs V. L., and Behar, E., *J. Quant. Spectrosc. Radiat. Transfer* **65**, 185 (2000).
17. Rosmej, F. B. et al., to be published.
18. Rosmej, F. B., these proceedings.
19. Trassl, R., Bräuning, H., Diemar, K. V., Melchert, F., Salzborn, E., and Hofmann, I., 12th APS Topical Conference on Atomic Processes in Plasmas, edited by R. C. Mancini and R. A. Phaneuf, AIP Conference Proceedings 547, New York, 2000, p. 157.
20. McLean, E. A., Stamper, J. A., Manka, C. K., Griem, H. R., Droemer, D. W., and Ripin, B. H., *Phys. Fluids* **27**, 1327 (1984).
21. Fujimoto, T., Proceedings of the 3rd Japan-US Workshop on Plasma Polarization Spectroscopy, edited by P. Beiersdorfer, Livermore, 2002, p. 1.
22. Kieffer, J. C., Matte, J. P., Pépin, H., Chaker, M., Beaudoin, Y., and Johnston, T. W., *Phys. Rev. Lett.* **64**, 480 (1992).

23. Oks, E., 13th International Conference on Spectral Line Shapes, edited by M. Zoppi and L. Ulivi, AIP Conference Proceedings 386, New York, 1997, p. 3.
24. Bell R., and Synakowski, E., 12th APS Topical Conference on Atomic Processes in Plasmas, edited by R. C. Mancini and R. A. Phaneuf, AIP Conference Proceedings 547, New York, 2000, p. 39.
25. Den Hertog, D. J., Craig, D., Fiksel, G. and the MST Group, Davydenko, V. I., Ivanov, A. A., and Lizunoy, A. A., Proceedings of the 3rd Japan-US Workshop on Plasma Polarization Spectroscopy, edited by P. Beierdorfer, Livermore, 2002, p. 205.

Benchmark Calculations for Electron Collisions with Complex Atoms

Klaus Bartschat

Department of Physics and Astronomy, Drake University
Des Moines, Iowa 50311, USA

Abstract. Advantages and disadvantages of various numerical methods to generate atomic data for electron–atom collisions are reviewed. Selected examples illustrate the current status regarding reliable and sufficiently complete data for such collisions, particularly for systems of interest for plasma processing and the lighting industry.

INTRODUCTION

In recent years, major progress has been achieved in calculating reliable cross-section data for electron scattering from atoms and ions. In particular, the "convergent close-coupling" (CCC) [1,2] and "R–matrix with pseudo-states" (RMPS) [3–5] methods have been extremely successful in describing elastic scattering as well as electron-impact excitation and ionization of light quasi-one and quasi-two electron targets, such as atomic hydrogen, helium, the alkalis, and the alkali-earth elements. The major advantage of the above methods, compared to standard close-coupling approaches with only discrete states included in the expansion of the scattering wavefunction, is their ability to account for the effect of the target continuum through a large number of pseudo-states. It is now known that this coupling can be very important for transitions between discrete states, particularly if these transitions are optically forbidden. In addition, total ionization cross sections can be extracted from these models by re-interpreting the excitation of positive-energy pseudo-states as a measure of ionization.

Despite the progress in the numerical treatment mentioned above, there are still significant challenges that need to be addressed. Specifically, the available data for electron collisions with noble gases other than helium (i.e., Ne, Ar, Kr, and Xe) are scarce and neither measurements nor theoretical predictions can realistically claim a high degree of accuracy. For other heavy targets, such as Cs, Ba, or Hg, some progress has been made recently, but the importance of relativistic effects has not been investigated in great detail, particularly in the near-threshold resonance region. Furthermore, the above systems present major challenges already in the structure description, due to strong term dependence in the one-electron orbitals.

In light of this structure problem, perturbative methods based on the Born-series expansion may still give results of comparable accuracy to those obtained in very sophisticated coupled-channel approaches. Finally, the field is wide open for truly complex targets such as Mo, which is a promising candidate for a mercury-free lighting concept [6], and other transition elements.

In the next section, the advantages and disadvantages of various numerical methods currently used to generate atomic data for electron collisions with atomic targets will be reviewed, with particular emphasis on the difficulties associated with heavy, complex, neutral targets. This is followed by a few examples to illustrate the current status regarding reliable and sufficiently complete data for such collisions, particularly for systems of interest for plasma processing and the lighting industry.

NUMERICAL METHODS

The numerical methods currently applied to calculate data for electron–atom collisions can be classified in two broad categories, perturbative and non-perturbative. The former are based on the Born-series expansion, with a very popular approach being the first-order distorted-wave method, while most of the latter are built on the close-coupling expansion, including recently developed time-dependent approaches. For a summary of recent developments in the numerical solution of the close-coupling equations, see ref. [5].

Distorted-Wave Methods

A semi-relativistic first-order distorted-wave approximation (DWBA) was described by Madison and Shelton [7] and later modified by Bartschat and Madison [8] to treat electron collisions with heavy noble gases and also with mercury. Other variants include the first-order many-body theory (FOMBT) of the Los Alamos group [9], the semi-relativistic version developed and described by Dasgupta et al. [10], and the relativistic distorted-wave method (RDW) of the Toronto group [11]. The basic idea behind all these methods is to calculate distorted waves for the projectile electron by solving the differential equation

$$\left[\frac{d^2}{dr^2} - \frac{l(l+1)}{r^2} - 2\{U(r) + V_r(r) - E\}\right]\chi_{E,l}(r) = 0. \qquad (1)$$

Here $U(r)$ is the static Coulomb potential, often modified by including terms to account for electron exchange and the charge distortion (polarization) of the target, while V_r contains relativistic effects. [Instead of (1), the full-relativistic RDW solves the corresponding Dirac equation.] The distorted waves are then used to calculate transition matrix elements, treating to first order the part of the projectile–target interaction not explicitly included in the distortion potentials.

Since the details of the methods can be found in the original references, we only mention here their principal advantages and disadvantages. The major problem with the distorted-wave approach is the fact that channel coupling is neglected. Consequently, the method cannot account for near-threshold Feshbach-type resonances, it is generally much less reliable for optically forbidden than for optically allowed transitions, and further problems can occur if the calculation is performed without explicit unitarization of the scattering matrix [12]. On the other hand, the method is fast and relatively easy to implement. It also allows for flexibility in the target description for the initial and final states since, in principle, the orbitals can be specifically optimized for the states of interest in a given transition.

Close-Coupling Methods

The close-coupling approximation has been a standard method of treating low-energy scattering, both elastic and inelastic, for many years. This non-perturbative method is based upon an expansion of the total wavefunction for a collision system in terms of a sum of products that are constructed from target states Φ_i and unknown functions $F_{E,i}$ describing the motion of the projectile for a total (target + projectile) collision energy E. If relativistic effects are neglected, the wavefunction for each total orbital angular momentum L, total spin S, and parity π is expanded as

$$\Psi^{LS\pi}(r_1,\ldots,r_{N+1}) = \mathcal{A} \sum_i\!\!\!\!\!\!\int \Phi_i^{LS\pi}(r_1,\ldots,r_N,\hat{r}) \frac{1}{r} F_{E,i}(r). \qquad (2)$$

Here $\sum\!\!\!\!\!\!\int$ denotes a sum over all discrete and an integral over all continuum states of the target, and $\mathcal{A}$ is the antisymmetrization operator that accounts for the indistinguishability of the projectile and the target electrons. Furthermore, the angular and spin coordinates of the projectile electron (collectively denoted by $\hat{r}$) have been coupled with the target states to produce the "channel functions" $\Phi_i^{LS\pi}(r_1,\ldots,r_N,\hat{r})$.

The unknown radial wavefunctions $F_{E,i}(r)$ are determined from the solution of a system of coupled integro-differential equations given by

$$\left[\frac{d^2}{dr^2} - \frac{\ell_i(\ell_i+1)}{r^2} + k^2\right] F_{E,i}(r) = 2 \sum_j\!\!\!\!\!\!\int V_{ij}(r) F_{E,j}(r) + 2 \sum_j\!\!\!\!\!\!\int W_{ij} F_{E,j}(r) \qquad (3)$$

with the direct coupling potentials $V_{ij}(r)$ and exchange terms $W_{ij} F_{E,j}(r)$.

The above coupled integro-differential equations may be generalized to a relativistic framework, and the collision problem essentially consists of finding the solution to this system for each total energy subject to the appropriate boundary conditions. This can be achieved by various iterative, noniterative, or algebraic methods. While the "convergent close-coupling" (CCC) method [1,13] can presently be applied to both quasi-one-electron and quasi-two-electron targets, the standard R-matrix program of the Belfast group [14] may also be used for complex atomic and ionic targets. The "R−matrix with pseudo-states" (RMPS) approach [3,4] has recently been implemented as well and extends the low-energy R−matrix method to higher energies. The principal advantage of both CCC and RMPS over standard discrete-state-only treatments lies in the fact that the coupling of high-lying discrete states and of the target continuum to the states involved in the transitions of interest is accounted for by including a large number of square-integrable pseudo-states in the close-coupling + correlation expansion (2). Also, re-interpreting the results for excitation of the pseudo-states as ionization allows for the treatment of such processes as well.

The principal advantages of close-coupling methods, including a time-dependent version described by Pindzola et al. [15], is their foundation on an expansion of the total wavefunction which, in theory, should converge to the correct answer. In practice, however, there remain significant problems to overcome. These include the treatment of the target continuum (with major progress being achieved through the use of a large number of pseudo-states) and the proper treatment of relativistic effects. Most importantly, however, recent work on heavy noble gases [12,16] and complex open-shell targets [17] has revealed potentially major difficulties associated with the target description itself. Since the standard implementations of the close-coupling method assume a fixed set of mutually orthogonal one-electron orbitals for *all states* included in the expansion of the total wavefunction, a strong term dependence can be problematic. Although pseudo-orbitals may increase the flexibility, this cure has its limits and, in fact, may cause additional problems through so-called pseudo-resonances.

Combination of Methods

In principle, perturbative and non-perturbative methods can complement each other to cover a wide range of incident energies. This was demonstrated successfully by Maloney et al. [18] for electron-impact excitation of Ar from the metastable $(3p^54s)^3P_{0,2}$ states, with one of the requirements being a consistent target description. A practical method was recently suggested by Kim [19]. It is based on a modification of first-order plane-wave Born results, using experimental oscillator strengths and an empirical recipe to shift the predicted position of the cross-section maximum. While lacking a justification based on "first principles", the method is very fast and apparently successful for strong, optically allowed transitions.

SELECTED RESULTS

Figure 1 shows results for the total (elastic + excitation + ionization) cross section for electron collisions with Cs atoms in their ground state $(6s)^2S_{1/2}$. Except for the data set measured by Brode [20], there is reasonable agreement between the other experimental data sets and theoretical predictions from a non-relativistic CCC calculation [24] and a semi-relativistic Breit-Pauli RMPS model [25]. For the

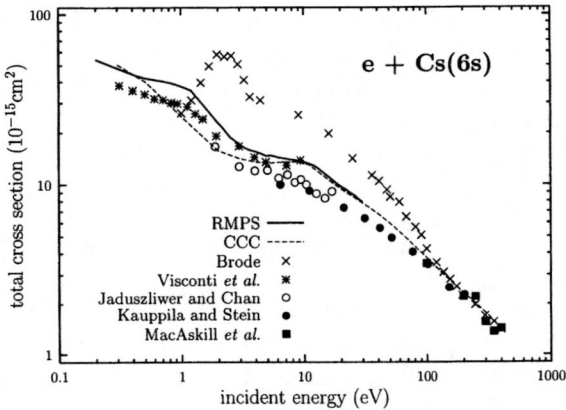

FIGURE 1. Total cross section for electron collisions with Cs atoms in their ground state $(6s)^2S_{1/2}$. Various experimental data sets [20–24] are compared with predictions from a non-relativistic CCC calculation by Bray [24] and a semi-relativistic RMPS model [25].

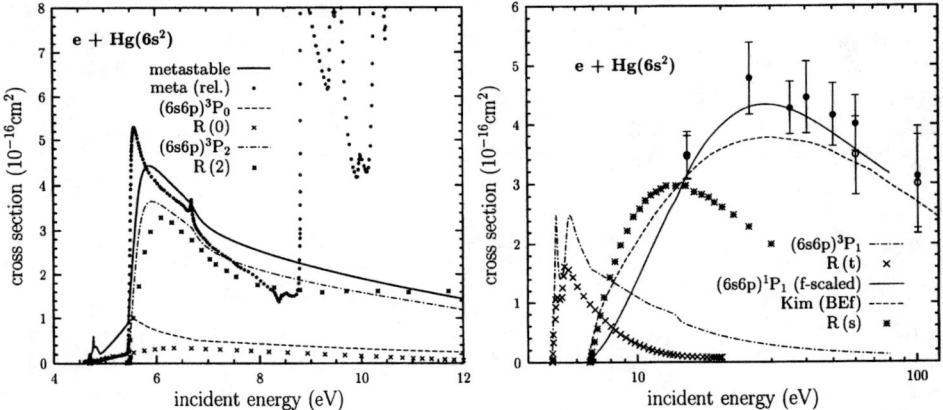

FIGURE 2. Cross sections for electron-impact excitation of the (6s6p) states of mercury. The left panel shows the individual results for the metastable states $^3P_0$ and $^3P_2$, as obtained in a five-state semi-relativistic Breit-Pauli R-matrix calculation, in comparison with a data set suggested by Rockwood [29] ("R") and the (relative) metastable yield determined by Newman et al. [26]. Results for excitation of the two $J=1$ states are shown on the right, together with Rockwood's suggestions for the triplet (t) and the singlet (s) states. For the $^1P_1$ state, Kim's "BEf" predictions [19] and two experimental data sets [27,28] are also presented.

total cross section in this particular system at the energies shown, it is apparently not necessary to account for relativistic effects. Note, however, that the predictions differ most significantly in the low-energy regime of incident energies below 3 eV. The reason for this discrepancy is currently under investigation.

Figure 2 shows state-selected results for electron-impact excitation of the (6s6p) states of mercury. Even a relatively simple 5-state Breit-Pauli R−matrix calculation is in remarkable agreement with the (relative) experimental data for the metastable yield measured by Newman et al. [26]. Note that the model does not include inner-shell excitation and the experimental data are likely affected by cascading for energies above $\approx 8\,\text{eV}$ incident energy. Furthermore, the theoretical predictions are in excellent agreement with two absolute cross-section measurements for the $^1P_1$ state — provided the results for this state are re-scaled to ensure the correct optical oscillator strength for the transition. Also shown are the cross section values suggested by Rockwood [29] that have been used to model discharge lamps containing mercury [30]. The figure clearly suggests that an *ab initio* collision calculation should provide more reliable data than the indirect approach based on transport data.

Results for the direct excitation cross sections of the states in the $(4p^55p)$ manifold from the initial metastable state $1s_5$ $(^3P_2)$ in Kr are presented in figure 3 as function of the incident-projectile energy, in comparison with the experimental data of Mityureva et al. [31] and of Kolokolov and Terekhova [32]. The predictions from four theoretical treatments [16], based upon two different distorted-wave approaches and two Breit-Pauli R−matrix models with 15 (BP15) and 51 (BP51) coupled states, respectively, is generally fair, while agreement with the experimental data is virtually non-existent. In order to even fit the experimental points on the graphs without extending the scale dramatically, the published values had to be reduced by one to two orders of magnitude ! However, based on previous experience for electron collisions with metastable argon atoms [33–35], this disagreement is not really surprising. Also, recent preliminary experimental results from the Wisconsin group [36] suggest much better agreement with the theoretical predictions.

Our last example deals with Mo, a truly complex open-shell transition element. A part of the MoI bound spectrum is shown in figure 4, only listing the LS-terms rather than the enormous number of individual fine-structure levels. Looking at this spectrum, it is immediately clear that both the structure and collision models will have to allow for significant compromises in terms of accuracy. Currently available computational resources make it impossible to account for all the term dependence to be expected for the one-electron orbitals, even in a purely nonrelativistic model.

Figure 5 exhibits a few predictions for electron-impact excitation of molybdenum from the $[4d^5(a^6S)5s]a^7S$ ground state. The left panel depicts results for elastic scattering and excitation of the two optically allowed transitions to the $[4d^5(a^6S)5p]z^7P^o$ and the $[4d^45s(a^6D)5p]y^7P^o$ states, while the right panel shows data for spin-changing transitions coupling the septet and quintet target states. In detail, we show results from 15-state, 29-state, and 67-state non-relativistic R−matrix (close-coupling) models, with the one-electron orbitals obtained from the atomic structure package CIV3 [37]. In addition, the 15-state calculation was performed with and without adjustment of the Hamiltonian matrix elements to ensure the correct experimental target thresholds. Furthermore, we show results obtained in a 56-state calculation with a target description generated by **SUPERSTRUCTURE** [38], and re-

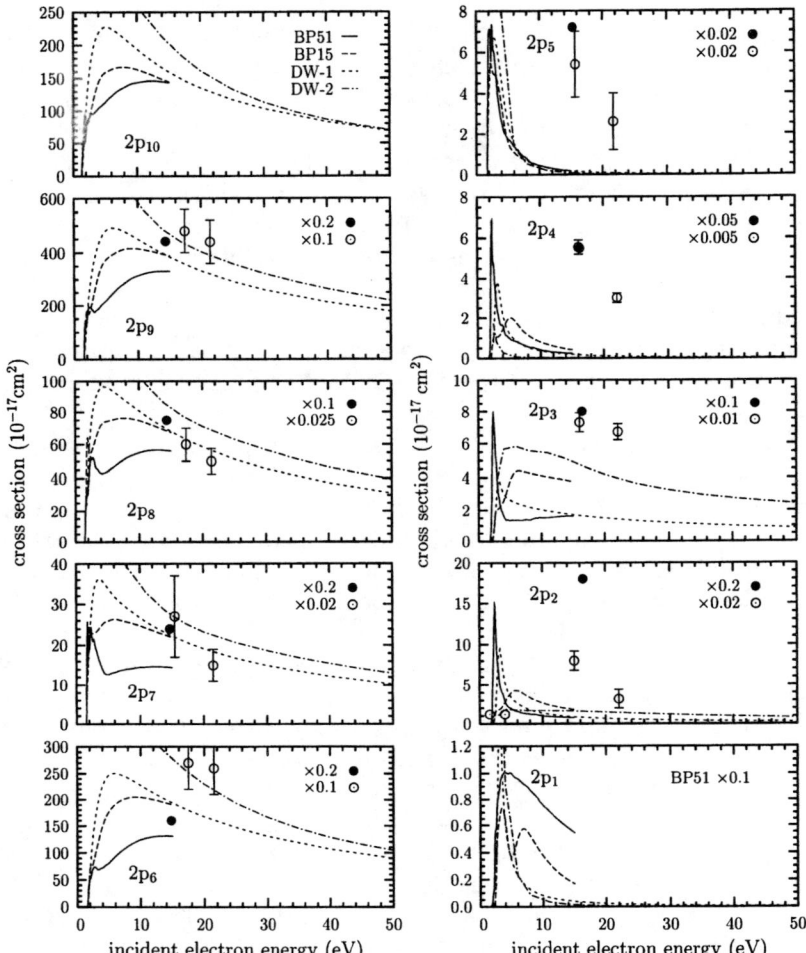

FIGURE 3. Cross sections for electron-impact excitation of krypton from the $1s_5$ ($J=2$) state to the $4p^5 5p$ manifold as a function of the collision energy. The experimental data of Kolokolov and Terekhova [32] (solid circles) and Mityureva et al. [31] (open circles) have been multiplied by the factors indicated.

sults published by Badnell et al. [39]. The agreement, or lack thereof, between the various predictions indicates the sensitivity of the results to the details of the target structure, to the amount of channel coupling included, and to the effect of pseudo-resonances. Despite the complexity seen in the figure, it is worth pointing out that the results from the presumably best model, namely the 67-state close-coupling expansion, have been used with some success in modeling a moly-oxide discharge lamp [40].

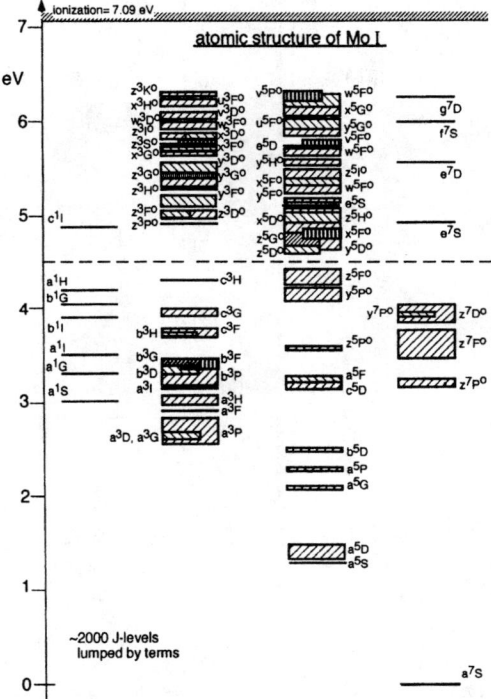

FIGURE 4. Simplified energy term spectrum of MoI. The shadowed areas indicate the range covered by the various fine-structure levels of the LS-term listed. The quintet and septet states below the dashed line are of primary interest for lighting applications.

CONCLUSIONS

Despite the progress in calculations of electron collisions with light quasi-one-electron and quasi-two-electron targets, there remain significant challenges to overcome in this field. For complex open-shell systems, term-dependence in the target structure is a major problem, and the need for a consistent treatment of both the N-electron target and the $(N+1)$-electron collision system complicates the situation. Given the difficulty and the cost of experimental efforts, a coordinated effort between experimental and theoretical data producers and the community of potential data users seems mandatory in order to fulfill the many data needs.

ACKNOWLEDGMENTS

I would like to thank I. Bray, A. Dasgupta, Y. Fang, A.N. Grum-Grzhimailo, D.H. Madison, and J.W. McConkey for their contributions. This work was supported by the US National Science Foundation under grant PHY-0088917.

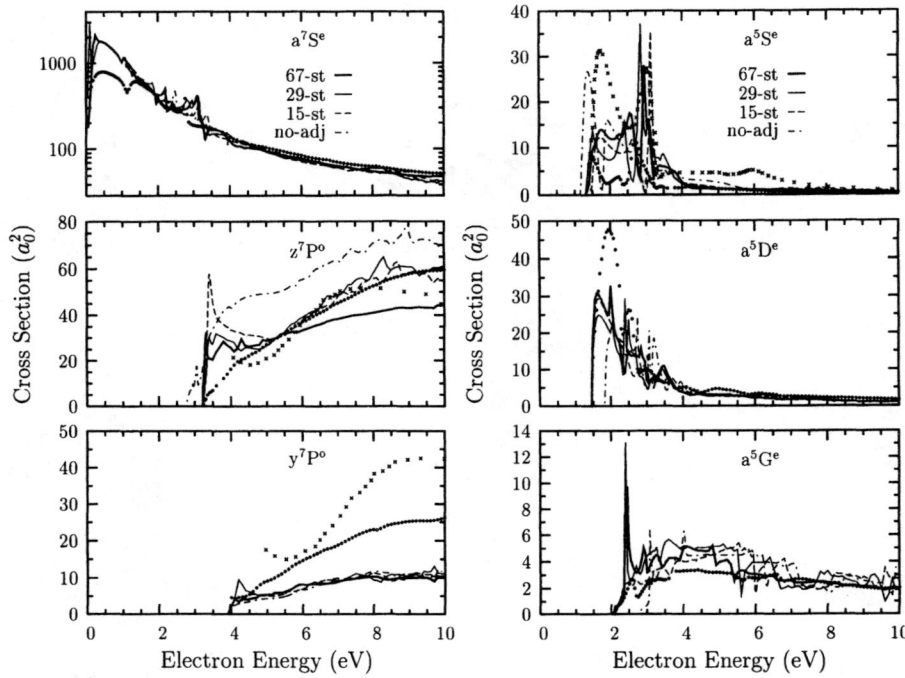

FIGURE 5. Cross sections for elastic electron scattering from the a^7S ground state of MoI, and for electron-impact excitation of the z^7P^o and y^7P^o states from the a^7S state [17]. The various models using CIV3 orbitals are described in the text. The dots represent the 56-state calculation with SUPERSTRUCTURE orbitals and the crosses the predictions of Badnell et al. [39].

REFERENCES

1. I. Bray and A.T. Stelbovics, Adv. At. Mol. Phys. **35**, 209 (1995).
2. I. Bray, in *Atomic and Molecular Data and Their Applications*, K.A. Berrington and K.L. Bell (eds.), AIP Conf. Proc. #543, 178 (2000).
3. K. Bartschat, E.T. Hudson, M.P. Scott, P.G. Burke, and V.M. Burke, J. Phys. B **29**, 115 (1996).
4. K. Bartschat, Comp. Phys. Commun. **114**, 168 (1998).
5. K. Bartschat, in *Atomic Processes in Plasmas*, E. Oaks and M.S. Pindzola (eds.), AIP Conf. Proc. #443, 121 (1998).
6. V.A. Shamamian, D.J. Vestyck, J.L. Giuliani, and J.E. Butler, U.S. Patent #6,157,133 (2000).
7. D.H. Madison and W.N. Shelton, Phys. Rev. A **7**, 499 (1973).
8. K. Bartschat and D.H. Madison, J. Phys. B **20**, 5839 (1987).
9. L. Machado, E.P. Leal, and G. Csanak, J. Phys. B **15**, 1773 (1982).
10. A. Dasgupta, M. Blaha, and J. L. Giuliani, Phys. Rev. A **61**, 012703 (2000).
11. T. Zuo, R.P. McEachran, and A.D. Stauffer, J. Phys. B **24**, 2853 (1991).

12. A. Dasgupta, K. Bartschat, D. Vaid, A.N. Grum-Grzhimailo, D.H. Madison, M. Blaha, and J.L. Giuliani, Phys. Rev. A **64**, 052710 (2001).
13. D.V. Fursa and I. Bray, Phys. Rev. A **52**, 1279 (1995).
14. K.A. Berrington, W. Eissner, and P.H. Norrington, Comp. Phys. Commun. **92**, 290 (1995).
15. M.S. Pindzola, F. Robicheaux, N.R. Badnell, and T.W. Gorczyca, Phys. Rev. A **56**, 1994 (1997).
16. A. Dasgupta, K. Bartschat, D. Vaid, A.N. Grum-Grzhimailo, D.H. Madison, M. Blaha, and J.L. Giuliani, Phys. Rev. A **65**, 042724 (2002).
17. K. Bartschat, A. Dasgupta, and J.L. Giuliani, J. Phys. B **35**, in press (2002).
18. C.M. Maloney, J.L. Peacher, K. Bartschat, and D.H. Madison, Phys. Rev. A **61**, 022701 (2000).
19. Y-K. Kim, Phys. Rev. A **64**, 032713 (2001).
20. R.B. Brode, Phys. Rev. **34**, 673 (1932).
21. P. Visconti, J. Slevin, and K. Rubin, Phys. Rev. A **3**, 1310 (1970).
22. B. Jaduszliwer and Y.C. Chan, Phys. Rev. A **45**, 197 (1992).
23. W.E. Kauppila and T.S. Stein, private communication (2001) [S.P. Parikh, *PhD Thesis*, Wayne State University, unpublished (1992).]
24. J.A. MacAskill, J. Domyslawska, W. Kedzierski, J.W. McConkey, and I. Bray, J. Electron Spectroc. and Rel. Phen., in press (2002).
25. K. Bartschat and Y. Fang, Phys. Rev. A **62**, 052719 (2000).
26. D.S. Newman, M. Zubek, and G.C. King, J. Phys. B **18**, 985 (1985).
27. F.J. Peitzmann and J. Kessler, J. Phys. B **23**, 2629 (1990).
28. R. Pajanatović, V. Pejčev, M. Konstantinović, D. Filipović, V. Bocvarski, and B. Marinković, J. Phys. B **26**, 1005 (1993).
29. S.D. Rockwood, Phys. Rev. A **8**, 2348 (1973).
30. R.B. Winkler, J. Wilhelm, and R. Winkler, Beitr. Plasmaphys. **22**, 401 (1981).
31. A.A. Mityureva, N.P. Penkin, and V.V. Smirnov, Opt. Spektrosk. **67**, 785 (1989) [Opt. Spectrosc. (USSR) **67**, 461 (1989)].
32. N.B. Kolokolov and O.V. Terekhova, Opt. Spektrosk. **86**, 547 (1999) [Opt. Spectrosc. (USSR) **86**, 481 (1999)].
33. J.B. Boffard, G.A. Piech, M.F. Gehrke, M.E. Lagus, L.W. Anderson, and C.C. Lin, J. Phys. B **29**, L795 (1996).
34. G.A. Piech, J.B. Boffard, M.F. Gehrke, L.W. Anderson, and C.C. Lin, Phys. Rev. Lett. **81**, 309 (1998).
35. K. Bartschat and V. Zeman, Phys. Rev. A **59** R2552 (1999).
36. J.B. Boffard, T. Stone, L.W. Anderson, and C.C. Lin, Bull. Am. Phys. Soc. **46**, Vol. 4, 9 (2001).
37. A. Hibbert, Comput. Phys. Commun. **9**, 141 (1975).
38. W. Eissner, M. Jones, and H. Nussbaumer, Comput. Phys. Commun. **8**, 270 (1974).
39. N.R. Badnell, T.W. Gorczyca, M.S. Pindzola, and H.P. Summers, J. Phys. B **29**, 3683 (1996).
40. J.L. Giuliani and G.M. Petrov, private communication (2002).

Experimental Data for Electron-Ion Collisions

A. Müller

Institut für Kernphysik, Strahlenzentrum, Justus-Liebig-Universität Giessen, D-35392 Giessen, Germany

Abstract. Electron-ion collisions are among the most important atomic processes in ionized gases. Experimental data on electron-ion collisions can be inferred from plasma observations, from ion trap measurements, from careful preparation and analysis of ion-atom collisions and from colliding beams experiments using table top size to large accelerator facilities. The different approaches have specific advantages and drawbacks which are briefly addressed in this paper. Some recent experimental results are discussed. References to the most recent experimental work are provided.

INTRODUCTION

Electron-ion collisions are atomic interactions of fundamental character. Studying details of these collision processes enhances our understanding of the quantum-physical basis of nature and provides knowledge about the structure and the dynamics of atomic systems. Besides their intrinsic relevance, however, electron-ion collisions are also most important in plasma applications. They determine the charge-state balance of atoms and the spectrum of electromagnetic radiation emitted by ionized gases. Understanding and diagnosing the state of a plasma, whether of astrophysical origin or man made, relies on information about cross sections and rate coefficients for electron-ion interactions. Accordingly, these interactions have attracted long-standing theoretical and experimental interest. The present overview summarizes the experimental approaches to the problem of determining cross sections and rate coefficients for electron-impact ionization, excitation, recombination and various scattering processes involving **atomic ions**.

EXPERIMENTAL ACCESS TO DATA

Plasma techniques

Well characterized Θ-pinch, tokamak and stellarator plasmas have been employed in numerous experiments to obtain plasma rate coefficients ([1, 2, 3, 4, 5, 6, 7, 8, 9] and references therein) for electron impact ionization, excitation and recombination. The concept of these measurements is to observe characteristic line emission from a plasma that is seeded with the desired atomic species, to model the observations by rate equations using measured electron densities and temperatures, and to obtain electron-ion collision rate coefficients for recombination, excitation or ionization by fitting the model to the observed quantities. The plasma has to be carefully diagnosed for this purpose and time dependences of electron densities n_e and temperatures T_e have to be recorded. Experiments with the Θ-pinch require diagnostics and line intensity measurements on a nanosecond time scale. Accordingly, the time-dependent development of densities n_q of seed ions in charge states q has to be described by a set of differential equations containing the information on collision rates of ions in different charge states q. Knowledge about the excitation of the observed line radiation from a given ion species is often taken from theoretical considerations. In the simplest approach a corona model can be applied to describe for example a phase of steady burn of a tokamak plasma. A set of rate equations such as

$$\frac{dn_q}{dt} = 0 = \alpha_{q-1}^{ion} n_e n_{q-1} - \alpha_q^{ion} n_e n_q + \alpha_{q+1}^{rec} n_e n_{q+1} - \alpha_q^{rec} n_e n_q \qquad (1)$$

with $q = 1, 2, 3, ..., Z$ relates the ionization rate coefficients α_q^{ion} and the recombination rate coefficients α_q^{rec} with the densities n_q of ions in different charge states q. These densities are inferred from the observed line radiation that is characteristic for the ion charge state of interest. All the rate coefficients can be treated like fit parameters that are determined by adjusting the model to the observations. Usually, additional assumptions have to be made in order to determine electron-ion collision rates. For example radiative recombination (RR) is usually neglected in comparison with dielectronic recombination (DR). Although uncertainties as low as 30 % are quoted for such measurements one has to admit that measured plasma rate coefficients can be off by as much as a factor of 2 to 4 [10, 11].

Traps

Trapped-ion techniques for the determination of electron-ion collision cross sections are based on the observation of spatially confined ions that are exposed to electron impact. In the first experiments of this kind [12] information was obtained on ionization of ions by the observation of sequential ionization in an ion source within which the target ions were trapped by the space charge of an intense electron beam. By applying suitable external electric potentials additional to the "internal" space charge potential it was possible to obtain trapping times of the order of one second and measure relative ionization functions [13] in such a device. A cylindrically symmetric hollow electron beam for ion trapping and primary ionization (target preparation) and a concentric variable-energy electron beam interacting with the target ions was developed by Hasted and collaborators [14, 15] for electron-ion collision studies. The method was employed for the investigation of electron-impact ionization of positive ions [14, 16], for electron-ion recombination [17], and for electron-impact dissociation of molecular ions [18]. After calibrating the apparatus to available crossed-beams data, uncertainties of measured cross sections were estimated to be in the vicinity of 30%.

The most productive use of traps for electron-ion collision studies was initiated by the development of the electron beam ion source EBIS [19] by Donets et al. employing a very dense magnetically guided electron beam for trapping and for sequential ionization of highly charged ions. From the observation of the time history of the charge state evolution of ions extracted from an EBIS after variable trapping times cross sections were deduced for ionization of ions over a wide range of charge states [20, 21, 22]. For the analysis rate equations similar to Eq.(1) were used. While Donets et al. neglected recombination processes this is not a good approximation when the electron energy in EBIS matches that of a recombination resonance. Ali et al. [23] demonstrated the substantial effect of dielectronic recombination (DR) on the charge state distribution of ions extracted from an EBIS and determined cross sections for DR [24].

Substantial extension of the technique had become possible with the development of a modified EBIS, the electron beam ion trap EBIT [25]. The geometry of an EBIT facilitates the observation and spectroscopy of electromagnetic radiation emitted by the trapped ions. Thus, two techniques can now be used to obtain cross sections for electron collisions with highly charged ions in EBIT devices. One is associated with the extraction of ions as described above. The other is based on the observation of photons emitted by the trap inventory during electron beam exposure. In both cases the electron energy is ramped over a selected range of interest providing signatures of electron-ion collisions either in the amount of extracted ions of a given charge state or in the differential photon yield at a fixed angle relative to the electron beam. The ion extraction technique has provided cross sections with experimental energy spreads as low as about 10 eV [26]. Such cross sections are relative in nature and have to be normalized, e.g. to theory. In contrast to that the X-ray technique [27] allows one to normalize the observed yields to the simultaneously recorded signal of radiative recombination (RR) and to make the well-founded assumption that cross sections for RR are theoretically well understood. The electron energy spread in this kind of measurements is typically above 50 eV [28]. Ambiguity arises from the presence of trapped ions of the investigated element in different charge states. All these ions emit similar X-ray spectra which can only be disentangled by relying on theory. Moreover, the lowest electron energies usable in the trap experiments are quite high because considerable electron current is necessary to trap the ions radially in the space charge potential well of the electron beam. Very high charge states are accessible in traps [29] but have to be produced by exposure of cold atoms to a dense electron beam of sufficiently high energy such as to overcome the ionization thresholds. By chopping the electron beam energy between ionization and measurement one can separate the phase of ion production from that of a cross section measurement. This phase is limited towards lower energies only by the fact that decent beams of electrons are required for a measurable signal. Such beams can only be produced in connection with strong magnetic guiding fields which are present in such devices anyway and at sufficiently high energies of the order of keV. Thus, measurements of low-energy electron collisions with highly charged ions are not accessible by the trapped-ion techniques.

By now there are numerous laboratories world wide where EBIT devices are available. EBITs are extensively used as sources of characteristic line emission and have provided a wealth of spectroscopic information [30]. They can deliver ions for external applications such as ion-surface interactions in high charge states up to completely stripped uranium. Their use in the measurements of electron-ion collision cross sections is continued (see e.g. [31, 28, 32]).

Quasi-free electrons

Experimental studies of two-body collisions require an arrangement in which the particles of interest can interact with each other. What immediately comes to mind is a scheme in which a beam of one species interacts with a target consisting of the other species. Targets of charged particles, however, are difficult to prepare [33] because of space-charge limitations and consequently the target densities are very low as compared to those possible with solid or gaseous targets. Therefore the investigation of interactions between free electrons and ions often suffers from low signal rates. The density of an electron target can be enhanced by compensating the electron space charge with positive ions [34]. The most efficient scheme of this kind is the use of electrons bound in atoms or solids. This scheme is particularly obvious for the quasi-free electrons in a crystal lattice which are assumed to move freely among the positive core ions. At sufficiently high projectile velocities ($v_{ion} \gg v_0$ with v_{ion} the velocity of the incident ion and v_0 the orbital electron velocity) also the electrons bound in target atoms can be treated like free particles, a concept that has been used with great success since almost a century to describe high-energy atomic collisions.

A number of experiments aiming at electron-impact ionization and recombination were carried out employing the channeling technique, i. e. the electron cloud in a channel of a single crystal was used as a target [35, 36, 37]. Numerous problems with the control of experimental parameters and conditions make the interpretation of the measured quantities difficult and lead to large uncertainties in the measured cross sections.

The importance of recombination and excitation measurements to test and to guide theory together with the appealing subject of observing resonance features in atomic collisions led to a very important development of accelerator based ion - neutral-target collision studies of resonance phenomena related to electron-ion collisions [38, 39]. The interaction of highly charged ions with quasi-free electrons had already been exploited to study the radiative-recombination equivalent in ion-atom collisions, the radiative electron capture (REC; [40]). The dielectronic-capture – related process of resonant transfer with excitation (RTE) was unambiguously demonstrated for the X-ray emission channel about a decade later [41, 38] and dielectronic recombination calculations could be directly tested by the equivalent RTE reaction, provided that the momentum distribution of the quasi-free (but still bound) electrons was taken into consideration [42]. Similarly, the ion-atom equivalent to resonant elastic electron scattering from ions was found [43] by observing the Auger decay channel of RTE resonances. This technique has been extended to measurements of differential cross sections also for inelastic and super-elastic scattering processes (see [44, 45, 46] and references therein).

The quasi-free electrons behave like an electron target with a very substantial energy spread. Even if the velocity condition $v_{ion} \gg v_0$ is fulfilled the measurement of cross sections appears like a low-resolution electron-ion collision experiment. However, this handicap can partly be overcome: if the photons or electrons emitted in the process are detected with high energy resolution, detailed information can be inferred from the emission spectrum. An important advantage of the ion-neutral versus direct electron-ion experiments is the comparatively high luminosity. Cold electron beams such as those applied for ion cooling in ion storage rings provide densities of usually less than 10^8 cm^{-3}. Targets of neutral gases can be easily produced with densities around

10^{14} cm^{-3}. Solid foil targets reach electron densities of the order of 10^{22} cm^{-3}. Accordingly, ion-neutral experiments enjoy much higher counting rates than interacting-beams experiments. Moreover, the higher density of neutral targets allows one to design luminous target spots where ions and neutrals collide in a small volume from which the photons or electrons are emitted. Thus the solid angle for the observation of radiation or electron emission can be optimized. Accordingly, the determination of differential cross sections for electron-ion collisions is relatively straightforward in ion-neutral target experiments [47, 48] and even (e,2e) scattering on ions is becoming accessible [49]. However, because of the condition $v_{ion} \gg v_o$ that has to be fulfilled when interpreting ion-atom collisions on the basis of electron-ion scattering mechanisms, these measurements are restricted to the regime of high electron-ion collision energies. Studying low-energy electron-ion collisions by RTE techniques is not possible.

Colliding-beams techniques

Clearly, the most direct access to unambiguous information about electron-ion collisions is to use the technique of interacting beams. The colliding-beams technique was first applied to electron impact ionization of He$^+$ ions [50]. Only a few years later the technique was extended to the observation of photons from excited ions [51], thus producing total cross sections for electron-impact excitation of ions. It took about two decades after the introduction of the crossed-beams technique to electron-ion interaction studies until the recombination of ions with electrons could be addressed by that technique [52, 53]. The cross sections for electron capture of an ion in a collision with an atom or molecule are very high at low ion velocities and typically increase with the charge state of the ion. Hence, background arising from slow collisions of multiply charged ions becomes excessive and almost inhibits the use of the crossed-beams technique for recombination measurements with multiply charged ions [54, 55]. Electron capture in residual-gas collisions can be greatly reduced by employing ions with velocities v_{ion} well above the orbital velocity v_0 of the bound target electrons since then the cross sections drop with high powers of v_{ion}. When ions from an accelerator are used background from residual gas collisions is greatly reduced and, moreover, fast ions can easily be stripped by passing them through solid foils so that high charge states become available for measurements. Pioneering experiments making use of accelerator-based merged beams arrangements for electron-ion recombination studies were reported by Dittner et al. [56] and Mitchell et al. [57].

A new observation channel in electron-ion collision studies was opened by Chutjian et al. [58] and Zapesochnyi et al. [59] measuring the electron-energy loss in a crossed-beams set-up thus providing the first angle-differential data for electron scattering from ions. While these measurements were successfully pursued ([60] and references therein) producing important information on doubly differential scattering cross sections (angle and energy of the outgoing electron) the applied data needs in plasma and astrophysics called for a new access to total cross sections for electron-impact excitation of ions. An electron-ion merged beams technique employing trochoidal analyzers for merging and de-merging the interacting beams was developed for the measurement of the inelastic energy loss of electrons scattered from ions [61, 62]. A drawback of this method is its restriction to the threshold energy region of excitation cross sections. There are difficulties with separating elastically and inelastically scattered electrons from each other. Collecting all scattered electrons even if they appear under backward angles in the electron-ion center-of-mass system is a challenge.

The accelerator based merged-beams technique introduced to recombination studies of multiply charged ions has undergone a remarkable technological development. By the construction of new cold electron targets and by the integration of the merged-beams technique in heavy-ion storage rings the signal rates and the energy resolution of electron-ion recombination experiments could be vastly improved. Storage ring merged beams experiments have become the method of choice to study low-energy electron-ion recombination. This implies the investigation of field effects on dielectronic recombination of multiply charged ions. Storage ring techniques were also applied to electron-impact ionization of highly charged ions. An important advantage of these measurements is in the accessibility of intense beams of highly charged ions provided the ring is attached to an accelerator that injects intense beam pulses of these ions which can then be accumulated and stored. With storage times in the range of seconds to hours electron-ion collision experiments can take advantage of the availability of pure ground-state ion beams for cross section measurements thus avoiding the often faced problems of unknown fractions of metastable states in beams of highly charged ions.

A drawback of accelerator based ionization measurements is the presence of high backgrounds produced by electron-stripping collisions of the fast parent ions with residual gas components -even at gas pressures in the range of 10^{-9} Pa typical for a storage ring vacuum. Therefore, experiments at low ion energies in the range of typically only 1 keV/u are still best suited for ionization studies on low to intermediate-high ion charge states that can be produced by an ion source with sufficient beam intensities. Better energy resolution and the observation of finer cross section details are accessible in dedicated low-energy ion beam arrangements as compared to storage-ring based electron-impact ionization experiments.

Storage ring electron-ion collision experiments have been limited so far to energies below several keV. This is not a principle limitation but can be overcome e.g. by employing a dedicated electron target besides the electron cooler. Preparations for such a target are underway at the Heidelberg storage ring.

STATUS AND RECENT RESULTS

The plasma technique has produced a relatively limited amount of electron-ion rate coefficients. Because of the inherent ambiguities and uncertainties and because of the development of the other techniques this method has not been very actively pursued during the last decade.

EBIT devices have an amazing potential of producing low-velocity ions in any given charge state. Thus they are unique sources of characteristic radiation providing sufficient intensities for detailed spectroscopic studies. Absolute cross sections for electron-impact ionization, excitation and recombination can be determined by normalizing intensities of

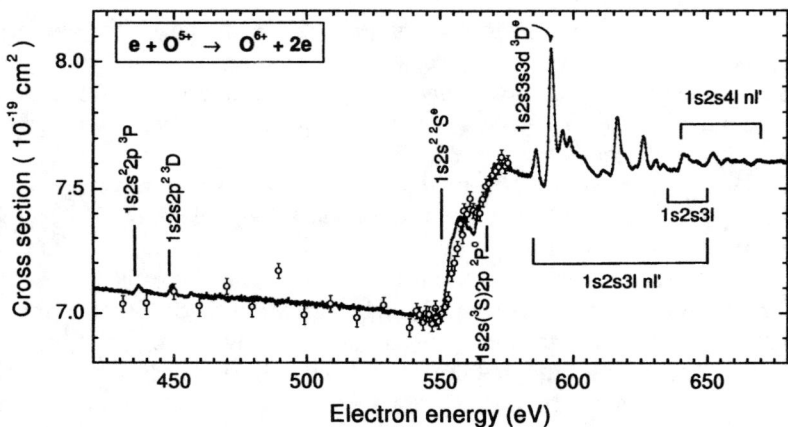

FIGURE 1. Experimental cross sections for electron-impact ionization of O^{5+} ions. The data are from Ref. [63]. Intermediate multiply excited states populated during the ionization process are indicated.

the observed radiation to the theoretical data on radiative recombination. Uncertainties are introduced by the presence of several ion charge states in the trap at a time and by the necessary modelling of the trap plasma. Low-energy electron-ion collisions are hardly accessible by EBIT techniques. For collision energies beyond several keV, however, EBIT has offered almost exclusive opportunities to investigate collisions of highly charged ions with electrons. More data can be expected from the available devices within the next years.

Studying electron-ion processes by careful observation of fast-ion – neutral-target collisions is an active field of research that produces unique information on differential cross sections. Further development of detection techniques combined with the exploitation of the relatively high luminosity of these experiments based on the possibilities to produce high-intensity ion beams up to the highest charge states and dense (neutral) quasi-electron targets will keep this field exciting and productive.

Direct experiments on electron-ion collisions employing interacting beams have reached a very high standard. Experimental arrangements employ low-energy merged and crossed beams of electrons and ions to study electron-impact ionization and excitation. Cross sections for single and multiple ionization of ions can be measured [64, 65] and data with very high precision and very good energy resolution can be obtained so that fine details of indirect processes are accessible [66, 63, 67, 68, 69, 70, 71, 72, 73]. An example for a very detailed ionization cross section measurement is shown in Fig. 1. Such studies are presently extended to metastable parent ions such as helium-like B^{3+} and C^{4+} ions in the $1s2s\ ^3S$ state [74].

The theoretical calculation of cross sections for electron-impact excitation is still very difficult and predictions can easily be off by substantial amounts. Experiments also remain to be difficult, however, there is steady progress in the field that is actively pursued employing the merged-beams technique [75, 76, 77, 78]. Eventually, the extension of the existing devices to accomodate more energetic ion beams would greatly enhance the ranges of collision energies and collision systems accessible to these experiments.

High-energy ion beams are now almost exclusively used in combination with ion storage rings. Merged-beams experiments are predominantly directed to electron-ion recombination but have also been employed for ionization studies. Dielectronic recombination of highly charged ions at low center-of-mass energies remains a field of particular interest because of the difficulties that theory has to face in the prediction of the energies and strengths of resonances near the threshold [11, 79, 80, 81, 82, 83, 84, 85, 86]. Such information is essential for the understanding of low-temperature (photon-driven) plasmas containing highly charged ions [87]. An aspect of special interest associated with this context is the influence of electric and magnetic fields on dielectronic recombination [88, 89, 90, 91]. Storage ring experiments continue to provide important new insights into this phenomenon [92, 93, 94, 95].

Recombination of completely stripped and very highly charged heavy ions are also receiving considerable attention [96, 96, 97, 98, 99, 100]. Although these data are not immediately relevant to applications in plasma physics they provide important test cases for state-of-the-art theory describing electron-ion collision phenomena. Special appeal is in the relative energy resolution of resonances in very highly charged ions where level splittings are maximum while the experimental energy spread is not much different from that obtained for low-charge ions.

REFERENCES

1. Hinnov, E., *J. Opt. Soc. Am.*, **56**, 1179 (1966).
2. Kunze, H. J., Gabriel, A. H., and Griem, H. R., *Phys. Rev.*, **165**, 267–276 (1968).
3. Kunze, H. J., *Phys. Rev. A*, **3**, 937–942 (1971).
4. Datla, R. U., Blaha, M., and Kunze, H. J., *Phys. Rev. A*, **12**, 1076–1083 (1975).
5. Källne, E., and Jones, L. A., *J. Phys. B*, **10**, 3637 (1977).
6. Breton, C., Michelis, C. D., Finkenthal, M., and Mattioli, M., *Phys. Rev. Lett.*, **41**, 110–113 (1978).
7. Brooks, R. L., Datla, R. U., and Griem, H. R., *Phys. Rev. Lett.*, **41**, 107–109 (1978).
8. Isler, R. C., Crume, E. C., and Arnurius, D. E., *Phys. Rev. A*, **26**, 2105–2116 (1982).
9. Wang, J.-S., Griem, H. R., Hess, R., and Rowan, W. L., *Phys. Rev. A*, **38**, 4761–4766 (1988).
10. Müller, A., *Int. J. Mass Spec.*, **192**, 9–22 (1999).
11. Müller, A., and Schippers, S., "Dielectronic Recombination: Experiment," in *The Challenge of High Resolution X-Ray Through Infrared Spectroscopy*, edited by G. Ferland and D. W. Savin, Astronomical Society of the Pacific, San Francisco, USA, 2001, vol. 247 of *ASP Conference Proceedings*, pp. 51–76.
12. Baker, F. A., and Hasted, J. B., *Phil. Trans. Roy. Soc. A*, **261**, 33 (1966).
13. Redhead, P. A., *Can. J. Phys.*, **49**, 3059 (1971).
14. Hasted, J. B., and Awad, G. L., *J. Phys. B*, **5**, 1719 (1972).
15. Hasted, J. B., "Confinement of Ions for Collision Studies," in *Physics of Ion-Ion and Electron-Ion Collisions*, edited by F. Brouillard and J. W. McGowan, Plenum Press, New York, 1983, vol. 83 of *NATO ASI Series B: Physics*, pp. 461–500.
16. Hamdan, M., Birkinshaw, K., and Hasted, J. B., *J. Phys. B*, **11**, 331 (1978).
17. Mathur, D., Khan, S. U., and Hasted, J. B., *J. Phys. B*, **11**, 3615 (1978).
18. Mathur, D., Hasted, J. B., and Khan, S. U., *J. Phys. B*, **12**, 2043 (1979).
19. Donets, E. D., Ilyushchenko, V. I., and Alpert, V. A., "Ultra High Vacuum Electron Beam Source of Highly Stripped Ions," in *Iere Conférerence Internationale sur Les Sources d'Ions*, Saclay, 1969, pp. 635–642.
20. Donets, E. D., and Pikin, A. I., *Sov. Phys. Tech. Phys.*, **20**, 1477 (1976).
21. Donets, E. D., and Ovsyannikov, V. P., *Sov. Phys. JETP*, **53**, 466 (1981).

22. Donets, E. D., *Phys. Scripta*, **T3**, 11 (1983).
23. Ali, R., Bhalla, C. P., Cocke, C. L., and Stöckli, M., *Phys. Rev. Lett.*, **64**, 633–636 (1990).
24. Ali, R., Bhalla, C. P., Cocke, C. L., Schulz, M., and Stöckli, M., "Electron-Ion Recombination Experiments on the KSU EBIS," in *Recombination of Atomic Ions*, edited by W. G. Graham, W. Fritsch, Y. Hahn, and J. A. Tanis, Plenum Press, New York, 1992, vol. 296 of *NATO ASI Series B: Physics*, pp. 193–202.
25. Marrs, R. E., Levine, M. A., Knapp, D. A., and Henderson, J. R., *Phys. Rev. Lett*, **60**, 1715 (1988).
26. DeWitt, D. R., Schneider, D., Chen, M. H., Schneider, M. B., Church, D., Weinberg, G., and Sakurai, M., *Phys. Rev. A*, **47**, R1597–R1600 (1993).
27. Knapp, D. A., Marrs, R. E., Levine, M. A., Bennet, C. L., Chen, M. H., Henderson, J. R., Schneider, M. B., and Scofield, J. H., *Phys. Rev. Lett.*, **62**, 2104 (1989).
28. Smith, A. J., Beiersdorfer, P., Widmann, K., Chen, M. H., and Scofield, J. H., *Phys. Rev. A*, **62**, 052717 (2000).
29. Marrs, R. E., Elliott, S. R., and Knapp, D. A., *Phys. Rev. Lett.*, **72**, 4082–4085 (1994).
30. Gillaspy, J. D., *J. Phys. B*, **34**, R93–R130 (2001).
31. Fuchs, T., Biedermann, C., Radtke, R., Behar, E., and Doron, R., *Phys. Rev. A*, **58**, 4518–4525 (1998).
32. O'Rourke, B., Currell, F. J., Kuramoto, H., Li, Y. M., Ohtani, S., Tong, X. M., and Watanabe, H., *J. Phys. B: At. Mol. Opt. Phys.*, **34**, 4003–4013 (2001).
33. Müller, A., *Nucl. Instrum. Methods A*, **282**, 80–86 (1989).
34. Müller, A., Hofmann, G., Tinschert, K., Sauer, R., Salzborn, E., and Becker, R., *Nucl. Instrum. Methods B*, **24/25**, 369 (1987).
35. Claytor, N., Feinberg, B., Gould, H., Bemis, C. E., Campo, J. G. D., Ludemann, C. A., and Vane, C. R., *Phys. Rev. Lett.*, **61**, 2081–2084 (1988).
36. Andriamonje, S., de Castro Faria, R. A. N. V., Chevallier, M., Cohen, C., Dural, J., Gaillard, M. J., Genre, R., Hage-Ali, M., Kirsch, R., L'Hoir, A., Farizon-Mazuy, B., Mory, J., Mouilin, J., C.Poizat, J., Quere, Y., Remillieux, J., Schmaus, D., and Toulemonde, M., *Phys.Rev.Lett.*, **63**, 1930–1933 (1989).
37. Datz, S., Vane, C. R., Dittner, P. F., Giese, J. P., Campo, J. G. D., Jones, N. L., Krause, H. F., Miller, P. D., Schulz, M., Schöne, H., and Rosseel, T. M., *Phys. Rev. Lett.*, **63**, 742 (1989).
38. Tanis, J. A., "Resonant Transfer Excitation Associated with Single X-Ray Emission," in *Recombination of Atomic Ions*, edited by W. G. Graham, W. Fritsch, Y. Hahn, and J. A. Tanis, Plenum Press, New York, 1992, vol. 296 of *NATO ASI Series B: Physics*, pp. 241–257.
39. Zouros, T. J. M., "Resonant Transfer Excitation Associated with Auger Electron Emission," in *Recombination of Atomic Ions*, edited by W. G. Graham, W. Fritsch, Y. Hahn, and J. A. Tanis, Plenum Press, New York, 1992, vol. 296 of *NATO ASI Series B: Physics*, pp. 271–300.
40. Schnopper, H. W., Betz, H. D., Delvaille, J. P., Kalata, K., Sohval, A. R., Jones, K. W., and Wegner, H. E., *Phys. Rev. Lett.*, **29**, 898–901 (1972).
41. Tanis, J. A., Bernstein, E. M., Graham, W. G., Stockli, M. P., Clark, M., McFarland, R. H., Morgan, T. J., Berkner, K. H., Schlachter, A. S., and Stearns, J. W., *Phys. Rev. Lett.*, **53**, 2551–2554 (1984).
42. Brandt, D., *Phys. Rev. A*, **27**, 1314–1318 (1983).
43. Itoh, A., Schneider, T., Schiwietz, G., Roller, Z., Platten, H., Nolte, G., Scheider, D., and Stolterfoht, N., *J. Phys. B*, **16**, 3965 (1983).
44. Richard, P., Bhalla, C., Hagmann, S., and Zavodsky, P., *Phys. Scripta*, **T80**, 87–92 (1999).
45. Zavodsky, P. A., Aliabadi, H., Bhalla, C., Richard, P., Tóth, G., and Tanis, J. A., *Phys. Rev. Lett.*, **87**, 033202 (2001).
46. Zamkov, M., Aliabadi, H., Benis, E. P., Richard, P., Tawara, H., and Zouros, T. J. M., *Phys. Rev. A*, **65**, 032705 (2002).
47. Kandler, T., Stöhlker, T., Mokler, P. H., Kozhuharov, C., Geissel, H., Scheidenberger, C., Rymuza, P., Stachura, Z., Warczak, A., Dunford, R. W., Eichler, J., Ichihara, A., and Shirai, T., *Z. Phys. D*, **35**, 15–18 (1995).
48. Benhenni, M., Shafroth, S. M., Swenson, J. K., Schulz, M., Giese, J. P., Schöne, H., Vane, C. R., Dittner, P. F., and Datz, S., *Phys. Rev. Lett.*, **65**, 1849–1852 (1990).
49. Kollmus, H., Moshammer, R., Olson, R. E., Hagmann, S., Schulz, M., and Ullrich, J., *Phys. Rev. Lett.*, **88**, 103202 (2002).
50. Dolder, K. T., Harrison, M. F. A., and Thonemann, P., *Proc. Roy. Soc. A*, **264**, 367–379 (1961).

51. Dance, D. F., Harrison, M. F. A., and Smith, A. C. H., *Proc. Roy. Soc. A*, **290**, 73 (1966).
52. Belić, D. S., Dunn, G. H., Morgan, T. J., Mueller, D. W., and Timmer, C., *Phys. Rev. Lett.*, **50**, 339–342 (1983).
53. Williams, J. F., *Phys. Rev. A*, **29**, 2936–2938 (1984).
54. Young, A. R., Gardner, L. D., Savin, D. W., Lafyatis, G. P., Chutjian, A., Bliman, S., and Kohl, J. L., *Phys. Rev. A*, **49**, 357–362 (1994).
55. Savin, D. W., Gardner, L. D., Reisenfeld, D. B., Young, A. R., and Kohl, J. L., *Phys. Rev. A*, **53**, 280–289 (1996).
56. Dittner, P. F., Datz, S., Miller, P. D., Moak, C. D., Stelson, P. H., Bottcher, C., Dress, W. B., Alton, G. D., Nešković, N., and Fou, C. M., *Phys. Rev. Lett.*, **51**, 31–34 (1983).
57. Mitchell, J. B. A., Ng, C. T., Forand, J. L., Levac, D. P., Mitchell, R. E., Sen, A., Miko, D. B., and McGowan, J. W., *Phys. Rev. Lett.*, **50**, 335–338 (1983).
58. Chutjian, A., Msezane, A. Z., and Henry, R. J. W., *Phys. Rev. Lett.*, **50**, 1357–1360 (1983).
59. Zapesochnyi, A. P., Imre, A. I., Aleksakhin, I. S., Zapesochnyi, I. P., and Zatsarinnyi, O. I., *Sov. Phys. JETP*, **63**, 1155–1160 (1986).
60. Williams, I. D., *Rep. Prog. Phys.*, **62**, 1431–1469 (1999).
61. Wåhlin, E. K., Thompson, J. S., Dunn, G. H., Phaneuf, R. A., Gregory, D. C., and Smith, A. C. H., *Phys. Rev. Lett.*, **66**, 157–160 (1991).
62. Smith, S. J., Man, K.-F., Mawhorter, R. J., Williams, I. D., and Chutjian, A., *Phys. Rev. Lett.*, **67**, 30–33 (1991).
63. Müller, A., Teng, H., Hofmann, G., Phaneuf, R. A., and Salzborn, E., *Phys. Rev. A*, **62**, 062720 (2000).
64. Khouilid, M., Cherkani-Hassani, S., Rachafi, S., Teng, H., and Defrance, P., *J. Phys. B: At. Mol. Opt. Phys.*, **34**, 1727–1744 (2001).
65. Khouilid, M., Cherkani-Hassani, S., Adimi, N., Rachafi, S., and Defrance, P., *J. Phys. B: At. Mol. Opt. Phys.*, **34**, 3239–3249 (2001).
66. Teng, H., Knopp, H., Ricz, S., Schippers, S., Berrington, K. A., and Müller, A., *Phys. Rev. A*, **61**, 060704(R) (2000).
67. Rejoub, R., and Phaneuf, R. A., *Phys. Rev. A*, **61**, 032706 (2000).
68. Aichele, K., Shi, W., Scheuermann, F., Teng, H., Salzborn, E., and Müller, A., *Phys. Rev. A*, **63**, 014701 (2000).
69. Shaw, J. A., Pindzola, M. S., Steidl, M., Aichele, K., Hartenfeller, U., Hathiramani, D., Scheuermann, F., Westermann, M., and Salzborn, E., *Phys. Rev. A*, **63**, 032709 (2001).
70. Aichele, K., Hathiramani, D., Scheuermann, F., Müller, A., Salzborn, E., Mitnik, D., Colgan, J., and Pindzola, M. S., *Phys. Rev. Lett.*, **86**, 620 (2001).
71. Aichele, K., Arnold, W., Hathiramani, D., Scheuermann, F., Salzborn, E., Mitnik, D. M., Griffin, D. C., Colgan, J., and Pindzola, M. S., *Phys. Rev. A*, **64**, 052706 (2001).
72. Aichele, K., Steidl, M., Hartenfeller, U., Hathiramani, D., Scheuermann, F., Westermann, M., Salzborn, E., and Pindzola, M. S., *J. Phys. B: At. Mol. Opt. Phys.*, **34**, 4113–4123 (2001).
73. Mitnik, D. M., Griffin, D. C., Colgan, J., Pindzola, M. S., Aichele, K., Arnold, W., Hathiramani, D., Scheuermann, F., and Salzborn, E., *Phys. Rev. A*, **64**, 062705 (2001).
74. Müller, A., Böhme, C., Jacobi, J., Knopp, H., and Schippers, S., "Electron-Impact Ionization of He-Like Metastable Ions," in *Proceedings of the XXIInd International Conference on Photonic, Electronic and Atomic Collisions*, edited by S. Datz, M. E. Bannister, H. F. Krause, L. H. Saddiq, D. Schultz, and C. R. Vane, Rinton Press, Princeton, New Jersey, 2001, p. 322.
75. Lozano, J. A., Niimura, M., Smith, S. J., Chutjian, A., and Tayal, S. S., *Phys. Rev. A*, **63**, 042713 (2001).
76. Smith, A. C. H., Bannister, M. E., Chung, Y.-S., Djuric, N., Dunn, G. H., Neau, A., Popovic, D., Stepanovic, M., and Wallbank, B., *J. Phys. B: At. Mol. Opt. Phys.*, **34**, L571–L577 (2001).
77. Popović, D. B., Bannister, M. E., Clark, R. E. H., Chung, Y.-S., Djurić, N., Meyer, F. W., Müller, A., Neau, A., Pindzola, M. S., Smith, A. C. H., Wallbank, B., , and Dunn, G. H., *Phys. Rev. A*, **65**, 034704 (2002).
78. Niimura, M., Smith, I. C. S. J., and Chutjian, A., *Phys. Rev. Lett.*, **88**, 103201 (2002).
79. Savin, D. W., Kahn, S. M., Linkemann, J., Saghiri, A. A., Schmitt, M., Grieser, M., Repnow, R., Schwalm, D., Wolf, A., Bartsch, T., Brandau, C., Hoffknecht, A., Müller, A., Schippers, S., Chen, M. H., and Badnell, N. R., *Astrophys. J. Suppl.*, **123**, 687–702 (1999).

80. Gwinner, G., Hoffknecht, A., Bartsch, T., Beutelspacher, M., Eklöw, N., Glans, P., Grieser, M., Krohn, S., Lindroth, E., Müller, A., Saghiri, A. A., Schippers, S., Schramm, U., Schwalm, D., Tokman, M., Wissler, G., and Wolf, A., *Phys. Rev. Lett.*, **84**, 4822–4825 (2000).
81. Schippers, S., Müller, A., Gwinner, G., Linkemann, J., Saghiri, A. A., and Wolf, A., *Astrophys. J.*, **555**, 1027–1037 (2001).
82. Glans, P., Lindroth, E., Badnell, N. R., Eklöw, N., Zong, W., Justiniano, E., and Schuch, R., *Phys. Rev. A*, **64**, 043609 (2001).
83. Madzunkov, S., Lindroth, E., Eklöw, N., Tokman, M., Paál, A., and Schuch, R., *Phys. Rev. A*, **65**, 032505 (2002).
84. Savin, D. W., Behar, E., Kahn, S. M., Gwinner, G., Saghiri, A. A., Schmitt, M., Grieser, M., Repnow, R., Schwalm, D., Wolf, A., Bartsch, T., Müller, A., Schippers, S., Badnell, N. R., Chen, M. H., and Gorczyca, T. W., *Astrophys. J. Suppl.*, **138**, 337–370 (2002).
85. Schippers, S., Kieslich, S., Müller, A., Gwinner, G., Schnell, M., Wolf, A., Bannister, M., Covington, A., and Zhao, L. B., *Phys. Rev. A*, **65**, 042723 (2002).
86. Tokman, M., Eklöw, N., Glans, P., Lindroth, E., Schuch, R., Gwinner, G., Schwalm, D., Wolf, A., Hoffknecht, A., Müller, A., and Schippers, S., *Phys. Rev. A* (2002), in print.
87. Savin, D. W., "Atomic Data Needs for Modeling Photoionized Plasmas," in *The Challenge of High Resolution X-Ray Through Infrared Spectroscopy*, edited by G. Ferland and D. W. Savin, Astronomical Society of the Pacific, San Francisco, USA, 2001, vol. 247 of *ASP Conference Proceedings*, pp. 165–171.
88. Müller, A., Belić, D. S., DePaola, B. D., Djurić, N., Dunn, G. H., Mueller, D. W., and Timmer, C., *Phys. Rev. Lett.*, **56**, 127–130 (1986).
89. Bartsch, T., Müller, A., Spies, W., Linkemann, J., Danared, H., DeWitt, D. R., Gao, H., Zong, W., Schuch, R., Wolf, A., Dunn, G. H., Pindzola, M. S., and Griffin, D. C., *Phys. Rev. Lett.*, **79**, 2233–2236 (1997).
90. Bartsch, T., Schippers, S., Müller, A., Brandau, C., Gwinner, G., Saghiri, A. A., Beutelspacher, M., Grieser, M., Schwalm, D., Wolf, A., Danared, H., and Dunn, G. H., *Phys. Rev. Lett.*, **82**, 3779–3781 (1999).
91. Müller, A., and Schippers, S., "Dielectronic Recombination and its Sensitivity to External Electromagnetic Fields," in *The Physics of Multiply and Highly Charged Ions*, edited by F. Currell, Kluwer Academic Publishers, Dordrecht, 2002, in print.
92. Bartsch, T., Schippers, S., Beutelspacher, M., Böhm, S., Grieser, M., Gwinner, G., Saghiri, A. A., Saathoff, G., Schuch, R., Schwalm, D., Wolf, A., and Müller, A., *J. Phys. B*, **33**, L453–L460 (2000).
93. Schippers, S., Bartsch, T., Brandau, C., Müller, A., Gwinner, G., Wissler, G., Beutelspacher, M., Grieser, M., Wolf, A., and Phaneuf, R. A., *Phys. Rev. A*, **62**, 022708 (2000).
94. Böhm, S., Schippers, S., Shi, W., Müller, A., Djurić, N., Dunn, G. H., Zong, W., Jelenković, B., Danared, H., Eklöw, N., Glans, P., and Schuch, R., *Phys. Rev. A*, **64**, 032707 (2001).
95. Böhm, S., Schippers, S., Shi, W., Müller, A., Eklöw, N., Schuch, R., Danared, H., Badnell, N. R., Mitnik, D., and Griffin, D. C., *Phys. Rev. A* (2002), in print.
96. Hoffknecht, A., Brandau, C., Bartsch, T., Böhme, C., Knopp, H., Schippers, S., Müller, A., Kozhuharov, C., Beckert, K., Bosch, F., Franzke, B., Krämer, A., Mokler, P. H., Steck, M., Stöhlker, T., and Stachura, Z., *Phys. Rev. A*, **63**, 012702 (2001).
97. Shi, W., Böhm, S., Böhme, C., Brandau, C., Hoffknecht, A., Kieslich, S., Schippers, S., Müller, A., Kozhuharov, C., Bosch, F., Franzke, B., Mokler, P. H., Steck, M., Stöhlker, T., and Stachura, Z., *Europhys. J. D*, **15**, 145–154 (2001).
98. Brandau, C., Bartsch, T., Böhme, C., Bosch, F., Dunn, G., Franzke, B., Hoffknecht, A., Knocke, C., Knopp, H., Kozhuharov, C., Krämer, A., Mokler, P., Müller, A., Nolden, F., Schippers, S., Stachura, Z., Steck, M., Stöhlker, T., Winkler, T., and A.Wolf, *Phys. Scr.*, **T80**, 318–320 (1999).
99. Lindroth, E., Danared, H., Glans, P., Pesic, Z., Tokman, M., Vikor, G., and Schuch, R., *Phys. Rev. Lett.*, **86**, 5027–5030 (2001).
100. Brandau, C., Bartsch, T., Hoffknecht, A., Knopp, H., Schippers, S., Müller, A., Steih, T., Grün, N., Scheid, W., Bosch, F., Franzke, B., Kozhuharov, C., Mokler, P. H., Nolden, F., Steck, M., Stöhlker, T., and Stachura, Z., *Phys. Rev. Lett.* (2002), submitted.

Electron-Molecule Collisions in the Static-Exchange Correlation-Polarization Approximation

Robert R. Lucchese[1] and F. A. Gianturco[2]

[1] *Department of Chemistry, Texas A&M University, College Station, TX, 77843-3255, USA*
[2] *Department of Chemistry, The University of Rome, Città Universitaria, 00185 Rome, Italy*

Abstract. The static-exchange correlation-polarization model for studying electron-molecule is presented. This model includes the important physical process needed to study elastic scattering. Applications of this approach to electron scattering from SF_6, C_6H_6, C_6F_6, and C_{20}, are discussed.

INTRODUCTION

There are several theoretical approaches for studying electron-molecule scattering. In general these methods can be very computationally challenging. The approach we will discuss here is based on a model potential, the static-exchange-correlation-polarization (SECP) potential, and can be fairly easily be applied to systems as large as $e-C_{60}$ scattering. The basic physics included in the model allows for the study of elastic scattering in the fixed nuclei approximation. A wide range of electron-molecule scattering systems have been studied using this approach [1-4].

Within the SECP model there are two interesting features of electron-molecule scattering that can be studied. They are shape resonances and virtual states. A shape resonance is a one-electron resonance, which occurs when an electron is temporarily trapped in the field of the molecule. In atoms, this trapping process can be easily understood by examining the effective radial potential for angular momenta with $l > 0$. In these systems, the attractive interaction between the electron and the atom combines with the $l(l+1)/(2r^2)$ effective radial potential due to the angular momentum, leading to an angular momentum barrier which can trap scattering states. These resonant scattering states can then only decay into the continuum by tunneling through the angular momentum barrier in the effective radial potential. In linear molecular systems, transforming the problem to confocal elliptic coordinates leads to a very similar picture [5]. For these resonances there are again effective radial potentials, in the radial confocal elliptic coordinate ξ, that have angular momentum barriers. Thus for linear systems, if one were to compute the resonant state, one could then analyze the nodal structure of the resonance and deduce the nature of the resonant state. Although we have not found a way to generalize the confocal elliptic coordinates to nonlinear molecular systems, we use insights gained from the linear systems to analyze shape resonances in the non-linear systems. In particular, the partial wave that traps a resonant state can be determined by analyzing the nodal structure of the resonant state on the periphery of the molecular system.

The second scattering feature that can be studied with the SECP model is a virtual state. A virtual state can occur in a scattering symmetry that includes the $l = 0$ partial wave. Consider the scattering with angular momentum $l > 0$ where one systematically makes the interaction potential more attractive. A shape resonance in the scattering will become a bound state when the potential becomes attractive enough. However for $l = 0$, there are no shape resonances. In this case, new bound states will also appear as the potential becomes more attractive. However, just before such a new state appears, there will be a virtual state. This state corresponds to a pole of the S matrix on the unphysical energy sheet at negative energy. In terms of the wave function, a virtual state is an eigenfunction of the electronic Hamiltonian that is purely exponentially growing at large r. This is in contrast to a bound state which is only exponentially decaying at large r. The presence of such a virtual state leads to negative scattering lengths and large cross sections near threshold. A virtual state has also been shown to provide a mechanism for low-energy attachment of electrons [6].

We will consider a number of systems to which the SECP model has been applied and discuss how the observables computed using this approach compare to available experimental data.

COMPUTATIONAL METHODS

The SECP Model

In the SECP model the electronic wave function of the molecular target is approximated by a Hartree-Fock wave function. The electron-molecule scattering equations are then approximated by using a one-electron optical potential that is the sum of the exact static and exchange potentials, V_{SE}, and a potential that approximately includes correlation effects, V_{CP}. The short-range effects of correlation between the target electrons and the projectile electron are treated using a local density-functional correlation potential, V_C. Unfortunately, the asymptotic behavior of such a potential is not correct. Asymptotically the polarization potential should be an attractive potential with a radial dependence of $1/r^4$ due to the static polarizability of the molecule, whereas local density-functional correlation potentials decay exponentially. In order to overcome this incorrect behavior, the correlation potential V_C is smoothly joined to a polarization potential, V_P, which has the correct asymptotic form. This combined potential is then the correlation-polarization potential, V_{CP}. Adding the V_{SE} and V_{CP} potentials together leads to the full SECP optical potential. The resulting non-local one-electron scattering problem is then solved using single-center expansion techniques [1-4]. This approach gives reasonably good agreement with elastic scattering experimental results as well as with the results of more sophisticated calculations [7].

The SECP model has several limitations. Most significantly, this model neglects electronic excitation, except to the extent that virtual excitations are the underlying physical processes that lead to the correlation and polarization potentials that are modeled by the V_{CP} potential. Although we have used the fixed-nuclei approximation, one could include the nuclear motion using the SECP model for the electronic part of the problem [8]; however, this has not yet been done for larger molecules.

A Local Model Potential For Studying Resonant States

In order to facilitate the analysis of the resonant states, we have developed a simpler local optical potential [9], the static-model-exchange-correlation-polarization (SMECP) model potential. In this potential we substitute the local Hara free-electron-gas exchange potential [10] for the non-local exchange potential. With this potential we can then solve the scattering equations at complex energies and directly locate the poles in the S matrix that are the sources of resonant behavior. We then plot the state at the complex resonance energy to determine its nodal structure. It was found that the SECP and SMECP potentials give very similar results, in particular they have the same resonances at nearly the same energies. However, the SECP and SMECP potentials yield total and differential cross sections that are quantitatively different, particularly at low energies, with the SECP results being generally in better agreement with experiment [1].

SELECTED APPLICATIONS OF THE SECP MODEL

Scattering From SF_6

We have investigated the convergence of the computed SECP cross sections in the e–SF_6 scattering system [11]. In particular, we considered the convergence of the cross section in the region of a narrow shape resonance with respect to partial wave expansion and with respect to the quality of the one-electron basis set used to describe the target. For all of the shape-resonances in this system, we found that going from the aug-cc-pVTZ to the aug-cc-pVQZ basis set [12-14] leads to shifts of no more than 0.13 eV. Thus at the aug-cc-pVQZ level of basis set, the resonance positions and cross sections are probably converged to within 0.2 eV with respect to the one-electron target basis. The convergence of scattering cross sections with respect to the partial wave expansion has been considered on several occasions. In the study of e–SF_6 we found that, in addition to the shift in the position of a resonance as the partial-wave expansion is increased, the shape of the a_{1g} resonant feature in the total cross section changed significantly as the expansion was increased. In this symmetry, the sensitivity to the expansion parameter was due to the interaction of the strong background scattering with the resonant scattering process.

One interesting result of our reexamination of the e–SF_6 system, is that we found a second scattering resonance of t_{1u} symmetry at 70.0 eV with a width of 15.2 eV. In electron scattering, this resonance is barely discernable, however the corresponding shape resonance in the photoionization of the $1s$ orbital of the S atom is quite pronounced [15] and occurs at a photoelectron kinetic energy of 53.1 eV with a width of 8.5 eV. This resonance is characterized by an $l = 9$ angular momentum barrier just outside the region of the F atoms. For most molecules there is a correspondence between the shape resonances seen in electron scattering and the shape resonances seen in photoionization. There is usually a significant shift in the position of such a resonance to lower electron kinetic energy in photoionization, which in the case of SF_6 is seen to be 16.9 eV. This shift is due to the additional attractive interaction between the electron and the positive charge of the residual molecular ion in the case of photoionization. In the C_6H_6 molecule, we have also found shifts of 10-12 eV for resonances seen in the valence ionization compared to electron-molecule scattering [16]. However for the photoionization of C_{60}, the corresponding shift is only about 2 eV [17]. This is presumably due to the much more delocalized charge in the C_{60}^+ valence ion.

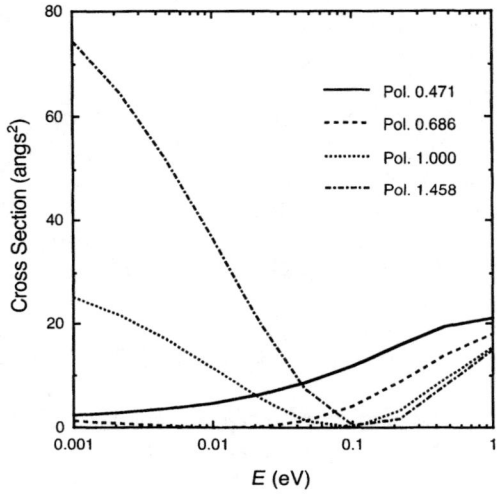

FIGURE 1. Cross section for e–SF_6 scattering in the a_{1g} scattering symmetry as a function of the assumed asymptotic polarization. The factor indicated in the legend is the ratio of static polarizability used in the potential relative to the experimental total static polarizability of 6.584 Å^3 [18].

As discussed above, one of the parameters that is required for the SECP model, as implemented here, is the static polarizability of the target molecule. The static polarizability can be divided into an electronic contribution and a nuclear contribution. For small molecules, the contribution to the polarizability from the motion of the nuclei is small compared to the electronic contribution. For CO_2, the nuclear contribution is approximately 9% of the total polarizability [19]. However in the case of SF_6 the nuclear contribution is 31% of the total [19]. In Fig. 1 we have plotted the low energy behavior of the a_{1g} scattering cross section in e–SF_6 scattering. The "Pol. 1.000" line corresponds to using the polarizability due to both electronic and nuclear motion, whereas the "Pol. 0.686" line corresponds to multiplying the static polarizability by 0.686 and thus corresponds to including only the electronic polarizability asymptotically. The large cross sections at low energy for some of the values of the polarizability are indicative of the presence of a virtual state. At very low kinetic energy one might expect that the scattering electron would feel both the nuclear and electronic polarizability. However as the velocity of the electron increases only the electronic part of the polarizability would be felt by the scattering electron. Thus as the low energy scattering is studied theoretically for more large systems, appropriate methods for including the nuclear polarizability will need to be developed.

Scattering From C_6H_6

We have used the SECP model to study the very low energy behavior of the e–C_6H_6 scattering cross sections. The SECP model does not include any nuclear motion except for that implicitly included in the static polarizability. Thus we are not able to study processes such as vibrational excitation thresholds. However as mentioned above, one can study virtual states. In earlier calculations on e–C_{60} scattering we found evidence for such a virtual state [20]. More recently we have considered the low energy scattering from C_6H_6. The low energy cross sections for e–C_6H_6 scattering show a low energy rise in the computed cross section that is due to a virtual state at –0.162 eV [21]. This result is very similar to that found in the experiments on the e–C_6D_6 system and thus suggests that the rise in the experimental cross section at low energy is due to a virtual state in the e–C_6D_6 system [21]. The behavior of the SECP cross section at low energy is substantially different from that found with a purely L^2 method [7] where the cross section goes rapidly to zero below 2 eV. One source of this difference may be the fact that the L^2 method implicitly truncates the interaction potential due to the relatively restricted range of the expansion functions used in the calculation [7]. This is particularly important for the part of the potential due to the quadrupole moment of the C_6H_6 molecule which only decays as $1/r^3$ at large r. In contrast to L^2 methods, in the SECP method, we numerically solve the scattering problem on a grid whose extent is a function of the asymptotic electron kinetic energy and the strength of the interaction potential. At the lowest energy studied with the SECP model, 0.0001 eV, the radial grid extended out to 11304 au.

Scattering From C_6F_6

We have considered electron scattering from C_6F_6 [22]. In this system we find twelve shape resonances between threshold and 30 eV. In Fig. 2, nine of these resonances are evident, with the strongest features being the narrow e_{2u} resonance at 1.6 eV, the a_{1g} resonance at 5.6 eV, the b_{2g} resonance at 8.8 eV, and the e_{2g} resonance at 12.8 eV. The general magnitude of the cross section is in reasonable agreement with the experimental data [23], however there is no evidence in the total scattering cross section data for the narrow shape resonances found in the calculations. Presumably these resonant features would be more prominent in vibrational excitation cross sections. The level of agreement is very similar to that obtained for the e–C_6H_6 system [7]. It is also interesting to compare the resonances in this system to those present in the e–C_6H_6 system. Nine of the resonances found in this energy region in e–C_6F_6 are also found in e–C_6H_6. There are also three additional σ symmetry resonances in the e–C_6F_6 system that presumably involve the relatively low-lying C-F σ* orbitals. Of the nine resonances found in both systems, the widths of the resonances are consistently narrower in the e–C_6F_6. An examination of the resonant states suggests that the presence of the F atoms in the molecule forces additional nodal surfaces into the wave functions. These nodal surfaces are required to ensure orthogonality to the bound orbitals localized on the F atoms. These additional nodal surfaces lead to more nodes asymptotically and thus to higher asymptotic angular momenta. The higher asymptotic angular momenta in turn imply higher angular momentum barriers and thus longer resonant lifetimes.

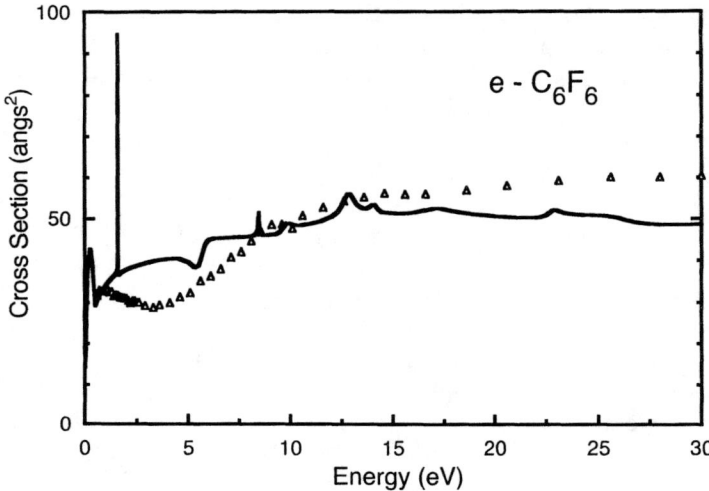

FIGURE 2. Total elastic scattering cross section for e–C_6F_6 scattering. Solid line is computed result using the SECP model. Triangles are experimental data from Ref. [23].

Scattering From C_{20}

Finally we have considered the nature of the shape resonances in electron scattering from C_{20} [24]. This carbon cluster is nearly spherical, however it has much lower symmetry than C_{60}. Due to the lower symmetry, one might expect the resonances in C_{20} to be somewhat broader than those found in C_{60}. This is indeed the case, however there still are a number of narrow resonances in this system in the low energy region.

In Fig. 3 we show a resonant state for scattering of A symmetry from the C_2 structure of C_{20} computed using the SMECP approximate local potential. This state has an energy of 1.838 eV and a width of 0.0851 eV. We can see that this state is a π state on the shell of the C_{20} molecule with a nodal structure that is characteristic of an $l = 3$ asymptotic state. At the low scattering energy of 1.838 eV, this is enough of an angular momentum barrier to make this a fairly narrow resonance.

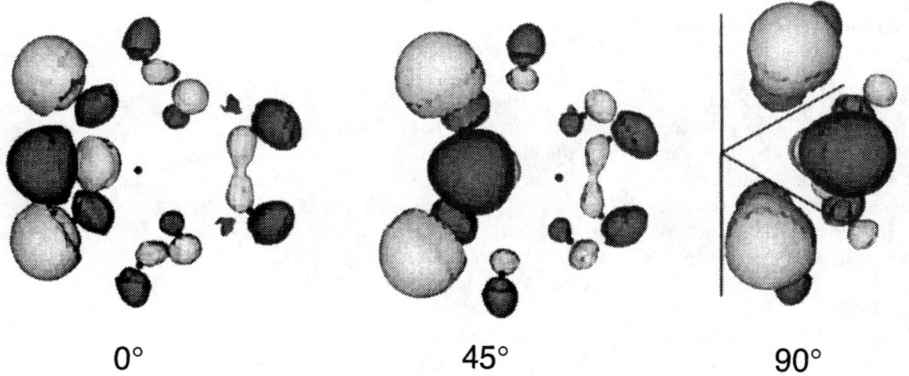

FIGURE 3. Three views of a resonant orbital in e–C_{20} scattering. Only half of the C atoms are shown and the angle label indicates the orientation of the half shell in each view. The solid lines in the 90° view indicate the location of the nodal planes characteristic of an $l = 3$ state.

CONCLUSIONS

The SECP model for electron molecule scattering is seen to give results that contain the major features seen in low-energy elastic scattering. In particular we note that it can be used to study the location and symmetry properties of shape resonances. In addition it can be used to study very low energy scattering from molecule where virtual states can strongly affect the scattering cross sections.

ACKNOWLEDGMENTS

We acknowledge the support of the NATO Scientific Affairs Division under grant CRG950552, the support of the Welch Foundation (Houston) under Grant No. A-1020, and the support of the Texas A&M University Supercomputing Facility.

REFERENCES

1. Gianturco, F. A. and Lucchese, R. R., *J. Chem. Phys.* **113**, 10044-10050 (2000).
2. Gianturco, F. A. and Lucchese, R. R., *J. Chem. Phys.* **111**, 6769-6786 (1999).
3. Gianturco, F. A. and Lucchese, R. R., *J. Chem. Phys.* **108**, 6144-6159 (1998).
4. Gianturco, F. A., Lucchese, R. R., Sanna, N., and Talamo, A., "A Generalized Single Center Approach for Treating Electron Scattering from Polyatomic Molecules," in *Electron Collisions with Molecules, Clusters, and Surfaces*, edited by H. Ehrhardt and L. A. Morgan, Plenum Press, New York, 1994, pp. 71-86.
5. Zurales, R. W., Ph. D. thesis, Texas A&M University, 1997.
6. Gauyacq, J. P. and Herzenberg, A., *J. Phys. B.* **17**, 1155-1171 (1984).
7. Bettega, M. H. F., Winstead, C., and McKoy, V., *J. Chem. Phys.* **112**, 8806-8812 (2000).
8. Morrison, M. A. and Sun, W., "How to Calculate Rotational and Vibrational Cross Sections for Low-Energy Electron Scattering from Diatomic Molecules Using Close-Coupling Techniques," in

Computational Methods for Electron-Molecule Collisions, edited by W. Huo and F. A. Gianturco, Plenum, New York, 1995, pp. 131-190.
9. Lucchese, R. R. and Gianturco, F. A., *Int. Rev. Phys. Chem.* **15**, 429-466 (1996).
10. Hara, S., *J. Phys. Soc. Japan* **22**, 710-718 (1967).
11. Gianturco, F. A. and Lucchese, R. R., *J. Chem. Phys.* **114**, 3429-3439 (2001).
12. Woon, D. E. and Dunning, T. H., Jr., *J. Chem. Phys.* **98**, 1358-1371 (1993).
13. Dunning, T. H., Jr., *J. Chem. Phys.* **90**, 1007-1023 (1989).
14. Kendall, R. A. and Dunning, T. H., Jr., *J. Chem. Phys.* **96**, 6796-6806 (1992).
15. Natalense, A. P. P. and Lucchese, R. R., *J. Chem. Phys.* **111**, 5344-5348 (1999).
16. Lin, P. and Lucchese, R. R., (in preparation).
17. Gianturco, F. A. and Lucchese, R. R., *Phys. Rev. A* **64**, 032706:1-032706:7 (2001).
18. Weast, R. C., editor, *CRC Handbook of Chemistry and Physics*, 66th edition, CRC Press, Boca Raton, 1986, p. E-69.
19. St-Arnaud, J. M. and Bose, T. K., *J. Chem. Phys.* **71**, 4951-4955 (1979).
20. Lucchese, R. R., Gianturco, F. A., and Sanna, N., *Chem. Phys. Lett.* **305**, 413-418 (1999).
21. Field, D., Ziesel, J.-P., Lunt, S. L., Parthasarathy, R., Suess, L., Hill, S. B., Dunning, F. B., Lucchese, R. R., and Gianturco, F. A., *J. Phys. B* **34**, 4371-4381 (2001).
22. Natalense, A. P. P. and Lucchese, R. R., *J. Chem. Phys.*, submitted for publication.
23. Kasperski, G., Mozejko, P., and Szmytkowski, C., *Z. Phys. D* **42**, 187-191 (1997).
24. Ginaturco, F. A., Kashenock, G. Yu., Lucchese, R. R., Sanna, N., *J. Chem. Phys.* **116**, 2811-2824 (2002).

Theoretical Atomic Data: Universality and Precision

Zenonas R. Rudzikas

State Institute of Theoretical Physics and Astronomy
Goštauto 12, Vilnius, LT-2600, Lithuania. E-mail: rudzikas@itpa.lt

Abstract. A universal method of finding the expressions for non-relativistic and relativistic matrix elements of any one- and two-electron operators for an arbitrary number of shells in an atomic configuration, requiring neither coefficients of fractional parentage nor unit tensors, is discussed. It is based on the second quantization method in the coupled tensorial form, angular momentum theory in three spaces (orbital, spin and quasispin) and generalized graphical technique. The method is implemented in a number of computer codes and has already demonstrated its high efficiency for generating various atomic data of high precision.

INTRODUCTION

With the advent of bio- and nanotechnologies the role of accurate data on the structure and properties of atoms and molecules is dramatically increasing. Many domains of contemporary physics and neighbouring branches of science and technology need such results. In some cases the data may be measured experimentally, but often their theoretical values are preferable. The latters may be obtained following measurements or observations, but they are most valuable when preceeding or replacing the experiments.

The theoretical methods must satisfy two often incompatible requirements - to be as universal as possible and to yield data of fairly high precision. In case of many-electron atoms and ions, highly charged ions including, this means that the method must combine the efficient way of accounting for the correlation and relativistic effects with a general method of the calculation of matrix elements of the relevant operators for complex electronic configurations, including matrix elements non-diagonal with respect to configurations. This paper is aimed at the review of several methods of such a kind.

For large variety of many-electron atoms and their ionization degrees the relativistic effects may be accurately accounted for as corrections of the order α^2 (α is fine structure constant) in the framework of the Hartree-Fock-Pauli approximation [1]. The most universal way to take into consideration the correlation effects is the method of superposition of configurations or multi-configurational approximation. The majority of difficulties when calculating the spin-angular parts of the relevant general matrix elements may be overcome by using the second quantization approach in the coupled tensorial form as well as the additional symmetry properties connected with the quasispin formalism. The relevant approach is described in [1,2]. It is already implemented in the universal computer programs [3]. A number of practical applications of this methodology has already demonstrated its high efficiency and general character. Similar approach is developed for the relativistic Hamiltonian and wave functions, too. It allows us to practically account for a very large amount of admixed configurations, to generate various spectroscopic data for heavy atoms and ions, atoms with open f-shells included.

For speeding up the convergency of the method of the superposition of configurations the use of the so-called transformed radial orbitals turned out to be efficient [4]. The relevant computer codes have already demonstrated their high efficiency and suitability to generate accurate data for the large variety of complex atoms and ions. Below we shall discuss the ideas mentioned above in more detail.

MANY-ELECTRON ENERGY OPERATOR

Numerous considerations and calculations have already demonstrated that the so-called Hartree-Fock-Pauli (HFP) approximation accounts with high accuracy for the relativistic effects in many-electron atoms and ions of wide domain of ionization degrees. The relevant Hamiltonian H for N-electron system may be presented in the form

$$H = H_0 + W, \tag{1}$$

where zero-order Hamiltonian

$$H_0 = T + P + Q = \frac{1}{2m} \sum_{i=1}^{N} \mathbf{p}_i^2 - Ze^2 \sum_{i=1}^{N} \frac{1}{r_i} + e^2 \sum_{i>j=1}^{N} \frac{1}{r_{ij}} \tag{2}$$

and the relativistic corrections of the order α^2

$$W = H_1 + H_2 + H_3 + H_4 + H_5, \tag{3}$$

where

$$H_1 = -\frac{1}{8m^3 c^2} \sum_i \mathbf{p}_i^4 \tag{4}$$

is the relativistic correction due to the dependence of the electron mass on the velocity,

$$H_2 = -\frac{e^2}{2m^2 c^2} \sum_{i>j} \frac{1}{r_{ij}} \left\{ (\mathbf{p}_i \cdot \mathbf{p}_j) + \frac{(\mathbf{r}_{ij} \cdot (\mathbf{r}_{ij} \cdot \mathbf{p}_i) \cdot \mathbf{p}_j)}{r_{ij}^2} \right\} \tag{5}$$

stands for the orbit-orbit interaction operator,

$$H_3 = H_3' + H_3'' = \frac{\pi e^2 \hbar^2 Z}{2m^2 c^2} \sum_i \delta(\mathbf{r}_i) - \frac{\pi e^2 \hbar^2}{m^2 c^2} \sum_{i>j} \delta(\mathbf{r}_{ij}) \tag{6}$$

describes the contact interactions,

$$H_4 = \frac{e^2 \hbar}{2m^2 c^2} \left(\left\{ \sum_i \frac{Z}{r_i^3} [\mathbf{r}_i \times \mathbf{p}_i] - \sum_{i>j} \frac{1}{r_{ij}^3} [\mathbf{r}_{ij} \times \mathbf{p}_i] + \sum_{i>j} \frac{2}{r_{ij}^3} [\mathbf{r}_{ij} \times \mathbf{p}_j] \right\} \cdot \mathbf{s}_i \right) \tag{7}$$

represents the spin-orbit interaction,

$$H_5 = H_5' + H_5'' = -\frac{8\pi e^2 \hbar^2}{3m^2 c^2} \sum_{i>j} (\mathbf{s}_i \cdot \mathbf{s}_j) \delta(\mathbf{r}_{ij}) +$$

$$+ \frac{e^2 \hbar^2}{m^2 c^2} \sum_{i>j} \frac{1}{r_{ij}^3} \left[(\mathbf{s}_i \cdot \mathbf{s}_j) - \frac{3(\mathbf{s}_i \cdot \mathbf{r}_{ij})(\mathbf{s}_j \cdot \mathbf{r}_{ij})}{r_{ij}^2} \right] \quad (8)$$

is in charge of spin-contact (H_5') and spin-spin (H_5'') interactions. In these formulas $\delta(\mathbf{r})$ is the Dirac δ-function of vectorial argument $\mathbf{r}$, the sums are from 1 up to N, where N denotes the number of electrons, symbol $(\mathbf{a} \cdot \mathbf{b})$ means the scalar product, and $[\mathbf{a} \times \mathbf{b}]$ is the vectorial product of $\mathbf{a}$ and $\mathbf{b}$.

One-particle operators H_1 and H_3' cause relativistic corrections to the total energy. Two-particle operators H_2, H_3'' and H_5' define more precisely the energy of each term, whereas H_4 and H_5'' describe their splitting (fine structure). These operators are also often called the describing magnetic interactions. Let us notice that all terms in (3) contain small parameter $1/c^2$, therefore they usually cause small corrections.

However, there are also numerous cases (highly charged ions, heavy atoms, studies of processes with the involvement of inner shells of many-electron atoms, etc.) when the relativistic effects are sufficiently large and therefore cannot be accounted for as corrections. Then one has to use relativistic wave functions and relativistic Hamiltonian, usually in the form of the so-called relativistic Breit operator. In the case of an N-electron atom the latter may be written as follows (in atomic units, in which the absolute value of electron charge e, its mass m and Planck constant $\hbar$ are equal to one, whereas the unit of length is equal to the radius of the first Bohr orbit of the hydrogen atom):

$$H = \sum_{i=1}^{N} (H_i^1 + H_i^2 + H_i^3) + \sum_{i>j=1}^{N} (H_{ij}^4 + H_{ij}^5 + H_{ij}^6). \quad (9)$$

Here

$$H_i^1 = c(\alpha_i^{(1)} \cdot \mathbf{p}_i^{(1)}), \quad (10)$$

$$H_i^2 = \beta_i c^2, \quad (11)$$

$$H_i^3 = V(r_i), \quad (12)$$

$$H_{ij}^4 = 1/r_{ij}, \quad (13)$$

$$H_{ij}^5 = -\frac{1}{2r_{ij}} (\alpha_i^{(1)} \cdot \alpha_j^{(1)}), \quad (14)$$

$$H^6_{ij} = -\frac{1}{2r^3_{ij}}(\alpha^{(1)}_i \cdot r^{(1)}_{ij})(\alpha^{(1)}_j \cdot r^{(1)}_{ij}). \tag{15}$$

In these formulas the Dirac matrices $\alpha^{(1)}$ and β are defined as follows:

$$\alpha^{(1)} = \begin{pmatrix} 0 & \sigma^{(1)} \\ \sigma^{(1)} & 0 \end{pmatrix}, \quad \beta = \begin{pmatrix} \hat{I} & 0 \\ 0 & \hat{I} \end{pmatrix}, \tag{16}$$

where $\sigma^{(1)}$ and $\hat{I}$ are Pauli and unit matrices of the second order, respectively. H^1, H^2 and H^3 are the one-electron operators. H^1 represents the electron kinetic energy and one-electron part of the spin-orbit interaction, H^2 describes the mass effect, and H^3 denotes the potential energy, in the Coulomb approximation being of the form $-Z/r_i$. Two-electron operator H^4 stands for the electrostatic (Coulomb) interaction, H^5 corresponds to the operator of the magnetic and a part of the retardation effects, whereas H^6 describes the remaining part of retardation interactions.

Wave function for the non-relativistic Hamiltonian (2) has a standard form

$$\psi_{nlsm_lm_s}(r,\vartheta,\varphi,\sigma) = R_{nl}(r) Y^{(l)}_{m_l}(\vartheta,\varphi) \chi^{(s)}_{m_s}(\sigma). \tag{17}$$

Here n, l, s represent the principal, orbital and spin quantum numbers of an electron, correspondingly, m_l and m_s describe the projections of l and s, whereas r, ϑ, φ are the variables of a spherical coordinate system and σ stands for a spin variable.

The relativistic analogue of one-electron wave function (17) is

$$\psi_{nljm}(r,\vartheta,\varphi) = \begin{pmatrix} \varphi_{nljm}(r,\vartheta,\varphi) \\ \chi_{nl'jm}(r,\vartheta,\varphi) \end{pmatrix} =$$

$$= \begin{pmatrix} |ljm|\vartheta,\varphi) f(nlj|r) \\ (-1)^\beta |l'jm|\vartheta,\varphi) f(nl'j|r) \end{pmatrix}. \tag{18}$$

Here $\beta = 1/2(1+l-l')$, $l' = 2j-l$, f and g are the so-called large and small components of relativistic radial wave function, respectively, and the spherical spinor $|ljm|\vartheta,\varphi)$ is

$$|ljm|\vartheta,\varphi) = \sum_{\sigma=\pm 1/2} \begin{bmatrix} l & s & j \\ m-\sigma & \sigma & m \end{bmatrix} Y^{(l)}_{m-\sigma}(\vartheta,\varphi) \Phi_\sigma, \tag{19}$$

where the quantity Φ_σ is a Pauli spin matrix defined by

$$\Phi_{1/2} = \begin{pmatrix} 1 \\ 0 \end{pmatrix}, \quad \Phi_{-1/2} = \begin{pmatrix} 0 \\ 1 \end{pmatrix}$$

and $\begin{bmatrix} l & s & j \\ m-\sigma & \sigma & m \end{bmatrix}$ is the Clebsch-Gordan coefficient.

It is worth mentioning that, as it has been shown in [5], instead of (19) in jj coupling, we can use the following function:

$$|ljm) = \left[\frac{2j+1}{8\pi}\right]^{1/2} \begin{pmatrix} D^{(j)}_{m\ 1/2} \\ (-1)^{l+1/2-j} D^{(j)}_{m\ -1/2} \end{pmatrix}, \tag{20}$$

where $D_{m\mu}^{(j)}$ is the so-called generalized spherical function.

The formula (20) is a new form of the angular part of the non-relativistic (in jj coupling) and relativistic wave function. Only its phase depends on the orbital quantum number l. The use of (20) considerably simplifies the finding of the matrix elements of physical operators and leads straightforwardly to their optimal expressions, having only radial integrals and phase multipliers, which depend on orbital quantum numbers.

GENERAL EXPRESSIONS FOR MATRIX ELEMENTS OF THE ENERGY OPERATOR

In order to find the expressions for matrix elements of any operator under consideration we must present its all terms in the form of the so-called irreducible tensors. The relevant expressions for the both Hamiltonians (1) and (9) may be found in [1]. More general irreducible forms of such operators are presented in [2, 6-9]. They are suitable for an arbitrary number of shells in a configuration and practically for any kind of matrix elements (diagonal or non-diagonal with respect to configurations). The relevant approach does not require either coefficients of fractional parentage or tensors composed of unit tensors. It is based on second quantization in the coupled tensorial form, angular momentum theory in three spaces (orbital, spin and quasispin) and generalized graphical technique.

Some expressions of two-electron matrix elements presented in [2, 6-9] may be simplified or written in the more compact form if to make use of the following recurrent relation for one-electron submatrix elements of spherical function:

$$\left(l||C^{(k-1)}||l'\right) = \left[\frac{(l+l'+k+2)(l+l'-k)(l-l'+k+1)(l'-l+k+1)}{(l+l'+k+1)(l+l'-k+1)(l-l'+k)(l'-l+k)}\right]^{\frac{1}{2}} \times$$
$$\times \left(l||C^{(k+1)}||l'\right). \quad (21)$$

Indeed, e.g., using (21) two terms (30) and (31) in [8] may be united or the term (31) [9] simplified.

Let us notice that of special interest is the term (31) [9] of the orbit-orbit interaction. Actually it accounts for the admixture of configurations of opposite parity, therefore it may be important for studies of the parity non-conservation in atoms. It is worth mentioning that all relativistic corrections to the Coulomb interaction may be easily accounted for as extra terms to radial integrals of electrostatic interaction.

The following general expression of two-particle operator satisfying the above-mentioned conditions was presented in [2]:

$$\widehat{G}^{(\kappa_1\kappa_2 k,\sigma_1\sigma_2 k)} = \sum_\alpha \sum_{\kappa_{12},\sigma_{12},\kappa'_{12},\sigma'_{12}} \Theta(\Xi)\left\{A_{p,-p}^{(kk)}(n_\alpha\lambda_\alpha,\Xi)\,\delta(u,1)\right.$$
$$+\sum_\beta \left[B^{(\kappa_{12}\sigma_{12})}(n_\alpha\lambda_\alpha,\Xi) \times C^{(\kappa'_{12}\sigma'_{12})}(n_\beta\lambda_\beta,\Xi)\right]_{p,-p}^{(kk)}\delta(u,2)$$

$$+ \sum_{\beta\gamma} \left[\left[D^{(l_\alpha s)} \times D^{(l_\beta s)} \right]^{(\kappa_{12}\sigma_{12})} \times E^{(\kappa'_{12}\sigma'_{12})}(n_\gamma\lambda_\gamma,\Xi) \right]^{(kk)}_{p,-p} \delta(u,3)$$

$$+ \sum_{\beta\gamma\delta} \left[\left[D^{(l_\alpha s)} \times D^{(l_\beta s)} \right]^{(\kappa_{12}\sigma_{12})} \times \left[D^{(l_\gamma s)} \times D^{(l_\delta s)} \right]^{(\kappa'_{12}\sigma'_{12})} \right]^{(kk)}_{p,-p} \delta(u,4) \right\}. \quad (22)$$

Here u stands for a number of shells acted upon by a given tensorial operator or their product. Quantity Ξ represents the whole array of parameters. The first term in (22) describes a two-electron operator acting upon the same shell $n_\alpha\lambda_\alpha$, the second term corresponds to operator acting upon two different shells $n_\alpha\lambda_\alpha$, $n_\beta\lambda_\beta$, etc.

The expression (22) is presented in the general form suitable for the both non-relativistic shells $n_i l_i^{N_i}$ and relativistic subshells $n_i l_i j_i^{N'_i}$.

The operators A, B, C, D and E, depending on the kind of shells they are acting upon, correspond to one of the following quantities (see details in [2]):

$$a^{(q\lambda)}_{m_q}, \quad (23)$$

$$\left[a^{(q\lambda)}_{m_{q1}} \times a^{(q\lambda)}_{m_{q2}} \right]^{(\kappa_1\sigma_1)}, \quad (24)$$

$$\left[a^{(q\lambda)}_{m_{q1}} \times \left[a^{(q\lambda)}_{m_{q2}} \times a^{(q\lambda)}_{m_{q3}} \right]^{(\kappa_1\sigma_1)} \right]^{(\kappa_2\sigma_2)}, \quad (25)$$

$$\left[\left[a^{(q\lambda)}_{m_{q1}} \times a^{(q\lambda)}_{m_{q2}} \right]^{(\kappa_1\sigma_1)} \times a^{(q\lambda)}_{m_{q3}} \right]^{(\kappa_2\sigma_2)}, \quad (26)$$

$$\left[\left[a^{(q\lambda)}_{m_{q1}} \times a^{(q\lambda)}_{m_{q2}} \right]^{(\kappa_1\sigma_1)} \times \left[a^{(q\lambda)}_{m_{q3}} \times a^{(q\lambda)}_{m_{q4}} \right]^{(\kappa_2\sigma_2)} \right]^{(kk)}. \quad (27)$$

Here the second quantization operator $a^{(q\lambda)}_{m_q m_\lambda}$ has the rank q in additional quasispin space and its projections $m_q = \pm\frac{1}{2}$, i.e. (in LS coupling)

$$a^{(q\lambda)}_{\frac{1}{2}m_\lambda} = a^{(ls)}_{m_l m_s} \quad (28)$$

and

$$a^{(q\lambda)}_{-\frac{1}{2}m_\lambda} = \tilde{a}^{(ls)}_{m_l m_s}. \quad (29)$$

Here a and $\tilde{a}$ stand for the electron creation and annihilation operators, respectively. The methodology of the second quantization in coupled tensorial form combined with the systematic use of the quasispin approach is described in more detail in [1].

The following relation between the coefficient of fractional parentage and completely reduced matrix element of the second quantization operator $a^{(qls)}$ gives the link of the approach under discussion to the traditional methods:

$$(l^N \alpha QLS \| l^{N-1}(\alpha_1 Q_1 L_1 S_1)l) = (-1)^{N-1}(N(2Q+1)(2L+1)(2S+1))^{-1/2} \times$$

$$\times \begin{bmatrix} Q_1 & 1/2 & Q \\ M_{Q_1} & 1/2 & M_Q \end{bmatrix} \left(l\alpha QLS \| \| a^{(qls)} \| \| l\alpha_1 Q_1 L_1 S_1 \right). \quad (30)$$

Here quasispin quantum number $Q = (2l+1-v)/2$ and its projection $M_Q = (N-2l-1)/2$. Numerical values of (||| |||) for $l=0$, 1 and 2 are presented in [10]. Their more complete tables (f^N shells including) may be found in [11]. The tables of the reduced coefficients of fractional parentage and reduced matrix elements of standard tensors in jj coupling, necessary for relativistic approach, are presented in [12].

A number of computer programs based on the above-mentioned methodology is written [13-15] allowing us to calculate very efficiently and accurately the energy spectra and the other spectral quantities of many-electron atoms and ions, those having open f-shells included.

CORRELATION EFFECTS VIA TRANSFORMED RADIAL ORBITALS

The methodology of the calculations of the matrix elements of one- and two-electron operators in the case of non-relativistic and relativistic wave functions considered above gives an effective way to take into account the correlation effects. The superposition of configurations method is one of the simplest ways to account for them. It is defined by

$$\Psi(K'\alpha J) = \sum_K a(K\alpha J)\Psi(K\alpha J), \quad (31)$$

showing that the actual electronic configuration is the superposition of a number (the larger the better) of configurations. This method, when combined with the methodology described in previous sections, is now applicable for practically any atom or ion.

The efficiency of the multi-configuration approach, based on the solution of the Hartree-Fock-Jucys equations [16], is also essentially increased if the spin-angular parts of the relevant matrix elements are calculated accounting for their symmetry properties in all three spaces. Many-body perturbation theory must be in such a case much more effective, too.

One of the most successful attempts to improve the convergence of the widely used superposition of configurations method was based on the so-called transformed Hartree-Fock radial orbitals (TRO) of the configuration under consideration for the description of admixed configurations [17]. They are obtained from the Hartree-Fock ones as follows:

$$P_{TRO}(n'l'|r) = Nf(r)P_{HF}(nl|r), \quad (32)$$

where N is a normalizing factor and $f(r)$ is a transforming function. In previous papers, e.g., [18] simple transforming function

$$f(r) = r^k \tag{33}$$

was used. It turned out that the following two kinds of such a function, namely, algebraic

$$f_a(r) = r^k/(\alpha + r^m) \tag{34}$$

and exponential

$$f_e(r) = r^k exp(-\alpha r^m) \tag{35}$$

lead to much better convergence and accuracy of the results obtained. The integer values of k and m as well as the real positive parameter α must be adjusted following the condition of maximum of the correlation correction to the energy, determined in second order of the perturbation theory [19]. A number of calculations shows that such an approach allows one to take into account up to 99% of the correlation correction.

Thus, transformed radial orbitals provide a fairly reliable and efficient replacement for the solutions of the multi-configurational Hartree-Fock equations. Such an approach combines successfully the simplicity, universality and high efficiency. The procedure converges quickly, allowing in this way the admixture of a great number of virtual configurations. Moreover, this speed-up is achieved without an increase in the order of the matrix to be diagonalized, just adding the extra terms to matrix elements of the usual single-configuration approximation. Therefore, this method, particularly if combined with the methodology of calculation of matrix elements discussed in previous sections, may be applied readily to study any many-electron atom or ion.

OPTIMIZATION OF CLASSIFICATION OF ENERGY LEVELS

The energy levels are identified and classified with the help of the sets of quantum numbers defining the so-called coupling scheme. However, these quantum numbers are exact only for pure couplings, which are actually more the exception than the rule. Therefore, one has to diagonalize the energy matrix, passing to the so-called intermediate coupling scheme. Then its wave function $\Psi(\beta J)$ looks as a linear combination of the functions of pure (usually LS) coupling scheme

$$\Psi(\beta J) = \sum_{\alpha_i L_i S_i} a(\alpha_i L_i S_i J) \Psi(\alpha_i L_i S_i J), \tag{36}$$

which is assumed to be closest to reality.

For finding the optimal coupling scheme we can make use of the following relationship between weights c_{jk} and a_{jr} of the wave functions φ_k and Ψ_r of two pure coupling schemes:

$$c_{jk} = \sum_r a_{jr}(\varphi_k|\Psi_r). \tag{37}$$

Here $(\varphi_k|\Psi_r)$ denotes the relevant transformation matrix. The method of finding its expressions for complex electronic configurations is described in [1]. Its main formula has the following recurrent form (here $[a,b] = (2a+1)(2b+1)$):

$$\left(l^N \alpha LSJ | l j_1^{N_1} j_2^{N_2} \beta_1 J_1 \beta_2 J_2 J\right) =$$

$$= \sqrt{[L,S]/N} \sum_{\alpha'L'S'} (l^N \alpha LS \| l^{N-1}(\alpha'L'S')l) \times$$

$$\times \sum_{J'} [J'] \left[\sqrt{[N_1[j_1,J_1]}\begin{Bmatrix} L' & l & L \\ S' & s & S \\ J' & j_1 & J \end{Bmatrix} \sum_{\beta_1'J_1'} (-1)^{j_1+J_1-J_2+J'} \times \right.$$

$$\times \begin{Bmatrix} J_2 & J_1' & J' \\ j_1 & J & J_1 \end{Bmatrix} \left(j_1^{N_1-1}(\beta_1'J_1')j_1 \| j_1^{N_1} \beta_1 J_1 \right) \times$$

$$\times \left(l^{N-1} \alpha'L'S'J' | l j_1^{N_1-1} j_2^{N_2} \beta_1'J_1' \beta_2 J_2 J' \right) + \sqrt{N_2[j_2,J_2]} \begin{Bmatrix} L' & l & L \\ S' & s & S \\ J' & j_2 & J \end{Bmatrix} \times$$

$$\times \sum_{\beta_2'J_2'} (-1)^{j_2+J_1+J_2'+J} \begin{Bmatrix} J_2' & J_1 & J' \\ J & j_2 & J_2 \end{Bmatrix} \times$$

$$\left. \times \left(j_2^{N_2-1}(\beta_2'J_2')j_2 \| j_2^{N_2} \beta_2 J_2 \right) \left(l^{N-1} \alpha'L'S'J' | l j_1^{N_1} j_2^{N_2-1} \beta_1 J_1 \beta_2'J_2'J' \right) \right]. \quad (38)$$

Its generalization for arbitrary number of shells may be presented as follows [20]:

$$\left(...((((\kappa_1^{-\bar{N}_1} \bar{v}_1 \bar{J}_1, \kappa_1^{+\bar{N}_1+} \overset{+}{v}_1 \overset{+}{J}_1)J_1, \kappa_2^{-\bar{N}_2} \bar{v}_2 \bar{J}_2)J'_{12}, \kappa_2^{+\bar{N}_2+} \overset{+}{v}_2 \overset{+}{J}_2)J_{12} \kappa_3^{-\bar{N}_3} \bar{v}_3 \bar{J}_3) \right.$$

$$J'_{123}, \kappa_3^{+\bar{N}_3+} \overset{+}{v}_3 \overset{+}{J}_3)J_{123}...) J \left| (...(((l_1^{N_1} \alpha_1 L_1 S_1, l_2^{N_2} \alpha_2 L_2 S_2)L_{12}S_{12}, l_3^{N_3} \alpha_3 L_3 S_3)L_{123}S_{123})...) J \right)$$

$$= \prod_{i=2}^{q} (-1)^{\bar{J}_i + \overset{+}{J}_i + J_{1...i-1} + J_{1...i}} \sqrt{[J'_{12...i}, J_{1...i-1}, L_{1...i}, S_{1...i}]} \sum_{J_i} \begin{Bmatrix} \bar{J}_i & \overset{+}{J}_i & J_i \\ J_{1...i} & J_{1...i-1} & J'_{1...i} \end{Bmatrix}$$

$$\times [J_i] \begin{Bmatrix} L_{1...i-1} & S_{1...i-1} & J_{1...i-1} \\ L_i & S_i & J_i \\ L_{1...i} & S_{1...i} & J_{1...i} \end{Bmatrix} \left(\kappa_i^{-\bar{N}_i} \bar{v}_i \bar{J}_i, \kappa_i^{+(N_i - \bar{N}_i)+} \overset{+}{v}_i \overset{+}{J}_i)J \left| l_i^{N_i} \alpha_i L_i S_i J_i \right). \quad (39)$$

The computer code based on the above-mentioned transformation matrices is in progress. Its practical use will allow one to find optimal coupling scheme for a large variety of configurations and to study the relative role of various intraatomic interactions.

MULTIPOLE ELECTRON TRANSITIONS

The wave functions of the intermediate coupling scheme obtained when accounting for correlation and relativistic effects may be used for calculation of the oscillator strengths or probabilities of the electric and magnetic multipole transitions. Let us present the relevant general expressions for non-relativistic wave functions. The following two alternative expressions are known for the probability of electric multipole radiation [1]:

$$W_{1\to2}^{Ek} = \frac{2(k+1)(2k+1)\omega^{2k+1}}{k[(2k+1)!!]^2 c^{2k+1}} \times$$

$$\times \left| \left(\psi_2 \left| \left[Q_{-q}^{(k)} + K\sqrt{\frac{k}{k+1}} \left\{ \frac{1}{\omega} Q'^{(k)}_{-q} - Q_{-q}^{(k)} \right\} \right] \right| \psi_1 \right) \right|^2, \quad (40)$$

$$W_{1\to2}^{Ek} = \frac{2(k+1)(2k+1)\omega^{2k-1}}{k[(2k+1)!!]^2 c^{2k+1}} \times$$

$$\times \left| \left(\psi_2 \left| \left[Q'^{(k)}_{-q} + K\sqrt{\frac{k}{k+1}} \left\{ Q'^{(k)}_{-q} - \omega Q_{-q}^{(k)} \right\} \right] \right| \psi_1 \right) \right|^2. \quad (41)$$

In these formulas

$$Q_{-q}^{(k)} = -r^k C_{-q}^{(k)} \quad (42)$$

denotes non-relativistic form of electric multipole transition operator corresponding, for $k=1$, to the length form of the electric dipole radiation operator. The quantity

$$Q'^{(k)}_{-q} = -r^{k-1}\left(kC_{-q}^{(k)}\frac{\partial}{\partial r} + \frac{i}{r}\sqrt{k(k+1)}\left[C^{(k)} \times L^{(1)}\right]_{-q}^{(k)}\right) \quad (43)$$

is a new form of the electric multipole radiation operator, which, for $k=1$, turns into the velocity form of the electric dipole transition operator.

The main peculiarity of the expressions (40) and (41) is the presence in them of an arbitrary parameter K, describing the gauge condition of the electromagnetic field potential. For exact wave functions these operators will not depend on K, because the multiplier in the curly brackets will be zero due to the commutation relations. Thus, this parameter may be used to evaluate the accuracy of the wave functions.

The probability of the magnetic multipole transition has the form

$$W_{1\to2}^{Mk} = \frac{2(k+1)(2k+1)}{k[(2k+1)!!]^2}\left(\frac{\omega}{c}\right)^{2k+1}\left|\left(\psi_2|mQ_{-q}^{(k)}|\psi_1\right)\right|^2. \quad (44)$$

Here the non-relativistic operator of magnetic multipole transitions is as follows:

$$mQ_{-q}^{(k)} = -i^k\frac{r^{k-1}}{c}\sqrt{k(2k-1)}\left\{\frac{1}{k+1}\left[C^{(k-1)} \times L^{(1)}\right]_{-q}^{(k)} + \right.$$

$$\left. + \left[C^{(k-1)} \times S^{(1)}\right]_{-q}^{(k)}\right\}. \quad (45)$$

Analogous general expressions for the relativistic electric and magnetic multipole radiation operators may be found in [1]. The expressions presented cover practically all cases of electronic transitions, two-electron transitions included. The latters become allowed when accounting for correlation effects.

CONCLUSION

This paper demonstrates that it is possible to formulate universal method of finding the expressions for non-relativistic and relativistic matrix elements of any one- and two-electron operators for an arbitrary number of shells in an atomic configuration, requiring neither coefficients of fractional parentage nor unit tensors. It is based on the second quantization in the coupled tensorial form, angular momentum theory in three spaces (orbital, spin and quasispin) and a generalized graphical technique. Both diagonal and non-diagonal, with respect to configurations, matrix elements are treated then uniformly.

For complex electronic configurations new programs lead to the speed-up of calculations of spin-angular parts of matrix elements of 3-8 or even 5-12 times. They allow one to efficiently account for correlation and relativistic effects practically for any atom or ion. Calculations of various atomic properties confirm the latter conclusion.

Finally let us also mention that the graphical technique of angular momentum was succesfully applied to the studies of polarization in photoionization of atoms [21,22] as well as of the Auger decay following photoionization of a polarized atom [23].

REFERENCES

1. Rudzikas, Z., *Theoretical Atomic Spectroscopy*, Cambridge University Press, Cambridge, 1997, pp. 1-424.
2. Gaigalas, G., Rudzikas, Z., Froese Fischer, Ch., *J. Phys. B.: At. Mol. Opt. Phys.*, **30**, 3747-3771 (1997).
3. Gaigalas, G., *Lithuanian J. Phys.*, **41**, 39-54 (2001).
4. Bogdanovich, P., Gaigalas, G., Momkauskaite, A., Rudzikas, Z., *Physica Scripta*, **56**, 230-239 (1997).
5. Rudzikas, Z.B., Kaniauskas, J.M., *Int. J. Quantum Chem.*, **10**, 837-852 (1976).
6. Gaigalas, G., Rudzikas, Z., *J. Phys. B.: At. Mol. Opt. Phys.*, **29**, 3303-3318 (1996).
7. Gaigalas, G., Bernotas, A., Rudzikas, Z., Froese Fischer, Ch., *Physica Scripta*, **57**, 207-212 (1998).
8. Gaigalas, G., *Lithuanian J. Phys.*, **39**, 79-105 (1999).
9. Gaigalas, G., *Lithuanian J. Phys.*, **40**, 395-405 (2000).
10. Rudzikas, Z.B., Kaniauskas, J.M., *Quasispin and Isospin in the Theory of Atom*, Mokslas Publishers, Vilnius, 1984, pp. 1-139 (in Russian).
11. Gaigalas, G., Rudzikas, Z., Froese Fischer, Ch., *Atomic Data and Nuclear Data Tables*, **70**, 1-39 (1998).
12. Gaigalas, G., Fritzsche, S., Rudzikas, Z., *Atomic Data and Nuclear Data Tables*, **76**, 235-269 (2000).
13. Gaigalas, G., Froese Fischer, Ch., *Computer Physics Communications*, **98**, 255-264 (1996).
14. Gaigalas, G., Fritzsche, S., *Computer Physics Communications*, 134, 86-96 (2001).
15. Gaigalas, G., Fritzsche, S., Grant, I.P., *Computer Physics Communications*, **139**, 263-278 (2001).
16. Rudzikas, Z., *Molecular Physics*, **98**, 1205-1212 (2000).
17. Bogdanovich, P., Zukauskas, G., *Lietuvos Fizikos Rinkinys*, **23**, No 3, 18-33 (1983), in Russian, English Translation: *Soviet Physics Collection*.

18. Bogdanovich, P., Tautvaisiene, G., Rudzikas, Z., Momkauskaite, A., *Mon. Not. R. Astron. Soc.*, **280**, 95-102 (1996).
19. Bogdanovich, P., Martinson, I., *Physica Scripta*, **61**, 142-145 (2000).
20. Gaigalas, G. (private communication).
21. Kupliauskiene, A., Rakstikas, N., Tutlys, V., *J. Phys. B: At. Mol. Opt. Phys.*, **34**, 1783-1803 (2001).
22. Tutlys, V., Kupliauskiene, A., *Lithuanian J. Phys.*, **41**, 31-37 (2001).
23. Kupliauskiene, A., Tutlys, V., *J. Phys. B.: At. Mol. Opt. Phys.*, **35**, 000-000 (2002), in press.

Electron Collision Data for Polyatomic Molecules in Plasma Processing and Environmental Processes

H. Tanaka, M. Kitajima, and H. Cho[#]

Department of Physics, Sophia University, Tokyo 102-8554, Japan
[#]*Physics Department, Chungnam National University, Daejeon 305-764, Korea*

Abstract. The experimental studies for electron-polyatomic molecule collision are reviewed in connection with the plasma processing and environmental issues. Recent developments in electron scattering experiments on the differential cross section measurements for various processes such as elastic scattering, vibrational, and electronic excitations are summarized from high to low energy regions (1-100 eV). The need for cross-section data for a broad variety of molecular species is also discussed because there is an urgent need to develop an international program to provide the scientific and technological communities with authoritative cross sections for electron-molecule interactions.

INTRODUCTION

The interaction of electrons with atoms and molecules is essential in many areas of modern science and technology, so called recently "nano-technology", and is pivotal to a greater understanding of aeronomy, astrophysics, gas and laser discharges, plasma processing, fusion plasmas, medical radiation studies and cellular biology. An example of applications is technology of plasma processing, which includes plasma-vapor deposition (CVD) and plasma etching [1, 2]. Similarly a detailed understanding of electron molecule collisions is fundamental in environmental processes, i.e., the chemistry of dissociated fluorocarbons to develop their use as replacements for the environmentally damaging chlro- and bromo-halocarbons, which is also notable as the global warming gases [3]. The current techniques of plasma diagnosis and modeling have elucidated plasma characteristics from the point of view of atomic, molecular, and optical physics, i.e., much more fundamentally scientific basis rather than the empiricism and intuition relied so far. For discharges utilized in industrial plasma processes, the most significant electron collisions occur in the electron energy range less than 100 eV. The generic primary processes are elastic and inelastic electron scattering, electron impact ionization, electron-impact dissociation, and attachment. Many possible excitation processes arising from the many degrees of freedom available within molecules, however, make the study of electron-molecule collisions extremely complex. The comprehensive set of electron-molecule cross section data for electron-molecule collisions is, therefore, only limited to the simplest diatomics (e.g. H_2,

N_2, O_2) [4] and a few polyatomics (e.g. CH_4, CF_4, SiH_4) [5]. The electron collision data, moreover, have provided the most stringent test of the theoretical methods.

Our research program is set out in terms of three major objectives determined experimentally by the electron-molecule collision mechanisms under: (1) Elastic Scattering, (2) Excitation Processes (vibrational and electronic), and (3) Resonant Electron Scattering. Three broad classes of polyatomic molecular targets have been studied (Table 1): Hydrocarbons, Fluorocarbons, and Linear Tri-atomic Molecules. A *systematic* measurement of Absolute Differential Cross Sections (DCSs) for electron scattering by those molecules have been performed with impact energy range from 1.5 to 100 eV and scattering angles between 15 and 130°.

Table 1. Molecules investigated in the author's laboratory.

CH_4, C_2H_6, C_3H_8, C_2H_4, C_3H_6, C_3H_4
CH_3F, CH_2F_2, CHF_3
CF_4, C_2F_6, C_3F_8, *cyclo*-C_4F_8, C_6F_6, C_2F_4
CF_3Cl, CF_3Br, CF_3I
CF_2Cl_2, $CFCl_3$
SiH_4, Si_2H_6, GeH_4, SiF_4
NF_3, H_2CO
CO_2, N_2O, COS, CS_2

Illustrative examples will be reviewed not only because of their relevance to applications, but also because it is possible in these cases to compare recent results obtained by different groups both with each other and to available theoretical data.

EXPERIMENTAL TECHNIQUES IN CURRENT USE

Electron energy loss spectroscopy (EELS) with a crossed beam geometry has been used to investigate the interaction of electron with polyatomic molecules, which provides detailed information on elastic scattering as well as vibrational and electronic excitation of the targets. Absolute cross sections may be obtained by careful measurement of all the experimental parameters. In practice, relative values of DCSs are usually determined and subsequently normalized by using the relative flow technique. The overall resolution of the spectrometer for these measurements was typically 15-30 meV with the electron beam current of 2-5 nA and the angular resolution of ±1.5°. Experimental errors are

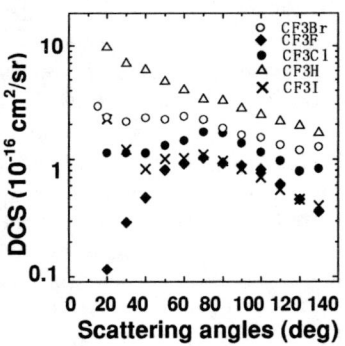

Fig. 1. Elastic DCSs of CF_3X(X=H, F, Cl, Br, I) at 1.5 eV impact energy.

estimated to be ~15% for elastic scattering and ~30%for inelastic one. The absolute energy scale is calibrated by observing the peak in the N_2^- $^2\Pi_g$ resonance at an energy of 2.198 eV for low impact energy region as well as the peak in the He- resonance at an energy 19.36 eV for higher impact energies.

ILLUSTRATIVE EXAMPLES

We briefly present a few examples that illustrate recent measurements from a large number of molecular species listed in Table 1. The main feed gases used by the plasma etching industry are perfluorocarbons (CF_4, C_2F_6, C_3F_8, CHF_3, and c-C_4F_8) however these are strong greenhouse gases and therefore, under the terms of Kyoto Protocol, must be replaced by alternative compounds that have low 'global warning potentials'. One possible replacement is CF_3I since, due to the weak C-I bond, it should be possible to produce high yields of CF_3 radical in any etching plasma by direct electron impact dissociation of CF_3I. CO_2 is the major greenhouse gas, accounting for about half of the total warming effect, because it powerfully absorbs IR. The CFCs and halocarbons have threatened stratospheric ozone.

Elastic scattering by CF_3X (X=H, F, Cl, Br, I), c-C_4F_8, and C_6F_6

The halogen-substituted methane molecules are of fundamental interest as simple tetrahedral-like systems for studying halogen substitutional effects and reactivities of halogen-containing fluorocarbons [6]. In Fig. 1, at 1.5 eV, DCS for CF_4 (non-polar molecule) shows a decreasing trend toward below 70º, while those for CF_3H (1.65D) and CF_3I (0.92D) rapidly continue to grow in small angles, resulting in a strong forward peak, a typical characteristics due to a long range dipole interaction of polar molecules. The intermediate features can be seen from the DCSs for CF_3Cl (0.5D) and CF_3Br (0.65D). In Fig. 2, an interesting feature of the angular distributions for c-C_4F_8 is appearance of maxima minima reminiscent of diffraction patterns as observed in scattering by the heavier atoms, such as Hg. The magnitude of the

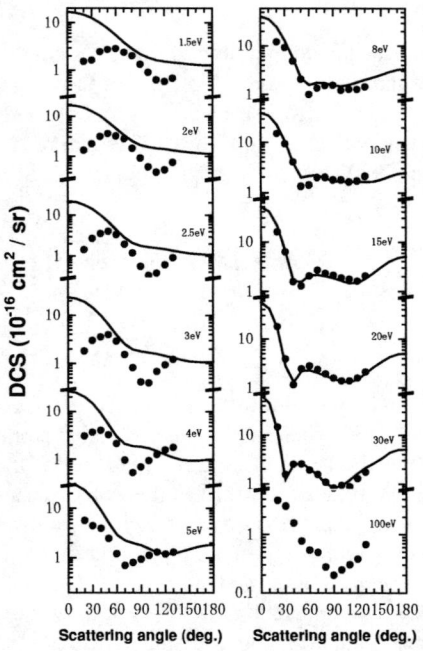

Fig. 2. Angular dependence in elastic DCSs for c-C_4F_8.

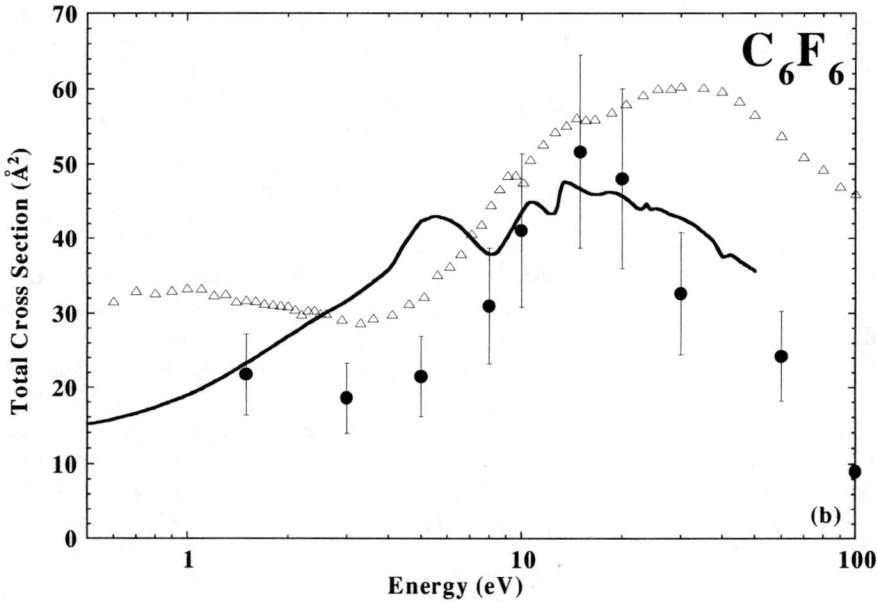

Fig. 3. Integral elastic cross section for C_6F_6.

cross sections is comparable to those for single bonded fluorocarbon molecules. These indicate that the size or complexity of the molecule affects neither the angular distributions nor the magnitude of cross sections. The integral elastic cross section for C_6F_6 have been evaluated from the differential cross section results by extrapolating to forward and backward angles, and integrating the resultant cross sections, as shown in Fig. 3 [7].

Electron Impact Excitation of CO_2 and CF_3I

2-1 Vibrational excitation

In Fig. 4, differential vibrationally inelastic cross sections for electron scattering from CO_2 [8] is presented in energy range from 1.5 to 30 eV over the scattering angles of 10 to 130º. To determine the intensity of the individual modes (010), (100), (001), and the overtone of (020) form the energy loss spectra, a deconvolution procedure was employed. The symmetric stretching shows a sharp increase in DCSs toward small-angle scattering due to the strong dipole interaction, which washes out resonance features seen at intermediate angles. DCSs for the symmetric and bending modes clearly show conspicuous features at a wide range of the scattering angle around 3.8 eV shape resonance. Fig. 5 displays results in the energy dependence of partially integrated cross sections [9] for all four vibrational modes separately. These results were obtained by integrating experimental differential cross sections over scattering angles

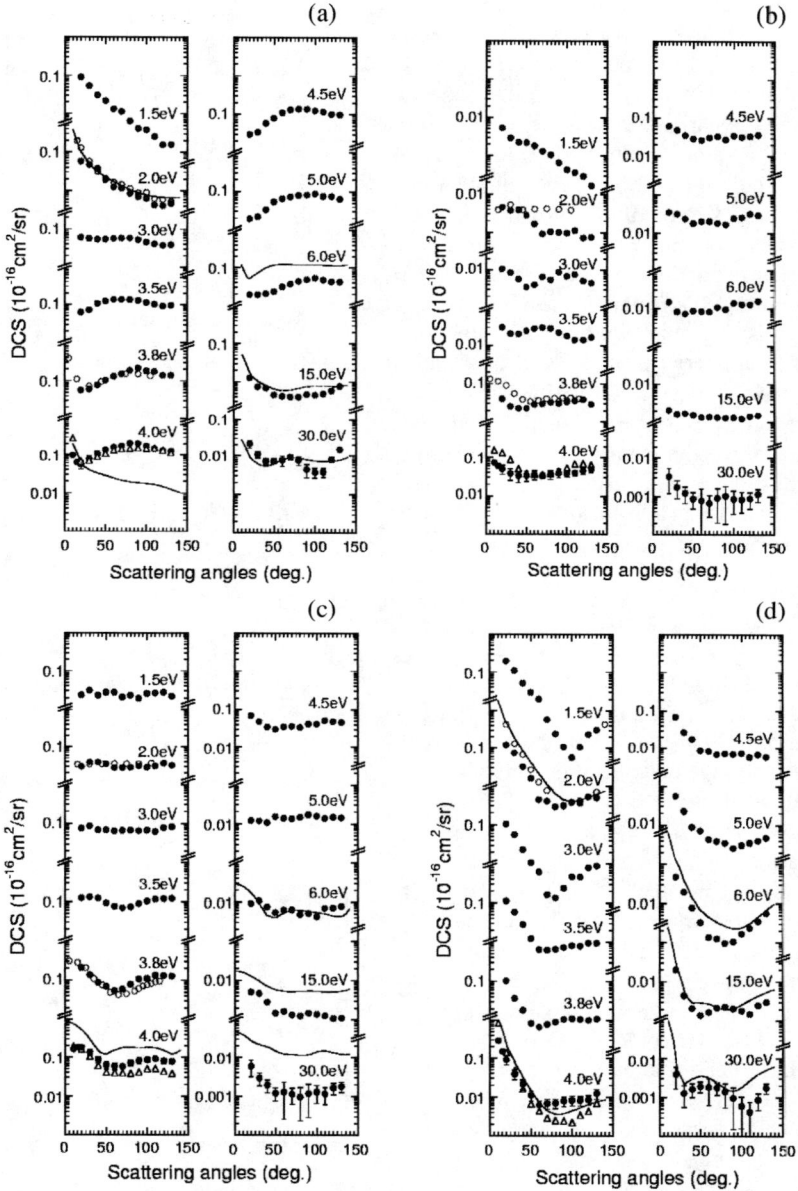

Fig. 4. Absolute DCSs of (a) (010), (b) (020), (c) (100), and (d) (001) vibrational excitation of CO_2.

in the angle region between 20° and 130°. From a systematic observation for both differential and integrated cross sections, vibrational excitation modes are found to show quite different dependence in the 3.8 eV $^2\Pi_u$ shape resonance peak in a CO_2

quite different dependence in the 3.8 eV $^2\Pi_u$ shape resonance peak in a CO_2 molecule. This feature arises from the different interaction scheme for excitation to each mode, i.e., the weak polarization interaction for excitation to symmetric stretching mode (100), weak dipole interaction to the bending mode (010), and strong dipole interaction to the asymmetric stretching mode (001). The vibrational excitation function for the 0.14 eV energy-loss peak (mainly the CF_3 stretching modes) is shown in Fig. 6 [6] as a function of the impact energy from 1.5 to 17 eV and for a scattering angle of 60°. Below 4 eV there is a steep increase in cross section and above this energy there are some overlapping structure, i.e. a peak at 5.5 eV, a shoulder at 8 eV, and a log tail up to 12 eV. These provide evidence for the presence of shape resonances.

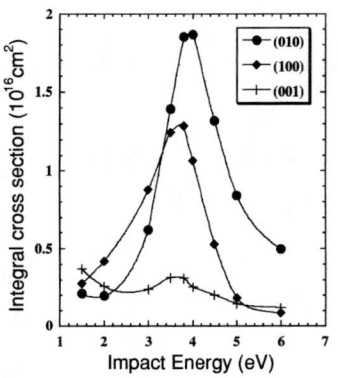

Fig. 5. Experimental integral cross sections for (010), (100), and (001) vibrational modes.

2-2 Electronic excitation

Electron impact excitation cross sections for the electronic states of CO_2 quantitatively measured by electron techniques are limited only at high-energy region over 300 eV. Absolute differential cross sections for electron impact excitation of electronic states of CO_2 in the 10.8-11.5 eV energy-loss range have been obtained at the incident electron energies 20, 30, 60, 100, and 200 eV and over the scattering angles from 3.5° to 90°. Generalized oscillator strengths [10] were derived from the DCSs and the extrapolated to zero momentum transfer, in order to determine the optical oscillator strengths as shown in Fig. 7. The electronic states of CF_3I [11] have been also investigated from 4 to 20 eV. Assignments have been suggested for each of observed absorption bands incorporating both valence and Rydberg transitions. At low energies and large scattering angles (the conditions for large momentum transfer to the target) forbidden transitions will be visible in the EELS spectra. Fig. 8 shows spectra recorded at 100 eV for 3 degrees compared

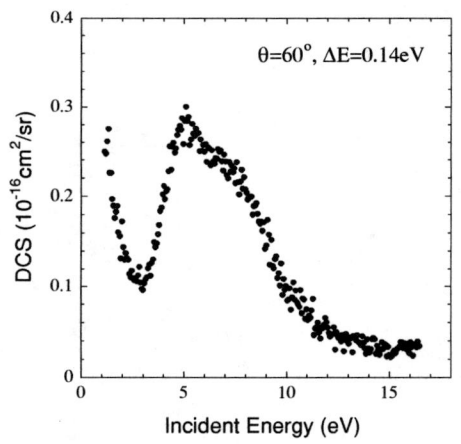

Fig. 6. Vibrational excitation function for energy loss peak of 0.14 eV in CF_3I.

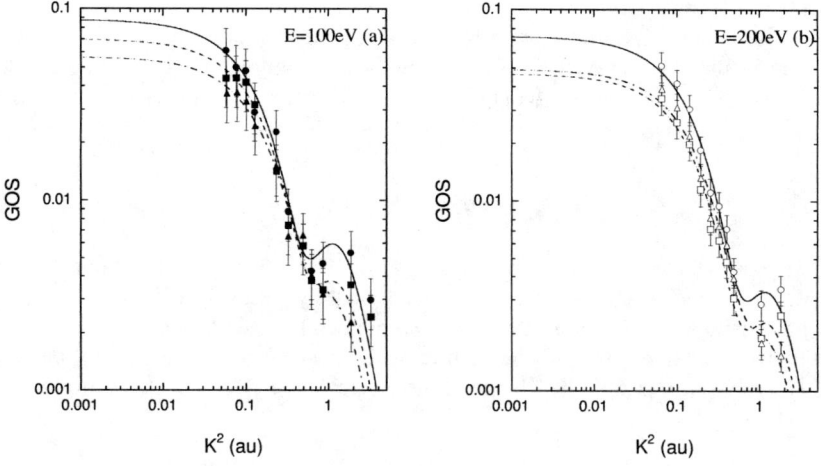

Fig. 7. Semiempirical formula fits to our GOS versus K^2 data at 100 eV and 200 eV.

with 30 eV and 15 degrees. Above 7 eV the spectra are in good agreement confirming that the transitions for Bands B and C are allowed. However notable differences are observed in Band A.

CONCLUSIONS

To provide electron-impact data relevant to a benchmark for experimental, theoretical, and model-calculation studies, these data are, however, needed to be processed as follows [5] (1) synthesis, assessment, and recommendation of electron collision data; (2) deduction of unavailable data and understanding from the assessed knowledge, including identification of new measurements and data needs; and (3) dissemination and updating of database. Hence, a quantitative analysis of the fundamental electron collision processes in terms of cross sections and rate coefficients is applied to understand discharges utilized in industrial plasma processes as well as environmental issues. The quality and reliability for cross section data set that are required to be comprehensive, absolute, and correct is evaluated by the data compilation and assessment through joint efforts involving

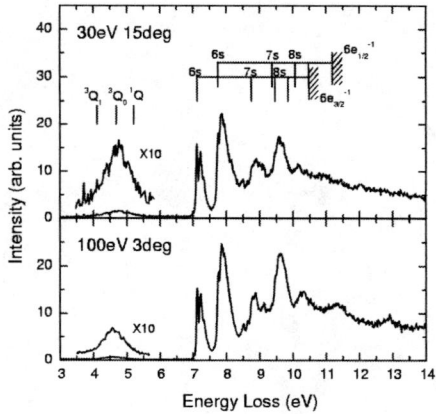

Fig. 8. Energy loss spectra of CF_3I obtained at 100 eV for 3° and 30 eV for 15°.

many knowledgeable works and international collaboration, as mentioned above. This is indeed now recognized, as seen in the ongoing program for data compilation at the NIST and in the International Conference on Atomic and Molecular Data and Their Applications.

ACKNOWLEDGMENTS

We are indebted to Professor S. J. Buckman (the Australian National University), Dr. N. J. Mason (University College London), and Dr. M. J. Brunger (Flinders University of South Australia) with whom the joint experimental research has been progressed. International exchange visits of the authors have been supported in part by the CUP (Core University Program, CR-01-02) of the Japan Society for the Promotion of Science.

REFERENCES

1. Tanaka, H., and Inokuti, M., *Adv. At. Mol. Opt. Phys.* **43**, 1-16 (2000).
2. Tanaka, H., and Sueoka, O., *Adv. At. Mol. Opt. Phys.* **44**, 1-31 (2001).
3. Hill, M. K., *Understanding Environmental Pollution*, Cambridge Univ. Press, London, 1998.
4. Brunger, M. J., and Buckman, S. J., *Phys. Rep.* **357**, 215-458 (2002).
5. Christophorou, L. G., and Olthoff, J. K., *Adv. At. Mol. Opt. Phys.* **44**, 59-96 (2001).
6. Kitajima, M. et al., accepted in *J. Phys. B*.
7. Cho, H. et al., *J. Phys. B* **34**, 1-20 (2001).
8. Kitajima, M. et al., *Phys. Rev. A* **61**, 060701(R)1-4 (2000).
9. Kitajima, M. et al., *J. Phys. B* **34**, 1929-1940 (2001).
10. Green, M. A. et al., *J. Phys. B* **35**, 567-587 (2002).
11. Mason, N. J. et al., submitted to *J. Electron Spectrosc. Relat. Phenom.*

Developing Cross Section Sets for Fluorocarbon Etchants

Carl Winstead* and Vincent McKoy*

A. A. Noyes Laboratory of Chemical Physics, California Institute of Technology, Pasadena, California 91125

Abstract. Successful modeling of plasmas used in materials processing depends on knowledge of a variety of collision cross sections and reaction rates, both within the plasma and at the surface. Electron–molecule collision cross sections are especially important, affecting both electron transport and the generation of reactive fragments by dissociation and ionization. Because the supply of cross section data is small and measurements are difficult, computational approaches may make a valuable contribution, provided they can cope with the significant challenges posed. In particular, a computational method must deal with the full complexity of low-energy electron–molecule interactions, must treat polyatomic molecules, and must be capable of computing cross sections for electronic excitation. These requirements imply that the method will be numerically intensive and thus must exploit high-performance computers to be practical. We have developed an *ab initio* computational method, the Schwinger multichannel (SMC) method, that possesses the characteristics just described, and we have applied it to compute cross sections for a variety of molecules, with particular emphasis on fluorocarbon and hydrofluorocarbon etchants used in the semiconductor industry. A key aspect of this work has been an awareness that cross section sets, validated when possible against swarm data, are more useful than individual cross sections. To develop such sets, cross section calculations must be integrated within a focused collaborative effort. Here we describe electron cross section calculations carried out within the context of such a focused effort, with emphasis on fluorinated hydrocarbons including CHF_3 (trifluoromethane), $c\text{-}C_4F_8$ (octafluorocyclobutane), and C_2F_4 (tetrafluoroethene).

INTRODUCTION

The critical importance of basic data to the understanding of plasma processes is well recognized (*e.g.*, [1]). In particular, it is apparent that numerical models of plasmas, regardless of their sophistication, cannot produce reliable output unless provided with reliable input. In materials-processing applications, many types of cross sections and reaction rates may be needed to understand the chemistry and physics within the plasma and at the surface, yet electron collision data have a special importance. Elastic electron cross sections are needed to model electron transport within the plasma, while inelastic cross sections, including those for electron-impact ionization and electronic excitation, are needed to understand the generation of reactive charged and neutral species. Yet the number of research groups engaged in the study of low-energy electron collisions in molecular gases is limited, and experimental measurement of neutral excitation and dissociation cross sections has proved very difficult. Computational approaches thus may play an important role.

Computational study of low-energy electron–molecule collisions is in itself a significant challenge, especially where polyatomic molecules and inelastic processes are concerned. Methods developed for spherical targets (atoms) may prove inapplicable or slowly convergent when applied to molecular targets having low or no symmetry. Likewise, potential-scattering approaches, in which the problem is approximated as that of a single electron moving in an electric potential field, are economical and often successful for elastic scattering but are not readily applicable to electronic excitation (or ionization). Indeed, a general, flexible approach to low-energy electron–molecule collisions should be based on many-electron wavefunctions in order to account for the full quantum-mechanical complexity of the problem. Such a many-electron treatment can be numerically demanding, with the size of the calculation increasing rapidly as the number of electrons in the molecule increases. Therefore, numerical efficiency and an ability to exploit powerful computers are important considerations.

We have developed a computational method, the Schwinger Multichannel or SMC method [2, 3], that is specifically intended to treat inelastic as well as elastic collisions of low-energy electrons with polyatomic molecules, and we have adapted the SMC method to run efficiently on parallel computers [4, 5, 6, 7], which provide the resources necessary to carry out large-scale computations. We have applied the parallel SMC method to obtain elastic and inelastic electron cross sections for a variety of molecules, with generally good results [8, 9, 10, 11]. Yet we have come increasingly to appreciate that isolated cross section calculations, no matter their basic scientific interest, are of sharply limited utility to consumers of collision data, who usually desire cross section *sets* describing all of the principal electron-collision processes for a given molecule. Indeed, the ideal is a *self-consistent cross section set* (wherein, for example, the total collision cross section equals the sum of the partial cross sections) that has been *validated* against swarm data—that is, the swarm parameters (ionization coefficient, diffusion coefficients, etc.) calculated from the cross section set are in agreement with swarm or electron-drift measurements. Such a cross section set is a sounder basis for modeling than a collection of cross sections from various sources, which may (and often do) prove mutually inconsistent.

Here we review our participation in a research program directed primarily at obtaining electron collision data for fluoro- and hydrofluorocarbon etchants used in semiconductor fabrication. Our role has been to apply advanced computers and computational methods to obtain elastic and electron-impact excitation cross sections critical to the construction of validated cross section sets. In that role, we have collaborated with experimentalists having expertise in the measurement of ionization cross sections and swarm parameters, as well as with experts in the construction and validation of cross section sets and plasma chemistries.

The following section briefly describes our computational procedures. We then survey some recent results for fluoro- and hydrofluorocarbon gases and provide a few closing remarks.

COMPUTATIONAL METHOD

Detailed descriptions of both the SMC method [2, 3] and its implementation for parallel computers [6, 7] can be found elsewhere. Here we point out a few features that may be useful in appreciating the nature of the calculations and their scaling.

The SMC method is a first-principles variational method for the scattering amplitude, the quantity whose square modulus determines the cross section. By first principles, we mean that the SMC method is based directly on the electronic Schrödinger equation (or rather its integral-equation equivalent, the Lippmann–Schwinger equation); by variational, we mean that the SMC method finds, within a given space of possible wavefunctions describing the collision process, the wavefunction that gives the "best" scattering amplitude in a well-defined sense.

As is typical in computational applications of variational methods, we employ a linear expansion of the wavefunction—that is, we write it as a sum of known functions χ_i with unknown coefficients x_i. Introducing that form into the SMC variational expression for the scattering amplitude produces a system of linear equations,

$$\mathbf{A}\mathbf{x} = \mathbf{b}, \tag{1}$$

where $\mathbf{A}$ is a known square matrix, $\mathbf{b}$ is a known vector, and $\mathbf{x}$ is the vector of unknowns x_i. Solving this equation by standard linear-algebraic techniques allows one to compute the variational approximation to the scattering amplitude.

In typical applications, the number of functions χ_i, and thus the dimension of the linear system of Eq. (1), ranges from about one hundred to a few thousand. Solving the linear system is thus not a major challenge. The real computational difficulty lies, rather, in the evaluation of the matrix $\mathbf{A}$ and the vector $\mathbf{b}$. Let us focus on the elements of $\mathbf{b}$, because the most difficult part of $\mathbf{A}$ involves elements of the same form; these elements are given by

$$b_i = \int d\tau \chi_i^* V \Phi(\vec{k}), \tag{2}$$

where the asterisk indicates complex conjugation, $d\tau$ indicates integration over all $3(N+1)$ coordinates of the N target electrons and the $(N+1)^{\text{th}}$ projectile electron, and V is the Coulombic interaction potential between the projectile electron and the electrons and nuclei of the target. $\Phi(\vec{k})$ is the product of an N-electron target wavefunction and a plane wave $\exp(i\vec{k} \cdot \vec{r}_{N+1})$ describing a free electron with momentum $\hbar k$. Borrowing standard techniques of computational chemistry, we can approximate the target wavefunction contained in $\Phi(\vec{k})$ as a spin-adapted, antisymmetrized product of N one-electron functions ϕ_m (molecular orbitals) that are mutually orthogonal and are optimized to describe the electronic structure of the molecule. The functions χ_i are expanded similarly, using $(N+1)$ rather than N molecular orbitals. As a last step, the ϕ_m are written as linear combinations of functions chosen specifically to make ultimate evaluation of the integrals easier, namely, Cartesian Gaussians $g(\vec{r}; \alpha, \ell, m, n)$,

$$g(\vec{r}; \alpha, \ell, m, n) = C_{\alpha \ell m n} x^\ell y^m z^n \exp(-\alpha r^2), \tag{3}$$

$C_{\alpha \ell m n}$ being a normalization constant.

When these successive expansions are introduced into (2), $\mathbf{b}_i$ reduces to a sum of integrals of two forms, the most numerous being

$$I_2 = \int d^3 r_1 \int d^3 r_2 \, g_\alpha(\vec{r}_1) g_\beta(\vec{r}_2) |\vec{r}_1 - \vec{r}_2|^{-1} g_\gamma(\vec{r}_1) \exp(i\vec{k} \cdot \vec{r}_2), \qquad (4)$$

where we have for simplicity written $g_\alpha(\vec{r})$ for $g(\vec{r}; \alpha, \ell, m, n)$, etc. These integrals may be evaluated efficiently to high accuracy. The computational challenge arises, first, from their sheer number: if N_g is the number of Gaussians and N_k the number of plane waves, there are approximately $N_g^3 N_k / 2$ unique integrals of this type. For a larger molecule we might have $N_g \sim 500$ and $N_k \sim 100,000$, implying $\sim 10^{12}$ such integrals. The other computationally intensive step is combining these raw integrals with the coefficients of expansion that describe how the g_α form the ϕ_m and the ϕ_m form the χ_i to obtain matrix elements of $\mathbf{b}$ and $\mathbf{A}$. One can show [6] that the number of arithmetic operations required is proportional to $N_g^4 N_k$. Because of the extra factor of N_g, transforming the raw integrals into final matrix elements will often, especially for larger molecules, require more computational effort than does evaluating them. Because total operation counts may lie in the range 10^{14}–10^{15}, single-processor calculations would be tedious even with gigaflop (10^9 arithmetic operations per second) processors.

Fortunately, the two main computational steps are amenable to parallelization. It is obvious that evaluation of the raw integrals of (4) can be done in parallel—one only need assign each processor a different batch of integrals to evaluate. More care is required in the second step, where data must flow among processors in the course of the calculation. However, this step can be formulated [4] as multiplication of large, dense, distributed matrices, and parallel matrix multiplication is well-studied and inherently efficient. The SMC method is thus able to make efficient use of parallel computers and workstation clusters having dozens or hundreds of processors.

SELECTED RESULTS

In this section, we give an overview of results that we have recently obtained for CHF_3 (trifluoromethane), C_2F_4 (tetrafluoroethene), and c-C_4F_8 (octafluorocyclobutane). We have also studied a number of other fluoro- and hydrofluorocarbons, including C_2HF_5 (pentafluoroethane) [12], C_2F_6 (hexafluoroethane), C_6F_6 (hexafluorobenzene), 1,3-C_4F_6 (1,3-hexafluorobutadiene), and c-C_5F_8 (octafluorocylopentene), as well as radicals such as CHF and CF_2 that may be present in plasmas at significant concentrations; however, the three examples we have chosen are fairly representative and are also of some current interest in plasma etching applications.

CHF_3

To aid in developing an electron cross section set for CHF_3 [13], we used the parallel SMC method described earlier to carry out calculations of the elastic cross section (differential, integral, and momentum transfer) as well as of the electron-impact excitation cross sections for several low-lying electronic states. In addition to these collision calculations, we carried out ancillary studies of fragmentation energetics and of the dissociative behavior of some of the excited states using standard quantum chemistry methodology. We found that each of the excited states whose excitation cross section we considered was dissociative; though not unexpected, this was a useful result in that a major goal was to obtain an estimate of the total dissociation cross section.

Comparison to other calculated [14, 15] and experimental [15] results confirmed that our elastic cross sections were reasonably good. For the inelastic cross sections, whose sum is shown in Fig. 1, there are no data to which we may directly compare. However, measured values of the total neutral dissociation cross section that are far smaller than our calculated excitation cross section have been reported [16, 17]. Given the limited nature of the calculations, we would not expect the cross section of Fig. 1 to be a highly accurate approximation to the total neutral dissociation cross section, but we would expect it to provide a good starting point for estimating that cross section. Such a discrepancy between calculated and laboratory-based results exemplifies the need for further consistency checks in the construction of a cross section set.

Such a consistency check was provided by using the calculated cross sections, together with measured electron-impact ionization cross sections [18], to carry out a swarm analysis, in which electron transport coefficients obtained by simulations based on the cross sections are compared to the electron transport coefficients for CHF_3 obtained in swarm measurements [19, 20]. If significant discrepancies exist, the cross sections are adjusted systematically [21] to bring the simulation results into better agreement with the measured swarm data.

The swarm analysis, performed by our collaborator, W. L. Morgan, resulted in a total dissociation cross section that was larger by a factor of 1.4 than the summed excitation cross section of Fig. 1, thus confirming our suspicion that the measured dissociation cross sections are (for unknown reasons) far too small. Similar conclusions were reached in an independent swarm analysis performed at about the same time by Kushner and Zhang [22], though there were significant differences in other respects between the cross section set of Kushner and Zhang and ours.

c-C_4F_8

Octafluorocyclobutane, c-C_4F_8, presents a significantly greater computational challenge than CHF_3, as is readily seen if we recall that two the principal computational steps increase in difficulty in proportion to N_g^3 and N_g^4, respectively, where N_g is the number of Cartesian Gaussian functions. Since N_g is in turn roughly proportional to the number of non-hydrogen atoms in the molecule, it is apparent that cross section calculations on c-C_4F_8 are inherently several dozen times more challenging than comparable calculations on CHF_3. Nonetheless, by exploiting multiprocessor computers, calculations of useful quality could still be carried out with reasonable turnaround times. We obtained [23] differential, integral and momentum transfer elastic cross sections as well as cross sections for excitation of the lowest singlet and triplet excited electronic states.

In Fig. 2, the computed differential elastic cross section is shown at several energies in

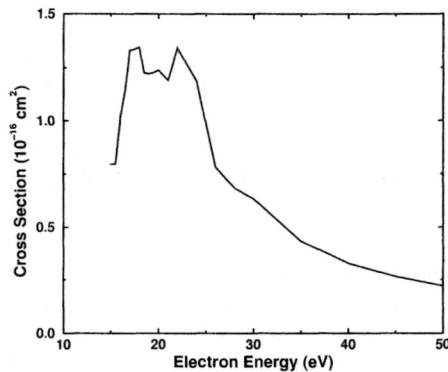

FIGURE 1. Summed cross sections for electronic excitation of CHF_3, from Ref. [13].

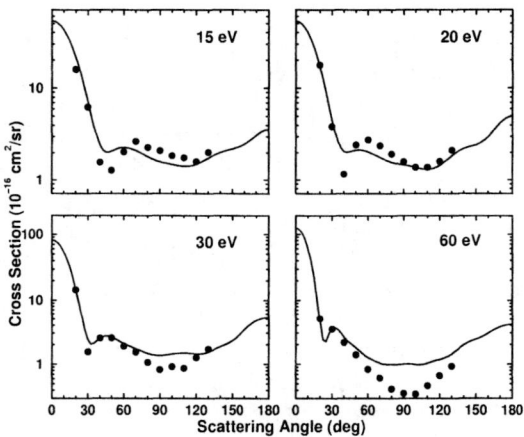

FIGURE 2. Differential elastic cross section for electron scattering by c-C_4F_8 at 15, 20, 30, and 60 eV. Experimental data (filled circles) are from Ref. [24]. Figure taken from Ref. [23].

comparison with experimental values obtained by Tanaka et al. [24]. The generally good agreement gives us confidence that our integral and momentum transfer cross sections (which are derived from the differential cross section) are reliable at these energies. At lower energies, agreement with the measured differential cross section is less good, for well-understood reasons. With additional computational effort, better low-energy results could be obtained; however, we were primarily interested in the energy range where excitation and ionization processes occur.

Font et al. [25] employed our calculated cross sections, along with ionization cross sections measured by Jiao et al. [26], to perform a swarm analysis similar to that previously described for CHF_3. They then used the resulting cross section set, together with a plasma chemistry that they also developed, in simulations of a plasma reactor, obtaining generally good agreement with diagnostic measurements for a variety of properties, including, for example, the densities of various radical fragments.

C_2F_4

Tetrafluoroethene (or tetrafluoroethylene), C_2F_4, has attracted considerable recent interest as an etching gas; as Samukawa et al. have demonstrated [27], adjusting the proportions of C_2F_4 and CF_3I in a mixture affords a degree of control over the relative densities of CF_2 and CF_3 radicals, allowing optimization of the etching process. From a chemical point of view, C_2F_4 also forms an interesting contrast to CHF_3 and c-C_4F_8. C_2F_4 has a double bond consisting of C–C σ and π orbitals, and there are low-lying and strongly excited electronic states arising from excitation out of the π bonding orbital into the conjugate π^* antibonding orbital. One also expects temporary trapping of the scattering electron in the π^* orbital to give rise to a prominent resonance (peak) in the low-energy elastic cross section.

The moderate size and high symmetry of C_2F_4 are advantages to computations. It was consequently possible to do an unusually thorough study [28]. For the elastic cross sections, we included polarization effects, which were completely omitted in the studies of CHF_3 and c-C_4F_8, in the description of resonant scattering, in order to obtain better resonance energies. For electron-impact excitation, we computed cross sections not only for the $\pi \to \pi^*$ triplet (T) and singlet (V) states but also for eight other states that had fairly low excitation thresholds; although we neither expected nor found any of these states to be individually as important as either $\pi \to \pi^*$ state, they did turn out collectively to be significant contributors to the total excitation cross section.

Our elastic calculation placed the π^* resonance at 3.6 eV; dissociative attachment [29, 30, 31] and derivative-mode transmission [32] measurements variously place the resonance at 2.8 to 3.6 eV. Given the limitations of the calculation (in particular, the neglect of vibrational motion), we consider the agreement reasonably good. We likewise found that higher-energy resonances in the elastic cross section correlated rather well with known dissociative attachment features. The only elastic cross section measurements available for C_2F_4 appear to be the 1976 relative differential cross sections of Coggiola et al. [33]. If those data are placed on an absolute scale by single-point normalization to our calculations, generally good agreement is again found [28].

The cross sections for the T and V ($\pi \to \pi^*$) states are shown in Fig. 3. The cross section for excitation of the T state is considerably smaller than the V-state cross section, but the low threshold for excitation gives the T excitation channel special importance in low-temperature plasmas, where the peak in the electron energy distribution may lie at few eV or less. The magnitude of the V-state cross section is explained by the strong optical transition moment of the singlet state, which promotes excitation to the V state even in distant collisions with relatively fast-moving electrons. In Fig. 4, cross sections for eight additional states are shown. Comparison to Fig. 3 verifies that, as stated earlier, none has the low threshold of the T-state cross section nor the large magnitude of the V-state cross section, but that collectively their contribution to the total cross section is significant in a certain energy range, roughly 10–30 eV.

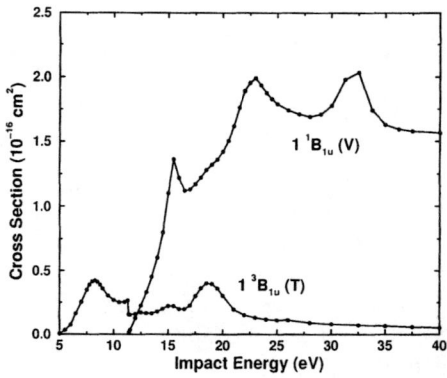

FIGURE 3. Integral cross sections for electron-impact excitation of the T ($1\ ^3B_{1u}$) and V ($1\ ^1B_{1u}$) states of C_2F_4, from Ref. [28].

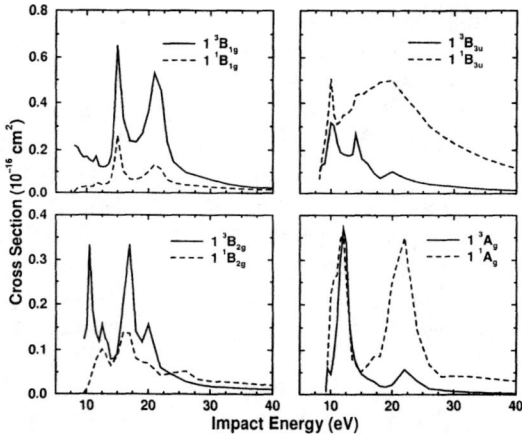

FIGURE 4. Integral cross sections for electron-impact excitation of eight low-lying electronic states of C_2F_4, from Ref. [28].

A cross section set for C_2F_4 was constructed [34] based on the calculated cross sections just described and measured ionization cross sections obtained by Haaland and Jiao [35]. At the time we began work on C_2F_4, no swarm data against which the cross section set might be validated were available, so the effort was extended to encompass swarm measurements by H. Tagashira, K. Yoshida, and S. Goto. W. L. Morgan once again carried out the swarm analysis necessary to produce a validated cross section set. One notable result of the swarm analysis was that it proved unnecessary to adjust the calculated cross sections in order to obtain good agreement with the measured swarm parameters. A second interesting observation was that the two-term Boltzmann analysis proved unreliable at higher values of E/N (the ratio of the applied field to the number density); in other words, the momentum transfer cross section does not capture sufficient information about the anisotropy of the elastic scattering. A Monte-Carlo analysis employing the calculated differential elastic cross sections, however, proved quite successful [34].

CONCLUDING REMARKS

We hope that the examples presented above demonstrate the potential of first-principles calculations to make useful and timely contributions to the supply of critical electron–molecule collision data. We are keenly aware that such calculations are still limited in some important ways—for example, electronic excitation remains a difficult problem, as does elastic scattering at very low energies (~ 1 eV or less)—and we are working to address those limitations. Nonetheless, there is much that can be accomplished with current computational technology, and given the persistent dearth of experimental data for technologically important molecules, computation will continue to have an important role to play.

At the same time, we hope to have made clear that cross section calculations are most useful as a component in a coherent program directed at obtaining self-consistent cross section sets (and, where necessary, swarm data against which those sets may be validated). It is possible to compute cross sections only for some of the (infinite) excitation processes in a given molecule, though fortunately it is often possible to identify *a priori* those that will be most important, and the accuracy currently attainable remains limited. Thus, while computations can provide a starting point for developing an excitation/dissociation cross section, in the form of the energy dependence and approximate magnitude of the cross section, validation against swarm measurements, and adjustment where necessary, remain essential steps. Likewise, some processes, such as electron-impact ionization of polyatomic molecules, remain out of reach of present first-principles computational techniques; despite the many successes of simple approximations such as the BEB model [36] and the modified additivity rule [37], measurements of these cross sections are the preferable alternative, especially as they can provide cross sections for individual fragment ions. As logic would suggest, critical data needs can be addressed most efficiently through a pragmatic approach that draws upon the best features of computation, experiment, and simulation.

ACKNOWLEDGMENTS

It is a pleasure to acknowledge collaboration with W. L. Morgan, P. D. Haaland, and their co-workers throughout the work described herein, as well as with H. Tagashira, K. Yoshida, and S. Goto in the development of a C_2F_4 cross section set. Work at Caltech was supported by Sematech, Inc., by the Department of Energy through the Office of Basic Energy Sciences, and by an equipment grant from Intel Corp. Calculations made use of the facilities of the Caltech Center for Advanced Computing Research and of the Jet Propulsion Laboratory's Supercomputing Project.

REFERENCES

1. Tanaka, H., and Inokuti, M., *Adv. At., Mol., Opt. Phys.* **43**, 1–16 (2000).
2. Takatsuka, K., and McKoy, V., *Phys. Rev. A* **24**, 2473–2480 (1981).
3. Takatsuka, K., and McKoy, V., *Phys. Rev. A* **30**, 1734–1740 (1984).
4. Hipes, P., Winstead, C., Lima, M., and McKoy, V., "Studies of Electron-Molecule Collisions on the Mark IIIfp Hypercube," in *Proceedings of the Fifth Distributed Memory Computing Conference, Charleston, SC, Vol. I: Applications*, edited by D. W. Walker and Q. F. Stout, IEEE Computer Society, Los Alamitos, California, 1990, pp. 498–503.
5. Winstead, C., Hipes, P. G., Lima, M. A. P., and McKoy, V., *J. Chem. Phys.* **94**, 5455–5461 (1991).
6. Winstead, C., and McKoy, V., *Adv. At., Mol., Opt. Phys.* **36**, 183–219 (1996).
7. Winstead, C., and McKoy, V., *Comput. Phys. Commun.* **128**, 386–398 (2000).
8. Winstead, C., and McKoy, V., *Adv. Chem. Phys.* **96**, 103–190 (1996).
9. Winstead, C., and McKoy, V., *Phys. Rev. A* **57**, 3589–3597 (1998).
10. Lee, C.-H., Winstead, C., and McKoy, V., *J. Chem. Phys.* **111**, 5056–5066 (1999).
11. Bettega, M. H. F., Winstead, C., and McKoy, V., *J. Chem. Phys.* **112**, 8806–8812 (2000).
12. Bettega, M. H. F., Winstead, C., and McKoy, V., *J. Chem. Phys.* **114**, 6672–6678 (2001).
13. Morgan, W. L., Winstead, C., and McKoy, V., *J. Appl. Phys* **90**, 2009–2016 (2001).
14. Natalense, A. P. P., Bettega, M. H. F., Ferreira, L. G., and Lima, M. A. P., *Phys. Rev. A* **59**, 879–881 (1999).
15. Varella, M. T. do N., Winstead, C., McKoy, V., Kitajima, M., and Tanaka, H., *Phys. Rev. A* **65**, 022702 (2002).
16. Goto, M., Nakamura, K., Toyoda, H., and Sugai, H., *Jpn. J. Appl. Phys.* **33**, 3602–3607 (1994).
17. Sugai, H., Toyoda, H., Nakano, T., and Goto, M., *Contrib. Plasma Phys.* **35**, 415–420 (1995).
18. Jiao, C. Q., Nagpal, R., and Haaland, P. D., *Chem. Phys. Lett.* **269**, 117–121 (1997).
19. Wang, Y., Christophorou, L. G., Olthoff, J. K., and Verbrugge, J. K., *Chem. Phys. Lett.* **304**, 303–308 (1999).
20. de Urquijo, J., Alvarez, I., and Cisneros, C., *Phys. Rev. E* **60**, 4990–4992 (1999).
21. Morgan, W. L., *J. Phys. D* **26**, 209–214 (1993).
22. Kushner, M. J., and Zhang, D., *J. Appl. Phys.* **86**, 3231–3234 (2000).
23. Winstead, C., and McKoy, V., *J. Chem. Phys.* **114**, 7407–7412 (2001).
24. Tanaka, H., private communication.
25. Font, G. I., Morgan, W. L., and Mennenga, G., *J. Appl. Phys.* **91**, 3530–3538 (2002).
26. Jiao, C. Q., Garscadden, A., and Haaland, P. D., *Chem. Phys. Lett.* **297**, 121–126 (1998).
27. Samukawa, S., Mukai, T., and Tsuda, K., *J. Vac. Sci. Technol. A* **17**, 2551–2556 (1999).
28. Winstead, C., and McKoy, V., *J. Chem. Phys.* **116**, 1380–1387 (2002).
29. Thynne, J. C. J., and MacNeil, K. A. G., *Int. J. Mass Spectrom. Ion Phys.* **5**, 329 (1970).
30. Heni, M., Illenberger, E., Baumgärtel, H., and Süzer, S., *Chem. Phys. Lett.* **87**, 244–248 (1982).
31. Illenberger, E., Baumgärtel, H., and Süzer, S., *J. Electron Spectrosc. Relat. Phenom.* **33**, 123–139 (1984).
32. Chiu, N. S., Burrow, P. D., and Jordan, K. D., *Chem. Phys. Lett.* **68**, 121–126 (1979).
33. Coggiola, M. J., Flicker, W. M., Mosher, O. A., and Kuppermann, A., *J. Chem. Phys.* **65**, 2655 (1976).
34. Yoshida, K., Goto, S., Tagashira, H., Winstead, C., McKoy, B. V., and Morgan, W. L., *J. Appl. Phys.* **91**, 2637–2647 (2002).
35. Haaland, P. D., private communication.
36. Hwang, W., Kim, Y.-K., and Rudd, M. E., *J. Chem. Phys.* **104**, 2956–2966 (1996).
37. Deutsch, H., Becker, K., and Märk, T. D., *Int. J. Mass Spectrom.* **167**, 503–517 (1997).

B. Atomic and Molecular Databases

Spectr-W³ Online Database On Atomic Properties Of Atoms And Ions

A.Ya. Faenov[1], A.I. Magunov[1], T.A. Pikuz[1], I.Yu. Skobelev[1],
P.A. Loboda[2], N.N. Bakshayev[2], S.V. Gagarin[2], V.V. Komosko[2],
K.S. Kuznetsov[2], S.A. Markelenkov[2], S.A. Petunin[2], V.V. Popova[2]

[1]*Multicharged Ions Spectra Data Center of VNIIFTRI, Mendeleevo, Moscow region, 141570, Russia*
[2]*Russian Federal Nuclear Center VNIITF, P.O.B. 245, Snezhinsk, 456770, Russia*

Abstract. Recent progress in the novel information technologies based on the World-Wide Web (WWW) gives a new possibility for a worldwide exchange of atomic spectral and collisional data. This facilitates joint efforts of the international scientific community in basic and applied research, promising technological developments, and university education programs. Special-purpose atomic databases (ADBs) are needed for an effective employment of large-scale datasets. The ADB SPECTR developed at MISDC of VNIIFTRI has been used during the last decade in several laboratories in the world, including RFNC-VNIITF. The DB SPECTR accumulates a considerable amount of atomic data (about 500,000 records). These data were extracted from publications on experimental and theoretical studies in atomic physics, astrophysics, and plasma spectroscopy during the last few decades. The information for atoms and ions comprises the ionization potentials, the energy levels, the wavelengths and transition probabilities, and, to a lesser extent, — also the autoionization rates, and the electron-ion collision cross-sections and rates. The data are supplied with source references and comments elucidating the details of computations or measurements. Our goal is to create an interactive WWW information resource based on the extended and updated Web-oriented database version SPECTR-W³ and its further integration into the family of specialized atomic databases on the Internet. The version will incorporate novel experimental and theoretical data. An appropriate revision of the previously accumulated data will be performed from the viewpoint of their consistency to the current state-of-the-art. We are particularly interested in cooperation for storing the atomic collision data. Presently, a software shell with the up-to-date Web-interface is being developed to work with the SPECTR-W³ database. The shell would include the subsystems of information retrieval, input, update, and output in/from the database and present the users a handful of capabilities to formulate the queries with various modes of the search prescriptions, to present the information in tabular, graphic, and alphanumeric form using the formats of the text and HTML documents. The SPECTR-W³ Website is being arranged now and is supposed to be freely accessible round-the-clock on a dedicated Web server at RFNC VNIITF. The Website is being created with the employment of the advanced Internet technologies and database development techniques by using the up-to-date software of the world leading software manufacturers. The SPECTR-W³ ADB FrontPage would also include a feedback channel for the user comments and proposals as well as the hyperlinks to the Websites of the other ADBs and research centers in Europe, the USA, the Middle and Far East, running the investigations in atomic physics, plasma spectroscopy, astrophysics, and in adjacent areas. The effort is being supported by the International Science and Technology Center under the project #1785-01.

INTRODUCTION AND OVERVIEW

Presently, the advances in the field of top technologies essentially depend on the availability of reliable and consistent data on the fundamental spectral properties of atoms and ions. The reason is that these properties are responsible for the interactions of matter with electromagnetic radiation and particle beams. Accordingly, the relevant data form the basis for upgrading high-precision diagnostic tools for the Inertial Confinement Fusion (ICF) studies, astrophysics as well as for the development of tabletop x-ray radiation sources for semiconductor micromachining, microlithography, diffractometry, studies of ultra-fast chemical reactions, and many other technological and scientific applications. However, this kind of data, resulting from costly calculations and experiments, are dissipated over a great number of original papers and are often published in an incomplete form. Therefore, the needs for systematization and adequate interpretation of spectral properties of atoms and ions along with the availability to effectively employ large-scale datasets for the applied research unambiguously assume the development of appropriate atomic databases (ADBs).

A tremendous progress in the novel information technologies achieved in the last few years on the basis of the World-Wide Web (WWW) has shown a very good potential for a worldwide exchange of scientific and technological information, including the atomic data. The most attractive features are friendly and understandable user interface, and powerful HTML language for creating and displaying hypertext documents able to support not only different alphanumeric formats but also graphics, including figures, formulas, tables, etc. Besides, WWW software is based on the standard operating systems of personal computers (PCs) and workstations and hence can readily be mastered by users.

A serious long-term effort made in 1988-95 at Multicharged-Ion Spectra Data Center (MISDC) of VNIIFTRI resulted in the development of a factual atomic databank SPECTR, which later served as a certified Databank of Russian State Service of Standard Reference Data. Previously, the relevant atomic database (SPECTR ADB), implemented using the FoxPro DBMS for IBM-compatible PCs, was available to users locally at MISDC of VNIIFTRI and was also distributed by the authors on magnetic tapes and diskettes upon the agreement with the interested organizations. The SPECTR ADB has been successfully used in a number of laboratories and universities worldwide, including RFNC-VNIITF. This database accumulates a considerable amount of factual information on the spectral properties of multicharged ions, low-ionized and neutral atoms (about 500,000 records), obtained at the leading research centers and university laboratories, which have been pursuing the studies in atomic physics, plasma spectroscopy and astrophysics during several decades. Nowadays the SPECTR ADB is still the largest factual database in the world, containing the information on the spectral properties of multicharged ions.

The information accumulated in the SPECTR ADB includes the experimental and theoretical data on ionization potentials, energy levels, wavelengths, radiation and autoionization transition probabilities, and parameters used in analytical expressions to approximate collisional cross-sections and electron transition rates in atoms and ions. The information is supplied with the references to the original sources and comments, elucidating the details of experimental measurements or calculations. SPECTR ADB accumulated practically all the experimental data for the x-ray spectral range published up to 1994. Publications also served as a source to obtain the theoretical data. It should be noted that a significant portion of theoretical data was calculated specially for the SPECTR ADB by the highly-qualified experts of a number of universities and institutes of Russia and the Former Soviet Union: the Institute of Theoretical Physics and Astronomy of Lithuanian Academy of Sciences, the Institute for Spectroscopy of Russian Academy of Sciences, Voronezh State University, MISDC of SRC VNIIFTRI, Uzhgorod State University (Ukraine), and RFNC-VNIITF. The data were calculated using various theoretical methods and were published in detail mostly in paper collections, preprints and reports of the relevant institutions. Therefore, the SPECTR ADB is the only resource for the world scientific community presenting a full-scale access to these data.

After the RFNC-VNIITF got a high-performance channel to access Internet in 1995, a wide experience in creating specialized Websites covering many directions of the research and spin-off activities of the RFNC-VNIITF has been gained by its specialists. This has resulted in a number of developments to create Web-versions of databases using the up-to-date commercial DBMS under Windows NT of IBM-compatible PCs like ORACLE.

The goal of this work is to create an online WWW information resource on atomic data based on the updated version of factual atomic database SPECTR. This resource can be used as a reference system for basic and applied research, promising technological developments, and university education programs. The main objectives of the project consist in an essential update of the SPECTR databank, development of a new Web-oriented database version SPECTR-W^3, and its further integration into the family of specialized atomic databases on the Internet.

ADB SPECTR-W^3

The atomic databank SPECTR is being updated by the inclusion of new experimental and theoretical information on the multicharged-ion spectra both published in literature and obtained in the participating organizations during the recent years, revision of the existing data, systematization of the referenced source data. Specifically, analysis of data on the state and transition properties stored in the SPECTR databank has shown that:
- contents of the SPECTR ADB adequately reflects the state of the art of experimental and theoretical research pursued in X-ray spectroscopy by early 90-ies of the 20th century;
- a huge amount of spectral data accumulated in the recent decade is poorly presented in the SPECTR ADB since during that time the SPECTR ADB was not actually updated;
- the SPECTR ADB contains a considerable amount of records with incomplete or erroneous data;
- designations of configurations and energy terms accepted in the SPECTR ADB with only capital letters, digits and brackets, that complied with the capabilities of the computer data processing of the past, do not meet the up-to-date practice

of handling spectroscopy data, thus, leading to complications or confusions in classifying these data.

The systematic analysis of the structures of data tables in the SPECTR ADB done with the tools of ORACLE 8 DBMS has shown that

- the reference tables on ions, atoms, and data acquisition methods should be added to the database structure;
- all the tables on spectroscopic data are to be supplemented with the fields for HTML-representation of configuration names enabling to display them in the form generally used in atomic spectroscopy;
- replacement of the symbolic identification of atoms and ions by a digital one, based on atomic numbers and weights of elements in the Periodic table, is desirable to eliminate a source of additional errors during data input;
- it is reasonable to combine tables of the upper and lower energy levels of radiative transitions into a single common one;
- it is reasonable to incorporate the tables of cross-sections and rates of dielectronic recombination, electronic excitation and ionization into a single integrated table, **SPE_CR**, in which the field **id_process** may take an appropriate values **DREC**, **EXC**, and **ION**.

Verifying calculations were done for the complicated spectra of [Be], [B], [N], and [Na]-ions with incomplete inner shells in order to properly classify terms when correcting the snippy or ambiguous records in their names. This work was done with the GRASP^2 package implementing multi-configuration Dirac-Fock method. The performed calculations were also used to develop a standardized term designation for the new version of the SPECTR ADB. The data on the parameters used to approximate cross-sections and rates of the elementary processes, given in the SPECTR ADB were also analyzed. The tasks are formulated to present the data on collision strength for ionization and excitation in the form needed for the SPECTR ADB. These data have been recently obtained by the Project participants who pursued the applied research in Julich (Germany). To make this presentation possible, a series of additional calculations were done with the ATOM code.

During the first half-year of the Project implementation, techniques to read new atomic data out of the appropriate electronic publications and recognize the data from graphic images of the scanned publication hardcopies were also developed during the first and second quarters. These techniques will be used in future to enter new information in the SPECTR ADB. To appropriately prepare the information for loading into the SPECTR ADB, a specialized interface was created using Microsoft Access database management software (DBMS) to handle these techniques. The scanned and recognized documents are dumped as tables in the rtf-format and then imported to the Microsoft Access database thru the delimited text format. Using a special visual control form, a table field containing transition representation in the text format, is transformed into HTML format and kept in the other field of the table. Normalized tables are stored and managed using the Microsoft ACCESS DBMS, and the processed original electronic copies or scanned images are moved to a special folder. The follow-on data transformation is to be powered by the Oracle DBMS. An algorithm is developed and tested to convert the names of atomic states originally written in different text-format representations (with digits, capital letters and various brackets) into the standardized representation in HTML-format including capital and small letters, subscripts, superscripts, and brackets.

Based on the analysis of the information in the SPECTR ADB, a new structure of database to be powered by Oracle 8.1 was designed (see Figure 1 and Table 1) in the environment of ERWin system and the kernel of the new ADB SPECTR version is being developed.

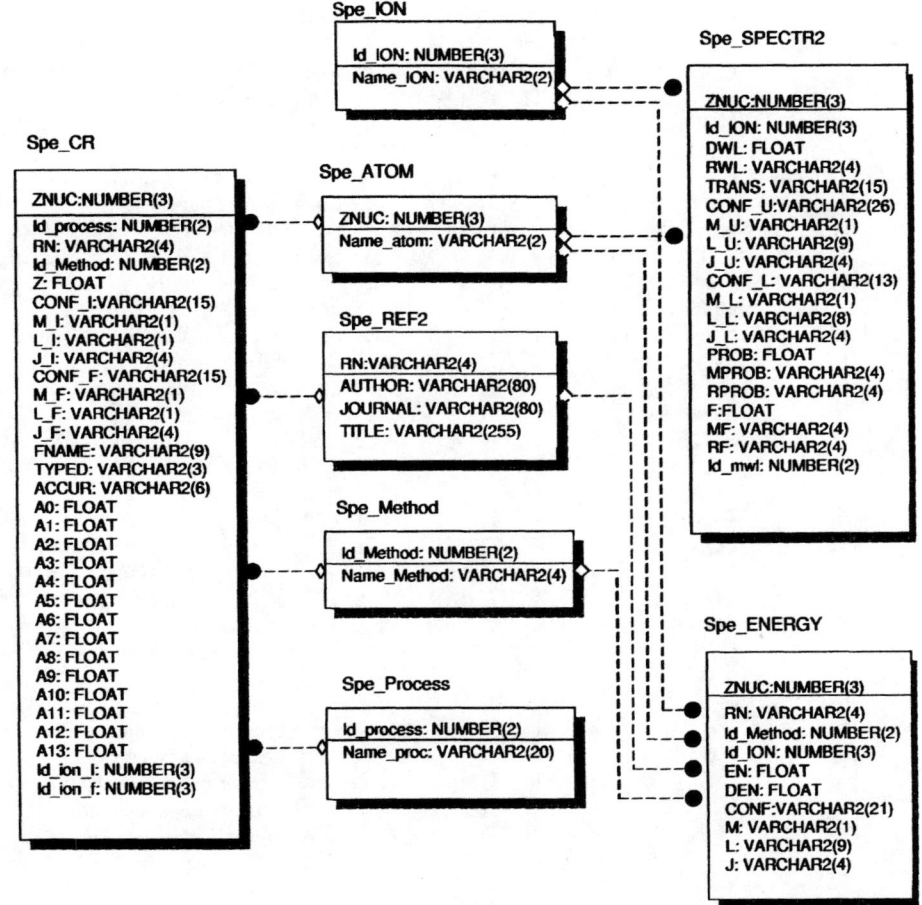

FIGURE 1. The structure of ADB Spectr-W^3.

In the renewed database
- records with snippy or erroneous data were corrected or deleted; and
- using HTML-format the energy level names were written in the standard form including capital and small letters, subscripts, and superscripts.

To optimize the development of the Web-version of the SPECTR ADB, **SPECTR-W^3**, Web-server *"Internet Information Server 5.0"* was selected which runs under Windows 2000 Professional operating system. After the database access system is developed, it will be moved on the freely accessible Apache-powered Web-server, running under Linux operating system.

TABLE 1. Field Description For ADB Spectr-W^3.

Name	Fields	Format	Field Description
Spe_ENERGY	CONF	TEXT(255)	Level configuration in text format
	CONF_HTML	TEXT(2000)	Level configuration in HTML-format
	DEN	FLOAT	Accuracy for the energy evaluation (1/cm)
	EN	FLOAT	Energy value(1/cm)
	J	TEXT(4)	Total angular momentum of the atomic state
	L	TEXT(9)	Total orbital momentum of the atomic state
	M	TEXT(1)	State multiplicity, M= 2S+1
	id_ION	INT(3)	Ion identifier
	id_Method	INT(2)	Method identifier
	RN	TEXT(4)	Bibliography reference identifier
	ZNUC	INT(3)	The element atomic number
Spe_ATOM	name_atom	TEXT(2)	The name of the element
	ZNUC	INT(3)	The element atomic number
Spe_ION	Name_ION	TEXT(2)	The name of the element
	id_ION	INT(3)	Ion identifier
Spe_Method	Name_Method	TEXT(4)	The name of the method of data obtaining
	id_Method	INT(2)	Method identifier
Spe_Process	id_process	INT(2)	Atomic process identifier
	Name_proc	TEXT(20)	The name of the process
Spe_REF2	AUTHOR	TEXT(80)	Author list
	JOURNAL	TEXT(80)	Publication name
	RN	TEXT(4)	Bibliography source identifier
	TITLE	TEXT(255)	Name of the paper
Spe_SPECTR2	CONF_L	TEXT(255)	Lower level configuration (text format)
	CONF_L_HTML	TEXT(2000)	Lower level configuration (HTML format)
	CONF_U	TEXT(255)	Upper level configuration (text format)
	CONF_U_HTML	TEXT(2000)	Upper level configuration (HTML format)
	DWL	FLOAT	Accuracy of the wavelength evaluation
	F	FLOAT	Oscillator strength
	id_mwl	INT(2)	Wavelength evaluation method identifier
	J_L	TEXT(4)	Lower level total angular momentum
	J_U	TEXT(4)	Upper level total angular momentum
	L_L	TEXT(8)	Lower level total orbital momentum
	L_U	TEXT(9)	Upper level total orbital momentum
	M_L	TEXT(1)	Lower level multiplicity
	M_U	TEXT(1)	Upper level multiplicity
	MPROB	TEXT(4)	Radiative probability evaluation method
	id_ION	INT(3)	Ion identifier
	PROB	FLOAT	Radiative probability value
	RF	TEXT(4)	Reference to the bibliography source for the oscillator strength evaluation
	RPROB	TEXT(4)	Reference to the bibliography source for the radiative probability evaluation
	RWL	TEXT(4)	Reference to the bibliography source for the wavelength evaluation
	TRANS	TEXT(15)	Optical transition specification
	WL	FLOAT	Value of the wavelength
	ZNUC	INT(3)	The element atomic number

To organize access to the Windows-based database, three search pages have been designed to-date using the well-proven Active Server Pages (ASP) technology:
- Ionization potentials search page
- Energy levels search page
- Bibliography search page

The example of "Energy levels search page" is presented in Figure 2.

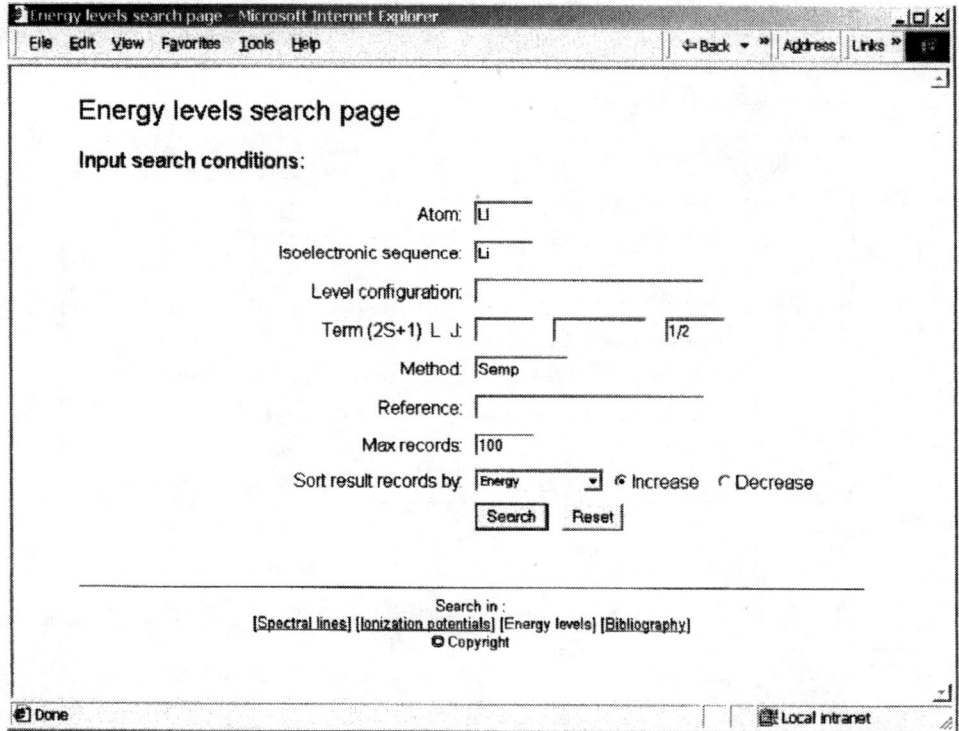

FIGURE 2. Energy level search page.

User specifies the data of interest, presses the Search button, and receives the result of inquiry in the form, presented in Figure 3. The result of inquiry may be sorted by any field of record. The page of inquiry results consists of three frames. In first of them displays the records of interest. In the second, the information on a data source is displayed when the user clicks on an appropriate reference in Reference field. In the third, navigation bookmarks are displayed. The frame size may be adjusted by the user to conveniently analyze the retrieved data. In the shown example, the frame size with navigation bookmarks is turned down to the minimal one.

Currently, the Spectral lines search page is under development.

In parallel with the development of the Web-site pages, the activities are being carried out to put the pages on Apache Web-server. Using the Web-programming technologies, an algorithm to temporarily dump previous information inquires sent by user to the **SPECTR-W^3** ADB is selected, which would be compatible both with

Unix/Linux and Windows platforms. The algorithm uses portions of information recorded automatically on the user's computer hard disk (the so called cookies). A version of data display at the user's request in the form of HTML-marked Web-table was implemented, maximum number of displayed records being specified by the user. Other version of data display in the plain text format and in the form of a Web-table to be emailed to the user is also being developed.

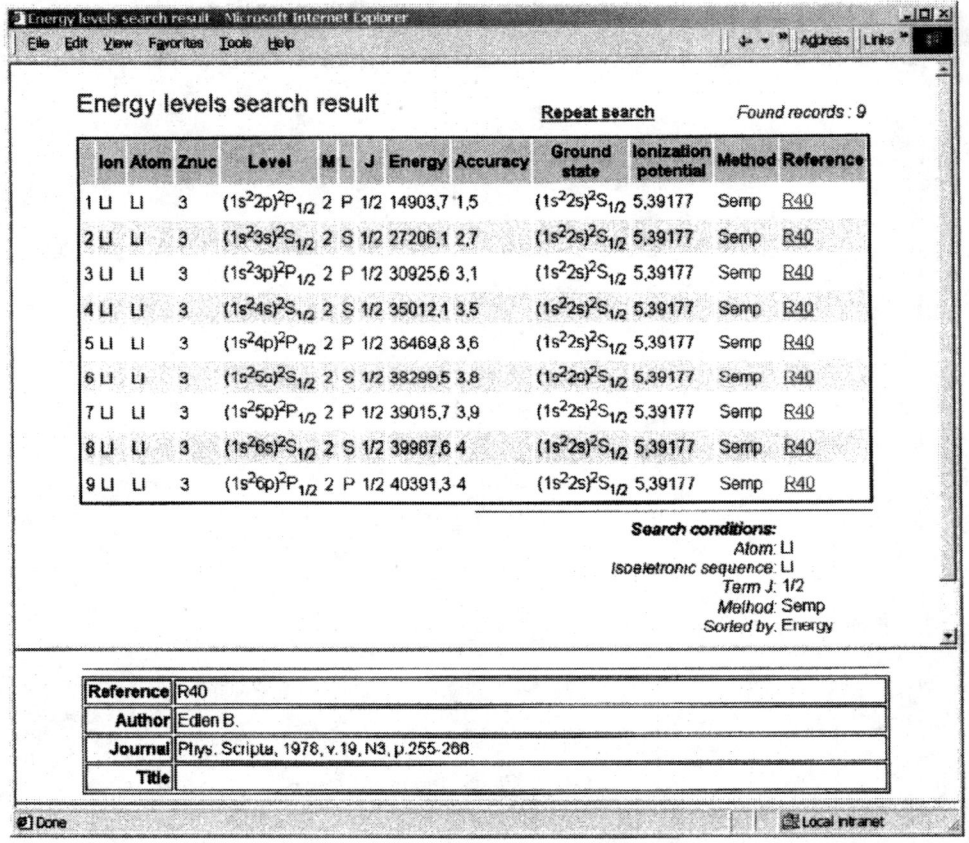

FIGURE 3. Energy levels search result.

The follow-on activities to update the SPECTR ADB will involve:
- the input of experimental data on the wavelengths of x-ray spectral lines and energy levels of multicharged ions (from the selected publications);
- the input of evaluated wavelengths and probabilities of radiative and autoionization transitions in multicharged ions (from the selected publications);
- the input of spectral data calculated by the Project foreign collaborators; and
- conduction of special-purpose experiments by the Project participants and foreign collaborators to obtain new high-resolution data to be incorporated in the SPECTR ADB, in particular, to obtain information on the structure of

Rydberg autoionization states of multicharged ions. It should be noted that during past years the Project participants have investigated a lot of emission spectra of multicharged ions in a high-temperature plasma (see Table 2), and many precise spectral data have been obtained.

TABLE 2. Precise Wavelength Data Produced By Project Participants During Past Years.

Atom	Isoelectronic sequence	Transitions	Wavelength region, Å	Accuracy, mÅ
		resonance transitions		
O	He-	$1snp - 1s^2$ (n = 4-10)	16.0 – 17.8	1.5 – 3.0
F	He-	$1snp - 1s^2$ (n = 4-9)	12.5 – 15.2	0.6 – 1.6
Mg	He-	$1snp - 1s^2$ (n = 4-9)	7.05 – 8.2	0.4 – 1.0
Al	He-	$1snp - 1s^2$ (n = 6-12)	5.7 – 6.1	0.3 – 0.4
F	He-	$2l2l' - 1s2l'$	13 - 17	1 – 1.2
Ni	O-, F-, Ne-	3 –2	10.5 -12.9	1.0
	Ne-	5 – 2,..., 15 – 2	7.8 – 9.3	0.5 – 2.5
Cu	O-, F-, Ne-	3 – 2	11.5 -14.1	1.0
	Ne-	4 – 2,..., 7 – 2	7.5 – 8.7	0.5 – 1.5
Zn	B-, C-, O-, F-	3 – 2, 4 – 2	6.5 -11.8	3.0
	Ne-	5 – 2,..., 9 – 2		3.0
Ge	Ne-	7 – 2,..., 9 – 2	8.7 – 9.3	0.5 – 1.5
Xe	Ni-	4 – 3	17 - 19	0.5
Cs, Ba	Ne-	3 – 2	2.6 -2.7	0.3 – 0.7
Ba	Cu-, Zn-, Ga-	6 – 3, 7 – 3	9.1 – 9.4	0.9
La	Cu-,..., As-	4 – 3,..., 7 – 3	7.4 – 11.9	1.0
Ce	Fe-,...,As-	4 – 3,..., 7 – 3	7.4 – 11.9	1.0
		satellite lines		
Si	He-	$2l2l' - 1s2l'$, $2l3l' - 1s3l'$	6.16 – 6.27	0.5 – 1.0
Mg - S	Li-,...F-	$1s2l2l' - 1s^22l'$	5.5 – 9.5	0.5 – 1.0
Mg	Li-	$1s2l3l' - 1s^23l'$, $1s2l2l'3l'' - 1s^22l'3l''$	9.16 – 9.38	1.0
	Be-			
Si	Li-	$1s2l3l' - 1s^22l'$, $1s3l3l' - 1s^23l'$	5.69 – 5.83	0.4 – 1.5
Ar	Li-	$1s2l2l' - 1s^22l'$, $1s2l3l' - 1s^22l'$, $1s2l4l' - 1s^22l'$	3.26 – 3.94	0.2 – 0.4
Mg, Si	Li-	$1snln'l'-1s^2nl$ (n=2,3, n'=3-6, n=2, n'<16)	7.4 – 8.2	0.5 - 1.0
			5.1 - 5.7	0.5 -0.7
Cr, Mn, Fe, Co	Na-	n=3 – n=2	12 -17	1.0 – 3.0
Se	Na-	n=3 – n=2	7.3 – 8.7	1.0 – 2.0
Ni, Cu, Zn, Kr	Na-	n=3 – n=2	6.3 -14.1	1.0
Y, Zr, Nb, Mo	Na-	n=3 – n=2	4.6 -5.9	2.0
Cu, Kr, Y	Na-	n=4 – n=2	4.1 -9.43	0.4 – 1.5
Y	Mg-	N=3 – n=2, n=4 – n=2	4.1 -5.9	1.5 – 2.0
Mo	Mg-	n=3 – n=2	4.6 -4.9	2.0

The activities under the Project are pursued in close contact with the foreign collaborators of the project (Dr. A.L. Osterheld, Dr. J. Nilsen - Lawrence Livermore National Laboratory, Dr. J. Abdallah, Jr. - Los Alamos National Laboratory, Prof. W.L. Wiese - National Institute of Standards & Technology, Dr. D. R. Schultz - Oak Ridge National Laboratory) and other participants of international scientific cooperation aimed at the creation and utilization of the global reference system on atomic and molecular data on the Web.

ACKNOWLEDGMENTS

This work is being supported by the International Science and Technology Center under the project #1785-01.

Electron-Molecule Cross Section Data for Hydrogen Plasma Applications

R. Celiberto*, A. Laricchiuta†, R.K. Janev** and M. Capitelli†‡

*Dipartimento di Ingegneria Civile ed Ambientale, Politecnico di Bari, Italy
†IMIP, Sezione Territoriale di Bari, CNR, Italy
**Macedonian Academy of Science and Arts, Skopje, Macedonia
‡Dipartimento di Chimica Università di Bari, Italy

Abstract. Electron-impact cross sections involving vibrationally and electronically excited molecules of hydrogen and its isotopes are presented, in particular reviewing the existing data and discussing some new results.

1. INTRODUCTION

Vibrationally excited diatomic hydrogen molecules play a role of fundamental importance in many plasma systems of technological interest. Typical examples are the negative ion sources [1] and divertor and edge plasmas, extensively studied in nuclear fusion research [2].

The negative ions, used for the ignition of nuclear fusion reaction by collision of high energy accelerated H^- (or D^-) beams on hydrogen gas target, are obtained in the so-called multicusp magnetic sources by thermoemitted electrons in a H_2 gas chamber, where they generate the plasma. In these conditions the main process for the production of H^- ions proceeds via dissociative attachment of an electron from a ground-state H_2 molecule:

$$H_2(X^1\Sigma_g^+, v_i) + e \rightarrow H_2^- \rightarrow H + H^-$$

It is well known that the efficiency of this process is greatly enhanced as increases the vibrational excitation of the molecules. As consequence, all the processes affecting the vibrational distribution, by destroying or creating vibrationally molecular states (dissociation, ionization, excitation, etc.) become important, and modelings of these systems aim at providing indications for the optimization of the source's operating conditions to enhance the concentration of vibrationally excited species [1].

The relatively low temperatures ($\sim 10^0 - 10^2$ eV) reached in tokamak divertor and edge plasmas, allow for the formation of electronic and vibrationally excited hydrogen molecules, which strongly affect the atomic physics of these systems. The dramatic reduction of the temperature, from hundreds eV close to the core plasma down to few eV (or sub eV) near the plates in the divertor chamber, for example, is due to the recombining processes in plasma volume mainly induced by the presence of vibrationally excited H_2 species (Molecular Assisted Recombination [3]).

In these systems the physics involving H_2 is rather complex and a role of primary importance is played by the electron-molecule collisions. In particular, vibrationally, as well as electronically, excited hydrogen molecules act in the plasma as individual chemical species, so that any impact event is characterized by a given cross section and rate constant. Thus, in this frame, the knowledge of state-to-state electron-impact cross sections, and related rate coefficients, becomes crucial in the applications, either in the construction of theoretical kinetics models of the plasma [4] that in experimental diagnostic methods [5].

Although the situation is far to be exhaustive, large sets of cross section data are now available for a number of processes, as a function of both incident electron energy and vibrational quantum numbers of the H_2 molecule, and in some case for the isotopic variants [12]. In this paper we will review the existing cross section data and present some new (preliminary) results.

2. ELECTRON-IMPACT CROSS SECTIONS

Cross section calculations for the processes discussed below have been performed by using mainly the resonant model for dissociative attachment and resonant vibrational excitation and dissociation, the classical Gryzinski method and the impact-parameter method, in combination with the Born approximation, for the excitation and ionization. Since we are not concerned here with the mathematical and numerical aspects of the theories, we refer the interested reader to the cited literature for details.

2.1. Resonant processes

In low-energy electron-molecule collisions a resonant molecular state can be formed by capture of the incident electron by the molecule initially in its ground electronic state and in a given vibrational level. The resultant negative ion H_2^- can evolve through the three following channels:

$$H_2(X^1\Sigma_g^+, v_i) + e \rightarrow H_2^- \rightarrow H + H^- \qquad \textit{dissociative attachment}$$

$$H_2(X^1\Sigma_g^+, v_i) + e \rightarrow H_2^- \rightarrow H_2(X^1\Sigma_g^+, \varepsilon; b^3\Sigma_u^+) \rightarrow 2H + e \qquad \textit{dissociation}$$

$$H_2(X^1\Sigma_g^+, v_i) + e \rightarrow H_2^- \rightarrow H_2(X^1\Sigma_g^+, v_f) + e \qquad \textit{vibrational excitation}$$

(v_i, v_f and $\varepsilon \equiv$ initial, final and continuum vibrational states respectively).

Early calculation for dissociative attachment cross sections have been performed by Bardsley and Wadehra [6] in the frame of the resonant model, and more recently by Wadehra and Atems for all the six hydrogen isotopes [12]. In figure 1 are displayed the cross section for H_2 molecule as a function of incident energy and for several vibrational levels.

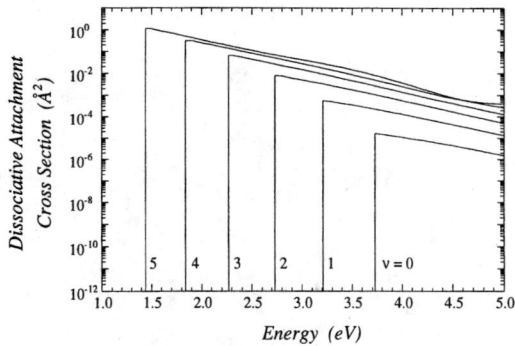

FIGURE 1. Cross section for the process $H_2(X^1\Sigma_g^+, v_i) + e \rightarrow H_2^-({}^2\Sigma_u^+) \rightarrow H + H^-$ as a function of energy for different vibrational levels.

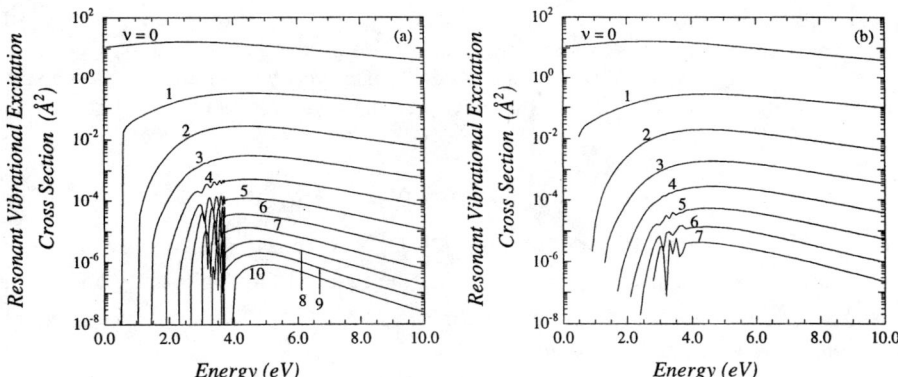

FIGURE 2. Cross section, as a function of energy for different final vibrational levels, for the process (a) $H_2(X^1\Sigma_g^+, v_i=0) + e \rightarrow H_2^- \rightarrow H_2(X^1\Sigma_g^+, v_f) + e$; (b) $HD(X^1\Sigma_g^+, v_i=0) + e \rightarrow HD^- \rightarrow HD(X^1\Sigma_g^+, v_f) + e$.

Resonant dissociation, occurring through the two $X^2\Sigma_u^+$ and $B^2\Sigma_g^+$ lower electronic states of H_2^- ion, has been recently discussed by Fabrikant et al. [7]. Cross sections have been reported for $v_i = 0, 3, 6, 9, 12$ and for an incident energy lower than 5 eV.

Resonant vibrational excitation cross sections have been firstly reported by Wadehra [6, 8, 9] for H_2. More recently, in our group, the calculations for $v_i \geq 0$ have been extended to all the molecular isotopes [10]. Preliminary results are shown in figure 2.

2.2. Electronic excitation and dissociation

Total singlet-singlet cross sections (summed over final bound and continuum vibrational states) for electronic excitation starting from the ground electronic state have been calculated, by using the impact-parameter method, for the following transitions:

$$H_2(X^1\Sigma_g^+, v_i) + e \rightarrow H_2(B^1\Sigma_u^+) + e \qquad (a)$$

$$H_2(X^1\Sigma_g^+, v_i) + e \rightarrow H_2(C^1\Pi_u) + e \qquad (b)$$

$$H_2(X^1\Sigma_g^+, v_i) + e \rightarrow H_2(B'^1\Sigma_u^+) + e \qquad (c)$$

$$H_2(X^1\Sigma_g^+, v_i) + e \rightarrow H_2(B'''^1\Sigma_u^+) + e \qquad (d)$$

$$H_2(X^1\Sigma_g^+, v_i) + e \rightarrow H_2(D^1\Pi_u) + e \qquad (e)$$

$$H_2(X^1\Sigma_g^+, v_i) + e \rightarrow H_2(D'^1\Pi_u) + e \qquad (f)$$

Cross section data for the (a) and (b) processes have been reported [11, 12] for hydrogen, deuterium, tritium and DT molecules, for all the vibrational levels.

A simple scaling law can be obtained for these transitions by manipulating the impact-parameter method equations. The resultant expression assumes the following form:

$$\sigma_{v_i}^{X \rightarrow \alpha_f}(x) = \tilde{\sigma}^{X \rightarrow \alpha_f}(x) \frac{1}{|\Delta E^{X \rightarrow \alpha_f}(R_{v_i})|^\gamma} \qquad (1)$$

where α_f indicates the B or C state and $\Delta E^{X \rightarrow \alpha_f}(R_{v_i})$ is the transition energy measured at the classical turning point R_{v_i}. $x = E/\Delta E$ is the reduced incident energy and γ is a constant (3 for $X \rightarrow B$ and 2 for $X \rightarrow C$ transition). $\tilde{\sigma}^{X \rightarrow \alpha_f}(x)$ is a "shape function" expressed in analytic form as [12]:

$$\tilde{\sigma}^{X \rightarrow \alpha_f}(x) = [c_1 e^{-c_2 \cdot x} + c_3 \ln(c_4 \cdot x)]/[x - c_5 x + c_6](\text{Å}^2 \cdot eV^\gamma) \qquad (2)$$

where c_1=4926.0, c_2=1.9586, c_3=2312.2, c_4=0.82802, and c_5=c_6=0 for $X \rightarrow B$ transition and c_1=-172.49, c_2=0.74392, c_3=27.212, c_4=0.014272, c_5=c_6=1 for $X \rightarrow C$ transition.

Eq.(1) relies on the fact that the vibrational cross sections are mainly determined by the inverse of the Franck-Condon transition energies. In figure 3 is shown a satisfactory comparison between scaled and impact-parameter cross sections for both $X \rightarrow B$ and $X \rightarrow C$ excitations.

In figure 4a are compared the $X \rightarrow B$ cross sections for different isotopes as a function of the vibrational quantum number and for a fixed incident energy. An isotopic effect is evident, characterized by a cross section shift towards higher quantum numbers as increases the molecular mass. This isotopic effect, however, is only apparent and completely disappears if we plot the cross sections in terms of the vibrational eigenvalues (figure 4b). This sort of collapse is due to cross section dependence, through eq. (1), on the transition energy which for different isotopes is practically the same for similar energy eigenvalues. This circumstance suggests that a mass scaling law for the cross

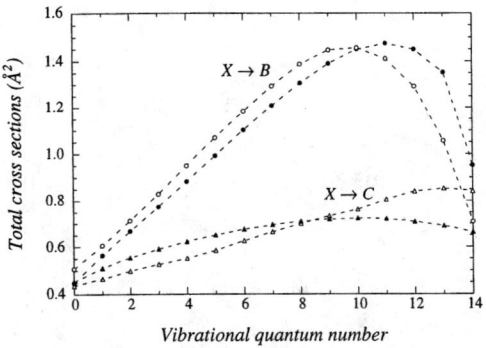

FIGURE 3. Cross sections as a function of initial vibrational quantum number, at a fixed energy of 40 eV, for the processes $H_2(X^1\Sigma_g^+, v_i) + e \to H_2(B^1\Sigma_u^+) + e$ and $H_2(X^1\Sigma_g^+, v_i) + e \to H_2(C^1\Pi_u) + e$ (open marks) compared with scaled values (filled marks).

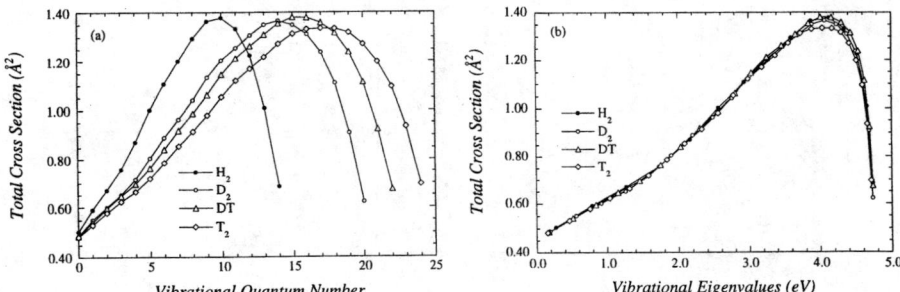

FIGURE 4. Cross section for the transition $(X^1\Sigma_g^+, v_i) \to (B^1\Sigma_u^+)$ for different isotopes, at a fixed energy of 40 eV, (a) as a function of initial vibrational quantum number; (b) as a function of vibrational eigenvalues.

sections can be obtained by simply inserting in eq. (1) the transition energy ΔE_v for the $v - th$ vibrational level of the considered isotope.

In figure 5 is shown a comparison between calculated and scaled cross sections (eqs. (1),(2)) for the heavier isotope T_2 molecule.

Dissociative cross sections for the above (a) and (b) processes are displayed in figure 6.

The upper electronic states involved in (c)-(f) processes are referred to in literature as Rydberg states. Recent experimental investigations have shown the importance of the manifold of high-lying electronic states in the extraordinarily efficient formation in particular experimental conditions of H^- negative ions by dissociative attachment [13]. Dissociative cross sections for (c)-(f) transitions are shown in figure 7a-d, while vibrational scaling laws for total cross sections can be found in literature [12].

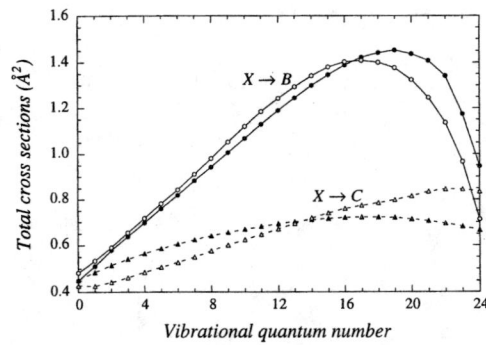

FIGURE 5. Cross sections as a function of initial vibrational quantum number, at a fixed energy of 40 eV, for the processes $T_2(X^1\Sigma_g^+, v_i) + e \rightarrow T_2(B^1\Sigma_u^+) + e$ and $T_2(X^1\Sigma_g^+, v_i) + e \rightarrow T_2(C^1\Pi_u) + e$ (open marks) compared with values obtained using mass scaling law (filled marks).

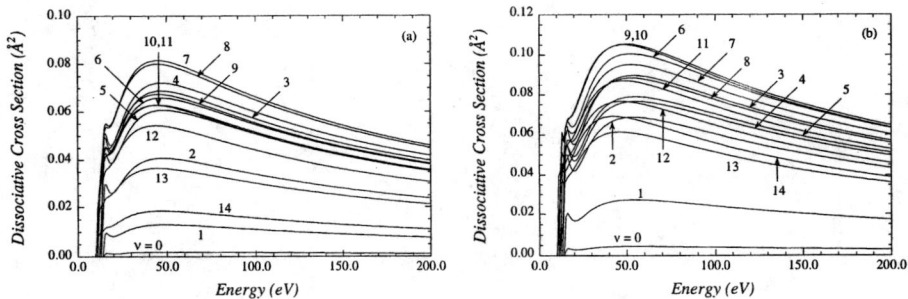

FIGURE 6. Cross section as a function of energy, for different initial vibrational levels, for the process: (a) $H_2(X^1\Sigma_g^+, v_i) + e \rightarrow H_2(B^1\Sigma_u^+) + e \rightarrow H + H + e$; (b) $H_2(X^1\Sigma_g^+, v_i) + e \rightarrow H_2(C^1\Pi_u) + e \rightarrow H + H + e$.

As a last example of singlet-singlet excitation cross section calculations we consider the process

$$H_2(B^1\Sigma_u^+, v_i) + e \rightarrow H_2(I^1\Pi_g) + e \tag{g}$$

This is, to our best knowledge, the only process involving two electronic excited states for which sets complete of vibrational cross sections are available. The related plots are shown in figure 8a-b. No analytic fits or scaling laws exist for these data, but numerical values are available in electronic form for total and dissociative cross sections [12].

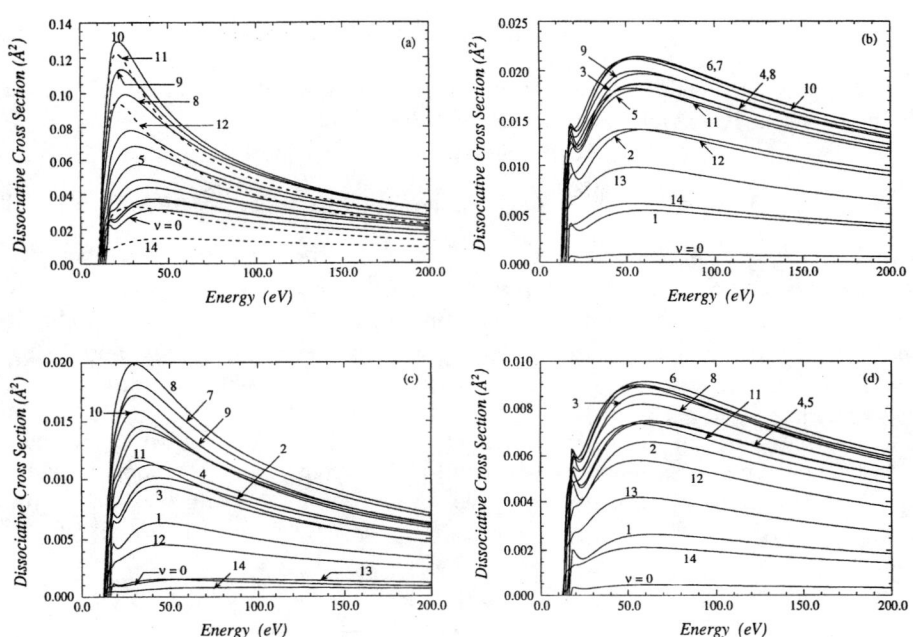

FIGURE 7. Cross section as a function of energy, for different initial vibrational levels, for the process: (a) $H_2(X^1\Sigma_g^+, v_i) + e \to H_2(B'^{\,1}\Sigma_u^+) + e \to H + H + e$; (b) $H_2(X^1\Sigma_g^+, v_i) + e \to H_2(D^1\Pi_u) + e \to H + H + e$; (c) $H_2(X^1\Sigma_g^+, v_i) + e \to H_2(B''^{\,1}\Sigma_u^+) + e \to H + H + e$; (d) $H_2(X^1\Sigma_g^+, v_i) + e \to H_2(D'^{\,1}\Pi_u) + e \to H + H + e$.

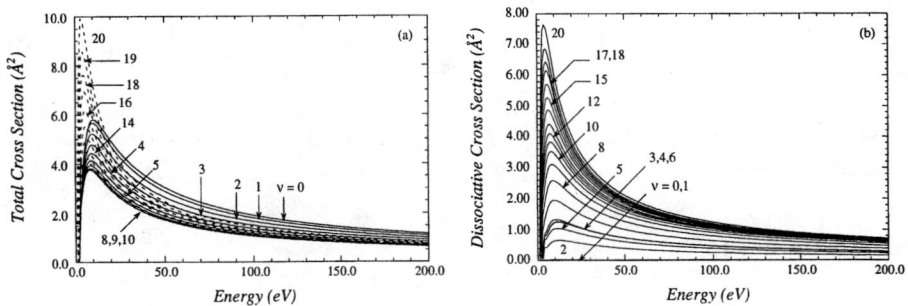

FIGURE 8. Cross sections as a function of energy for the process (a) $H_2(B^1\Sigma_u^+, v_i) + e \to H_2(I^1\Pi_g) + e$; (b) $H_2(B^1\Sigma_u^+, v_i) + e \to H_2(I^1\Pi_g) + e \to H + H + e$.

2.3. Radiative decay

Probably the most efficient way to populate the high vibrational levels of the ground electronic states of H_2 molecule is the so-called $E - V$ process

$$H_2(X^1\Sigma_g^+, v_i) + e \to H_2^*(excited\ singlet\ states) \to H_2(X^1\Sigma_g^+, v_f\ or\ \varepsilon) + e + h\nu \quad (h)$$

The first steps of this two-steps process is the electron-impact excitation of the molecule to singlet states, then followed by radiative decay back to the ground electronic state either on the bound vibrational manifold (v_f) or on the continuum spectrum (ε). In the former case a global vibrational excitation results, populating significantly the high levels; in last case dissociation occurs with the production of two 1s hydrogen atoms.

Cross section for this process (involving the $B^1\Sigma_u^+$, and $C^1\Pi_u$, singlet states) have been calculated for all the vibrational ladder of H_2 molecule and only for some v_i values for deuterium. In figure 9 is displayed a comparison between vibrational excitation and dissociative cross sections for hydrogen molecule as a function of initial vibrational quantum number.

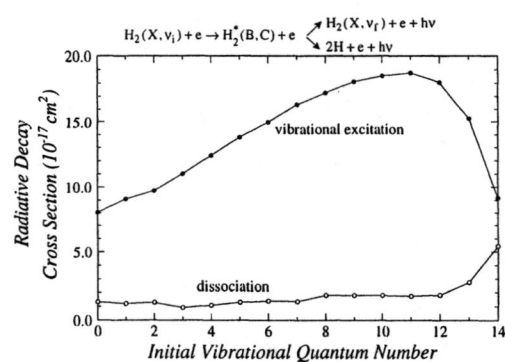

FIGURE 9. Cross section, as a function of initial vibrational quantum number, for the processes: $H_2(X^1\Sigma_g^+, v_i) + e \to H_2^*(excited\ singlet\ states)$ (filled circles) $\to H_2(X^1\Sigma_g^+, v_f) + e + h\nu$; (open circles) $H_2(X^1\Sigma_g^+, \varepsilon) + e + h\nu$.

2.4. Direct excitation to $b^3\Sigma_u^+$ state

The $X^1\Sigma_g^+ \to b^3\Sigma_u^+$ transition is the most important channel leading to dissociation, due to the completely repulsive nature of the $b^3\Sigma_u^+$ state. The related cross sections are shown in figure 10. The calculation have been performed by using the classical Gryzinski method, which is known to give usual only qualitative results. A comparison with more accurate calculations [14] however, is at least questionable due to the evident discrepancy among the different sets of results, as shown in figure 11.

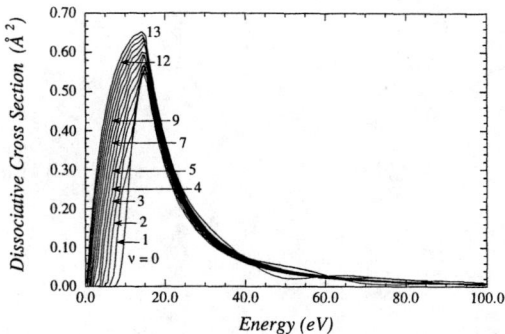

FIGURE 10. Cross sections as a function of energy for the process $H_2(X^1\Sigma_g^+, v_i) + e \to H_2(b^3\Sigma_u^+) \to H + H + e$, for different initial vibrational levels.

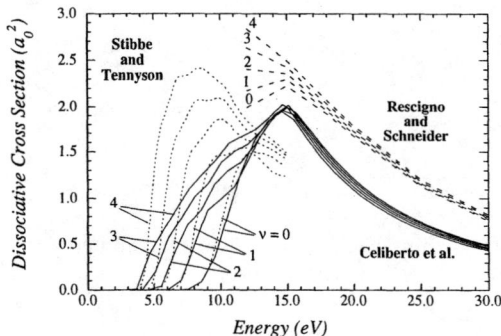

FIGURE 11. Cross section for the process of figure 10 for $v_i = 0$ - 4 compared with results of literature. Solid lines: Celiberto et al.; dashed lines: Rescigno and Schneider; dotted lines: Stibbe and Tennyson.

2.5. Ionization

Gryzinski method has also been used for the following H_2 (and D_2) ionization processes

$$H_2(X^1\Sigma_g^+, v_i) + e \to H_2^+(X^2\Sigma_g^+, v_f) + 2e \qquad \text{ionization} \qquad \text{(i)}$$

$$H_2(X^1\Sigma_g^+, v_i) + e \to H_2^+(X^2\Sigma_g^+) + 2e \to H + H^+ + 2e \qquad \text{dissociative ionization} \qquad \text{(j)}$$

$$H_2(X^1\Sigma_g^+, v_i) + e \to H_2^+(^2\Sigma_u^+) + 2e \to H + H^+ + 2e \qquad \text{dissociative ionization} \qquad \text{(k)}$$

In figure 12a,b are shown the cross sections for hydrogen molecule. For the direct dissociative case (k) a scaling law of the same nature of eq. (1) holds, with $\gamma = 2.5$, and the shape function given by [15]:

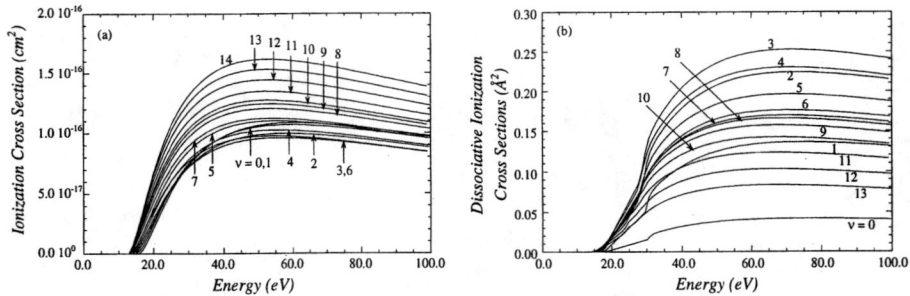

FIGURE 12. Cross section as a function of energy for the process: (a) $H_2(X^1\Sigma_g^+, v_i) + e \rightarrow H_2^+(X^2\Sigma_g^+, v_f) + 2e$; (b) $H_2(X^1\Sigma_g^+, v_i) + e \rightarrow H_2^+(X^2\Sigma_g^+) + 2e \rightarrow H + H^+ + 2e$.

$$\widetilde{\sigma}^{X \rightarrow ^2\Sigma_u^+}(x) = 20013 \left(\frac{x-1}{x}\right)^{2.63323} (1 + 0.57363 \ln x)]/x (\overset{\circ}{A}^2 \cdot eV^{\gamma}) \qquad (3)$$

2.6. Triplet-triplet transitions

Triplet-triplet cross sections have been recently calculated for the following transitions

$$H_2(a^3\Sigma_g^+, v_i) + e \rightarrow H_2(d^3\Pi_u) + e \qquad (l)$$

$$H_2(c^3\Pi_u, v_i) + e \rightarrow H_2(g^3\Sigma_g^+) + e \qquad (m)$$

$$H_2(c^3\Pi_u, v_i) + e \rightarrow H_2(h^3\Sigma_g^+) + e \qquad (n)$$

and a brief and preliminary account will be given here.

Potential curves for the electronic states $a^3\Sigma_g^+$, $c^3\Pi_u$, $d^3\Pi_u$, $g^3\Sigma_g^+$ and $h^3\Sigma_g^+$ are shown in figure 13.

Process (l) is known as Fulcher transition and is involved in spectroscopic diagnostic methods which provide information about vibrational and rotational population of hydrogen plasmas [5]. Total and dissociative cross sections for this transition as a function of energy are shown in figure 14a and b. These figures show that the dissociation gives a small contribution to the total total cross section. Same comments hold for the $c^3\Pi_u \rightarrow g^3\Sigma_g^+$ cross sections shown in figure 15a and b. The potential curve for the state $h^3\Sigma_g^+$ shows a barrier above the dissociation limit. This implies the existence of the so-called quasi-bound vibrational states which can lead to dissociation via tunneling effect.

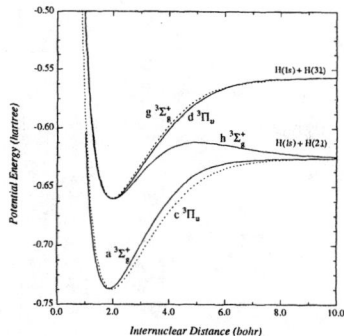

FIGURE 13. Energy diagram for triplet excited states of H_2 molecules.

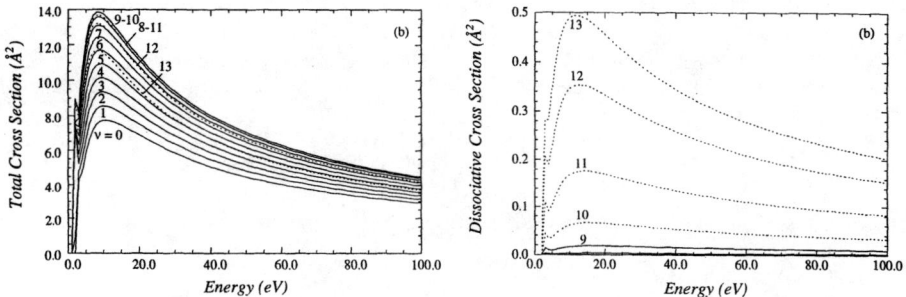

FIGURE 14. Cross sections as a function of energy for the process (a) $H_2(a^3\Sigma_g^+, v_i) + e \to H_2(d^3\Pi_u) + e$; (b) $H_2(a^3\Sigma_g^+, v_i) + e \to H_2(d^3\Pi_u) + e \to H + H + e$.

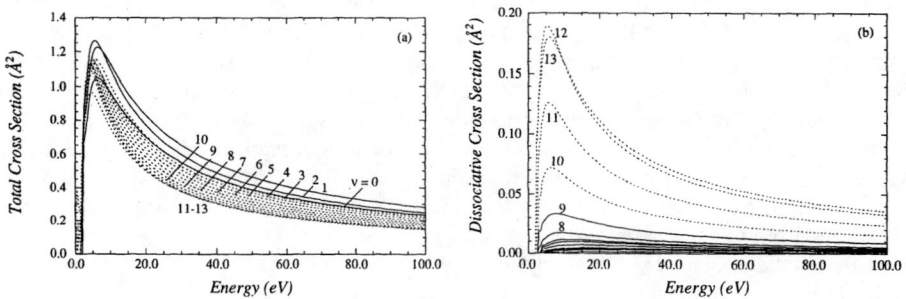

FIGURE 15. Cross sections as a function of energy for the process (a) $H_2(c^3\Pi_u, v_i) + e \to H_2(g^3\Sigma_g^+) + e$; (b) $H_2(c^3\Pi_u, v_i) + e \to H_2(g^3\Sigma_g^+) + e \to H + H + e$.

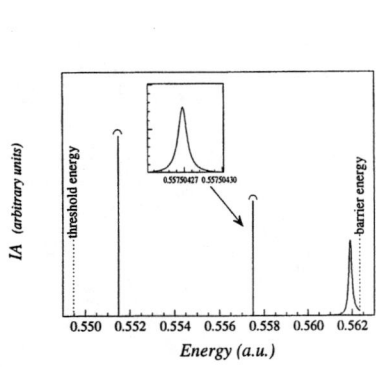 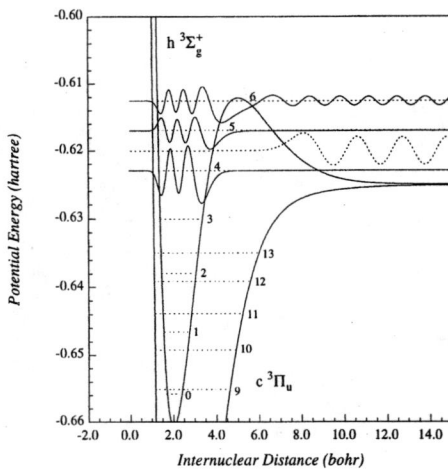

FIGURE 16. (a) Internal amplitude (4) as a function of continuum energy (the second peak is magnified in the inserted picture); (b) potential energy curves for $c^3\Pi_u$ and $h^3\Sigma_g^+$ states of H_2, wavefunctions for quasi-bound states are also displayed.

These states have been searched by calculating the "Internal Amplitude", defined as

$$IA(\varepsilon) = \int_{R_a}^{R_b} \frac{|\Psi_\varepsilon(R)|^2 \, dR}{[R_b - R_a]} \qquad (4)$$

(R_a and R_b are the two classical turning points of the vibrational ε energy level). The IA is shown in figure 16a as a function of the potential energy ε. The peaks indicate three regions of resonances which, except for the last one very close to the top of the barrier, present a very sharp intensity. At a closer resolution becomes evident the energy enlargement, as shown in the inserted picture in figure 16a. In the next figure (16b) are shown the vibrational wavefunctions of three quasi-bound states, at a selected energy corresponding to the maximum of the related IA. In the same plot the dotted line represent, as a comparison, the wavefunction of completely dissociative state. The cross sections for these states depend on the ratio between the tunneling and radiative times. The calculated values (obtained by evaluating the tunneling probabilities and radiative Einstein coefficients) are shown in figure 17. As expected, the last resonance, located at the maximum of the barrier, shows a strong dissociative character, being the contrary for the first state, placed inside the potential well, which behaves practically as a bound state. A hybrid character is displayed by the second resonance.

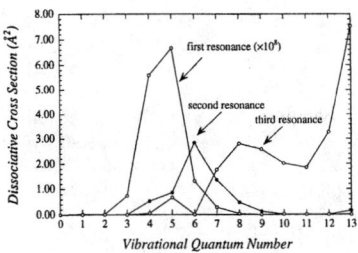

FIGURE 17. Quasi-bound state dissociative cross section as a function of initial vibrational quantum number.

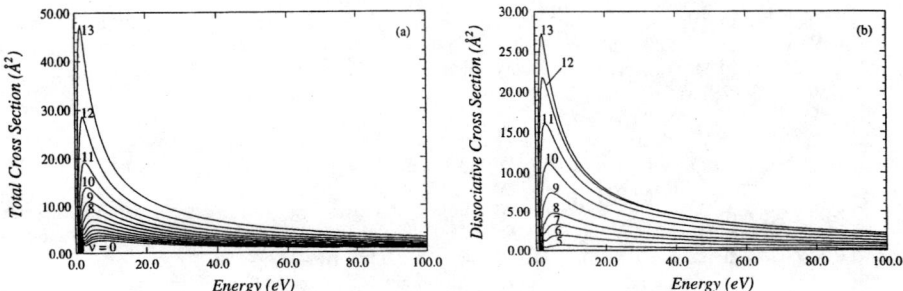

FIGURE 18. Cross sections as a function of energy for the process (a) $H_2(c^3\Pi_u, v_i) + e \rightarrow H_2(h^3\Sigma_g^+) + e$; (b) $H_2(c^3\Pi_u, v_i) + e \rightarrow H_2(h^3\Sigma_g^+) + e \rightarrow H + H + e$.

Finally, figure 18 gives the total and dissociative cross sections for this transition, as a function of the energy and the initial vibrational quantum number [16].

ACKNOWLEDGMENTS

The present work has been supported by ASI (I|R|038|01) and MIUR (cof.2001 Project n. 2001031223_009).

REFERENCES

1. M. Capitelli, R. Celiberto, and M. Cacciatore, in "Advanced in Atomic, Molecular and Optical Physics: Cross Section Data", edited by M. Inokuti, Academic Press, N.Y. and London, **33**, (1994) 321-372.
2. R.K. Janev, in "Atomic and Molecular Processes in Fusion Edge Plasmas", edited by R.K. Janev, Plenum Press, N.Y. and London, (1995) 1-13.

3. R.K. Janev, in "Atomic and Molecular Processes in Divertor Plasma Volume Recombination", edited by R.K. Janev and D.R. Schultz, Physica Scripta, **T96**, (2002) 94-101.
4. C. Gorse, R. Celiberto, M. Cacciatore, A. Laganà and M. Capitelli, "From dynamics to modeling of plasma complex systems: negative ion (H^-) sources", Chemical Physics, **161**, (1992) 211-227.
   C. Gorse, M. Bacal, R. Celiberto and M. Capitelli, "Nonequilibrium plasma kinetics of D_2 in magnetic multicusp plasmas for (D^-) production", Chemical Physics Letters, **192**, (1992) 161-165.
   M. Capitelli, R. Celiberto, F. Esposito, A Laricchiuta, K. Hassouni and S. Longo, "Elementary processes and kinetics of H_2 plasmas for different technological applications", Plasma Sources and Plasma Technology, submitted (2001).
5. U. Fantz, B. Heger and D. Wünderlich, "Using the radiation of hydrogen molecules for electron temperature diagnostics of divertor plasmas", Plasma Physics Controlled Fusion, **43**, (2001) 1-12.
6. J.N. Bardsley and J.M. Wadehra, "Dissociative attachment and vibrational excitation in low-energy collisions of electrons with H_2 and D_2", Physical Review A, **20**, (1979) 1398-1405.
7. I.I. Fabrikant, J.M. Wadehra and Y. Xu, in: "Atomic and Molecular Processes in Divertor Plasma Volume Recombination", edited by R.K. Janev and D.R. Schultz, Physica Scripta, **T96**, (2002) 45-51.
8. J.M. Wadehra in "Nonequilibrium Vibrational Kinetics", edited by M. Capitelli, Springer-Verlag, New York and London, (1986).
9. D.E. Atems and J.M. Wadehra, "Non local effects in dissociative electron attachment to H_2", Physical Review A, **42**, (1990) 5201-5207.
10. R. Celiberto et al., unpublished results.
11. R. Celiberto and T.N. Rescigno, "Dependence of electron-impact excitation cross sections on the initial vibrational quantum number in H_2 and D_2 molecules: $X^1\Sigma_g^+ \to B^1\Sigma_u^+$ and $X^1\Sigma_g^+ \to C^1\Pi_u$ transitions", Physical Review A, **47**, (1993) 1939-1945.
12. R. Celiberto, R.K. Janev, A. Laricchiuta, M. Capitelli, J.M. Wadehra and D.E. Atems, "Cross section data for electron-impact inelastic processes of vibrationally excited molecules of hydrogen and its isotopes", Atomic Data and Nuclear Data Tables, **77**, (2002) 161-213.
    R. Celiberto, R.K. Janev and A. Laricchiuta, "Total and dissociative electron-impact cross sections for $X^1\Sigma_g^+ \to B^1\Sigma_u^+$ and $X^1\Sigma_g^+ \to C^1\Pi_u$ transitions of vibrationally excited tritium and deuterium-tritium molecules", Physica Scripta, **64**, (2001) 26-33.
13. L.A. Pinnaduwage, W.X. Ding, D.L. McCorkle, S.H. Lin, A.M. Mebel and A. Garscadden, "Enhanced electron attachment to Rydberg states in molecular hydrogen volume discharge", Journal of Applied Physics **85**, (1999) 7064-7069.
14. T.N. Rescigno and B.I. Schneider, "Electron-impact excitation of the $b^3\Sigma_u^+$ state of H_2 using the Complex Kohn Method: R-dependence of the cross section", Journal of Physics B, **21**, (1988) L691.
    D.T. Stibbe and J. Tennyson, "Near-threshold electron impact dissociation of H_2", New Journal of Physics, **1**, (1998) 21-29.
15. R. Celiberto, M. Capitelli, and R.K. Janev, "Scaling of electron-impact dissociative ionization cross sections of vibrationally excited H_2 and D_2 molecules", Chemical Physics Letters, **278**, (1997) 154-158.
16. State-to-state cross sections for all the processes discussed here have not been published but they are available from the authors on request (a.laricchiuta@area.ba.cnr.it, r.celiberto@poliba.it).

Atomic and Molecular Databases for Fusion Divertor Plasmas

P.S. Krstić and D.R Schultz

Controlled Fusion Atomic Data Center, Physics Division, Oak Ridge National Laboratory, Oak Ridge, TN 37831

Abstract. We describe our recent activities regarding the production of atomic and molecular data that are needed for modeling divertor plasmas in fusion tokamaks. The transport of particles and, in particular, the exchange of momentum in such plasmas can be dominantly influenced by elastic scattering and resonant charge transfer among hydrogen ions, atoms, and molecules. We have undertaken a comprehensive calculation of these processes, in all isotopic combinations of hydrogen, and highlights of that study, with intercomparison of various integral and differential cross sections, are shown here. Also, in the formation of the detached plasma layers the processes of charge transfer between hydrogen ions and vibrationally excited molecules might play a crucial role in the chain of reactions called Molecule Assisted Recombination (MAR). All other inelastic processes involving hydrogen molecules are also of interest in the colder parts of the divertor plasma. We have studied all processes that involve hydrogen ions/atoms and vibrationally excited molecules/molecular ions (charge transfer, excitation, dissociation, association), in the range of Center-of-Mass (CM) collision energies 0.5-10 eV. Here we discuss integral cross sections for these reactions, resolved in both initial and final vibrational states. The full set of the cross sections for both elastic and inelastic processes which involve molecules can be obtained through the ORNL Controlled Fusion Atomic Data Center's website.

1. INTRODUCTION

The Oak Ridge national Laboratory (ORNL) Controlled Fusion Atomic Data Center (CFADC) has multiple goals defined by the needs of the fusion community, which can be organized into several categories: the production of atomic collision data, compilation of numerical and bibliographic data, dissemination of the data to the fusion community, and theoretical support for ORNL atomic collision experiments. These goals are addressed by the development of new theoretical methods or utilization of existing ones as well as by compilation of bibliographic data by a network of consultants. Besides the need for improved understanding of the relevant atomic processes, the production of data is characterized by the need for comprehensive calculations of the cross sections over a sufficiently wide energy range to determine the rate coefficients. The data are disseminated by publication in scientific journals and on the CFADC www site (*www-cfadc.phy.ornl.gov*).

It is widely recognized in the fusion community that the divertor region is of crucial importance for the successful engineering, design, and operation of tokamak fusion reactors, while the atomic collision data needed to model and diagnose this region are often unknown or only poorly considered in the extant literature. The high density, low temperature divertor plasma is characterized by similar densities of ions and neutrals, consisting predominantly of hydrogen atoms, ions and molecules in various isotopic combinations. The transport of particles and, in particular, the exchange of momentum in this plasma can be dominantly influenced by elastic scattering and resonant charge transfer among hydrogen ions, atoms, and molecules. We have recently undertaken a very large set of fully quantal calculations to remedy this need. Four types of systems, ion-atom (P^++Q), atom-atom (P+Q), ion-molecule (P^++QR) and atom-molecule (P+QR) have been studied [1-4], where P, Q, and R are any of the hydrogen isotopes (H, D, or T). In addition, P^++He [1,2], P^++C [5] and P^++Ar systems were studied, as well. Several thousands of accurate differential and integral cross sections were calculated, spanning the center of mass (CM) collision energy range of 0.1-100 eV.

In addition, formation of the divertor plasma detachment layer might critically depend on Molecule Assisted Recombination (MAR), which involves ion conversion via electron capture by a proton from a vibrationally and rotationally excited hydrogen molecule, followed by dissociative recombination of the molecular ion with plasma electrons. Such collisions require accurate data for the relevant processes, particularly for vibrationally resolved charge transfer from H_2 to H^+. We report on a comprehensive study of scattering of hydrogen ions on vibrationally excited hydrogen molecules as well as of hydrogen atoms on vibrationally excited hydrogen molecular ions in the range of center of mass energies 0.5 - 10 eV. Total and partial, initial and final vibrational state resolved cross sections for excitation, charge transfer, dissociation (including dissociative energy spectra), and association have been calculated "on the same footing" using a fully-quantal, coupled-channel approach. An extensive vibrational basis set of several hundred states, including discretized dissociative continua in a large configuration space (to include nuclear particle arrangements) was employed, while the rotational dynamics of H_2 and H_2^+ were treated within the sudden approximation. Concerning the details of the treatment of the reactive dynamics, the data summarized here are an improvement over these previously published [6]. Still, limitations of the sudden approximation require extension of the method to fully include rotational dynamics if isotopic variations of hydrogen are employed. These will be the subject of future research.

In Section 2 we discuss the elastic processes. We describe our study of inelastic processes involving molecules and molecular ions in Section 3.

2. ELASTIC AND TRANSPORT COLLISION PROCESSES IN A COLD HYDROGEN PLASMA

Elastic scattering among ions, atoms, and molecules can cause dissipation of the divertor plasma momentum. Neutrals, produced through recombination processes, can carry off a part of the collision energy, but are not controlled by the magnetic field and thus diffuse to the walls unless re-ionized. In the detached plasma regime, diffusion of neutrals is crucial.

We have performed a study of elastic processes for the H^++H, $H+H$, $H+H_2$, and $H+H_2^+$ systems, in all combinations of the hydrogen isotopes for both projectiles and targets, in the CM collision energy range 0.1-100 eV, with ten energy values per decade. In addition, the systems H^++He, H^++C, and H^++Ar were studied for all isotopic combinations of H. The calculation resulted in over 3000 differential and more than 250 integral cross sections, describing elastic scattering (EL), and transport of momentum (momentum transfer, MT) and energy (viscosity, VI), charge transfer (CT), and spin exchange (SE). These can all be found at the CFADC www site (*www-cfadc.phy.ornl.gov*).

Details of the study can be found in recent publications [1-5] so here we summarize only the most important points.

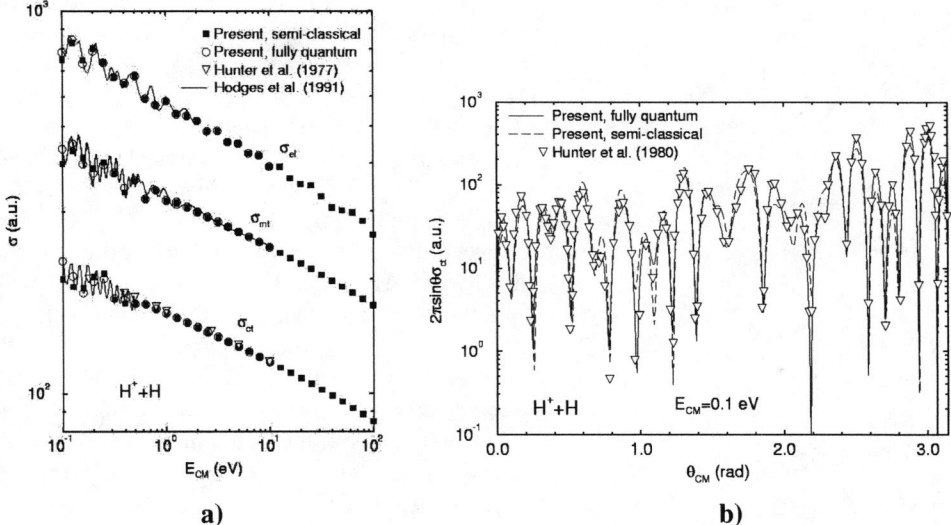

FIGURE 1. a) H^++H integral elastic, momentum, and charge-transfer cross sections, compared with calculations of Hunter et al. (1977) [7] and Hodges et al. (1991) [8]; b) Differential cross section H^++H at low energies compared with the results of Hunter et al. (1980) [9].

In studying symmetric ion-atom and atom-atom collisions (H^++H, D^++D, T^++T, as well as H+H, D+D and T+T) an important problem arises from the indistinguishability of the projectile and the target particles [1-3]: in the ion-atom cases one cannot distinguish at the exit channel the elastically scattered projectile (A) from the nucleus coming from charge transfer (B) unless these particles are somehow labeled. While classical mechanics allows such labeling, quantum mechanical wave packets are not well localized at the lowest energies, allowing for overlap and thus, interference of channels A and B. Only tracking of the nuclear spin of these particles can be used for positive identification of the elastically scattered and charge transfer channels, which is neither typical for common experiments nor plasma environments. One rather averages over the spins in the input and sums them in the output channel which produces a quantity called the elastic cross section, although it is made of a coherent sum of elastically scattered and charge transferred ions, which we denote the IP (Indistinguishable Particle) elastic cross section in our www database. The integral moments (momentum transfer and viscosity) (Fig. 1a) calculated from the IP differential cross sections (Fig. 1b) define the transport due to both elastic scattering and charge transfer. This needs to be explicitly taken into account in particle transport modeling to avoid double counting. On the other hand, for CM collision energies above 1 eV, most of the charge transfer differential cross section can be distinguished by its strong peak in the backward direction, with weak overlap with the elastically forward scattered particles, i.e. one can label the scattered particles A and B by their different angular distributions, which allows for "standard" distinguishable particle (DP) definitions of the transport moments. For modelers whose codes require such distinguishability, we provide also "labeled particle" elastic cross sections and moments (denoted by DP in our www database). Still, the DP cross sections introduce an error of about 10% at energies below 1 eV, due to neglect of interference effects [3].

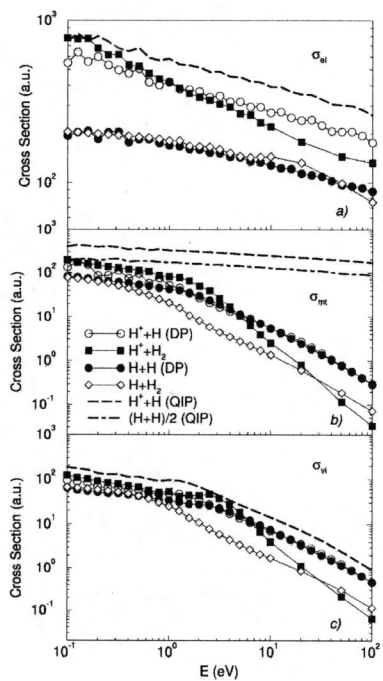

Figure 2. Effects of the symmetry of the colliding hydrogen ions, atoms, and molecules.

In the case of symmetric atom-atom collisions, a similar problem emerges due to indistinguishability of the elastically scattered and recoiled particles, which disappears at higher energies due to the decreasing overlap between the elastic (forward) and recoil channel (backward angles) [1,2]. Due to the full symmetry of the problem, the integral elastic cross section follows exactly its "classical" counterpart by simple division of the total integral cross section by a factor of 2. On the other hand, we recognize that, after correct averaging-summing over the nuclear and electron spins, the elastic cross section is identical to the momentum transfer cross section in symmetric atom-atom scattering [1,2,10].

The behaviors discussed are illustrated in Fig. 2 for hydrogen. For ion/atom-molecule scattering, as well as for ion/atom-atom scattering with isotopically different particles, the problems arising from the symmetry of the particles are not present, yielding the standard elastic and charge transfer cross sections.

Another important issue is the accuracy of the data. By choice of highly accurate intermolecular potentials as well as by controlling the accuracy of the computational procedures, 7-digit numerical accuracy was reached for ion-atom collisions [1,2]. Still, the physical accuracy was far worse, caused by neglect of inelastic processes, mainly at the higher end of the energy range considered. Thus we found that inclusion of inelastic charge transfer processes in ion-atom collisions changed the integral cross sections by 3% for viscosity, 2% for momentum and charge transfer, and 1% for "pure" elastic scattering. For the atom-atom scattering, the contribution of neglected inelastic channels, electronic excitation of the projectile or target and dissociation, is significantly smaller in the energy range considered. Concerning the cross sections for ion/atom-molecule cases, up to nine vibrationally excited states were taken simultaneously into account in the calculation of the elastic channels, while charge transfer (of importance at higher energies) was neglected. In addition, the rotational motion of the target was frozen during the collision, and the cross sections averaged over different orientations of the molecule (Infinite Order Sudden Approximation, IOSA [11-13]), which might produce some error at the lowest energies considered. Having in mind all these effects, the overall physical uncertainty of the integral cross sections over the whole energy range (0.1-100 eV) is estimated to be less than 1% for ion-atom and atom-atom scattering, and less than 10% for the ion/atom-molecule systems considered. In Fig. 3a, we show a comparison of our differential cross sections for vibrational excitation of H_2 by proton impact (dashed line) with the experimental results of Herman et al. [14]. The solid lines in the same figure are for vibrational excitation of H_2 by impact of hydrogen.

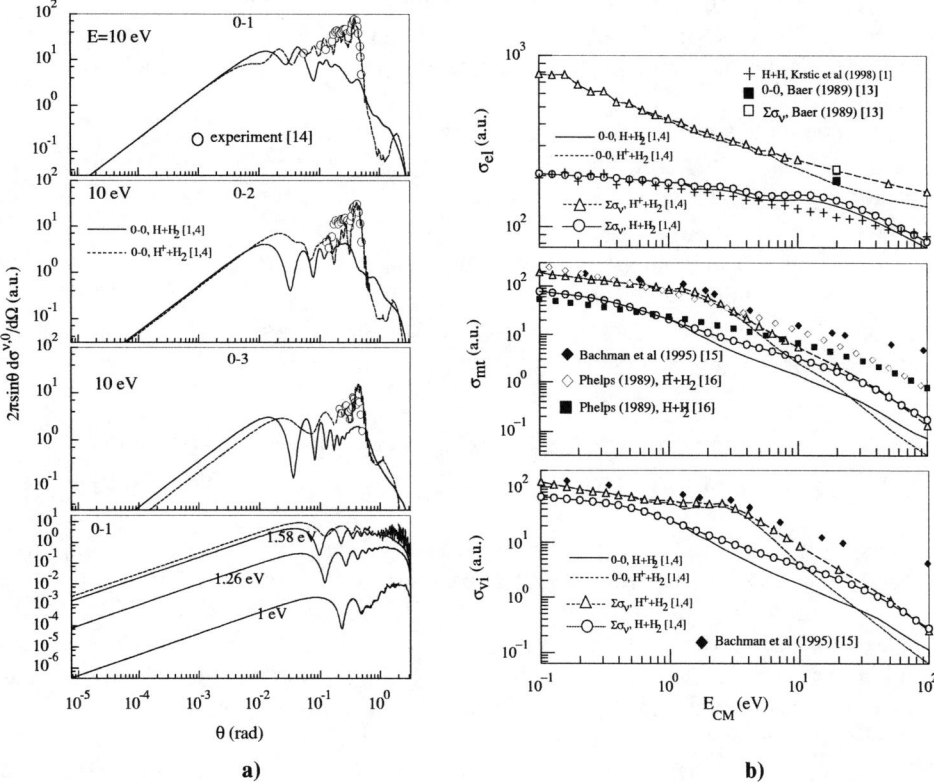

Figure 3. a) Differential elastic cross sections for vibrational excitation in H^+, $H+H_2$ scattering; b) Elastic, momentum transfer, and viscosity cross sections for the ground vibrational state of H_2.

Next important issue in ion/atom-molecule scattering is the depletion of the ground vibrational state due to vibrational excitation. The effect is visible in Fig. 3b, where the elastic, momentum transfer, and viscosity cross sections are compared with evaluated cross sections of Phelps [16]. The latter were obtained by interpolation of the measured data at about 1 eV, and calculated data at several hundred eV collision energy. In the intermediate, interpolated region of energies, where the vibrational processes are the most intensive, the deviation of Phelps' data, induced by neglecting the depletion of the ground vibrational state reaches an order of magnitude. On the other hand, the agreement of our data with that of Phelps is very good in the region of the measured data.

To summarize, we list the highlights of the database of elastic cross sections for the hydrogen collision systems:

- Benchmarks were set for both differential and integral cross sections (elastic, momentum transfer, viscosity and charge transfer) for ion-atom and atom-atom collisions involving hydrogen atoms in all isotopic combinations.
- Scaling with isotopic mass was found for the integral cross sections and their moments for all cases studied (including molecules) [1-5].
- Momentum transfer and viscosity cross sections were shown to be independent of isotopic composition, to a large extent [1-5].
- New definitions of the elastic cross sections were introduced to deal with inconsistencies in the modeler's implementations of the data in symmetric systems [3].
- Inadequate transport formulas in the literature were corrected [1,3,10].
- By inclusion of vibrational excitations, the cross sections for molecular targets were significantly improved [1,4], in the vibrationally intensive region of energies, over the previous published ones [16].
- All differential and integral cross sections were fitted to analytical expressions. These were published in Ref. [1].
- Finally, all raw data are available in both tabular and graphical form through the CFADC web site *www-cfadc.phy.ornl.gov*.

3. INELASTIC PROCESSES IN SLOW COLLISIONS OF IONS AND ATOMS WITH VIBRATIONALLY EXCITED MOLECULES

We have also produced a comprehensive database for inelastic, vibrationally resolved processes on the two lowest adiabatic electronic surfaces of the H_3^+ complex, from all vibrationally excited states of H_2 and H_2^+ for collisions of H^++H_2 and $H+H_2^+$, in the CM energy range 0.5-10 eV. As described in the introduction, the principal motivation for this study was to improve the very incomplete knowledge base regarding the potentially important process of MAR in diverter plasmas.

Only one comprehensive data set exists for the charge transfer from excited vibrational states of H_2, calculated by the Trajectory Surface Hopping (TSH) method [17]. Due to the classical nature of the TSH method, this approach might not give reliable data in the eV energy range. We performed a fully quantal calculation of the vibrationally resolved processes [6,18], where electronic and vibrational dynamics are completely taken into account, while the rotational motion of the diatom was treated within IOSA, as in Section 2. The rearrangements channels were included implicitly by doing calculations in a large configuration space of nuclear motion (encompassing 40 a.u. in all directions.) The role of dissociative continua (of both H_2 and H_2^+) was taken into account by discrete continuum states obtained by constraining the system motion to the configuration box.

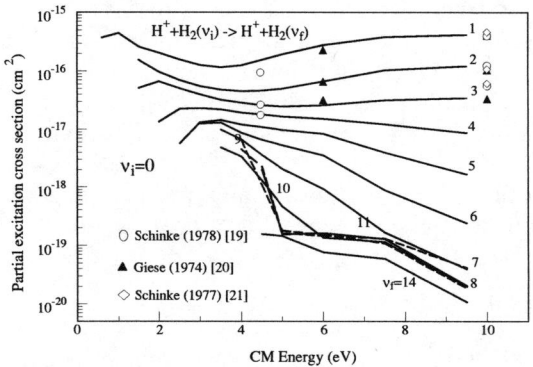

Figure 4. Vibrational excitation cross sections for collisions of protons with H_2 in the ground state.

In addition to a limited number of bound vibrational states (totaling 35 on both H_2 and H_2^+), several hundred continuum states were needed (including the closed channels) to achieve convergence of the cross sections in the energy range considered.

Fig. 4 shows cross sections for excitation by proton impact of the ground state of H_2 to all bound vibrational states. The results are compared with cross sections from the literature. The role of the particle rearrangements in the excitation process is of particular importance at the lowest energies. The previous quantum calculations shown in Fig. 4, do not include the latter channels, underestimating the cross sections at lowest energies. Our data contain coherent sums over the direct and rearrangement channels.

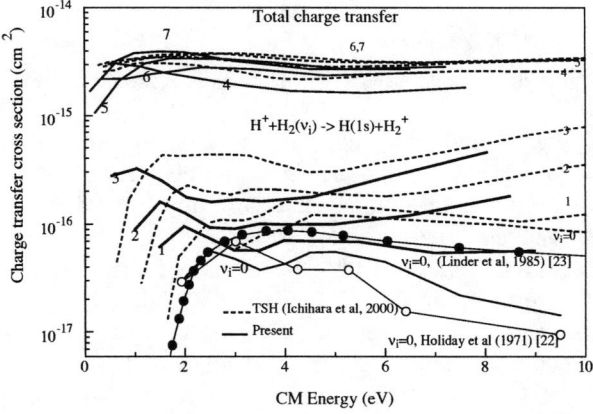

Figure 5. Charge transfer from vibrationally excited states of H_2.

The charge transfer cross sections (summed over final vibrational states) from initially excited vibrational states of H_2 in collisions with protons are shown in Fig 5. Comparison with experimental data of Holiday [22] and Linder [23] shows good agreement at lowest energies, but lies between these two data sets at the higher energies. Among all the cross sections shown, charge transfer from the first three excited states constitutes a separate group. The reason is that CT from the higher excited states is dominantly exoergic and often quasi-resonant with the vibrational states of H_2^+. The corresponding cross sections therefore are large even at the lowest energies. Comparison of our data with the TSH ones of Ichihara [17] shows good agreement for the higher initial vibrational states, as could be expected from the correspondence principle. It is interesting to note that our quantum and the TSH calculations show similarly large contributions of the particle re-arrangement channels at low energies.

As a contrast to Fig. 5, Fig. 6 shows charge transfer cross sections from excited states of H_2^+ in collision with H. Since this process is exoergic from all vibrational

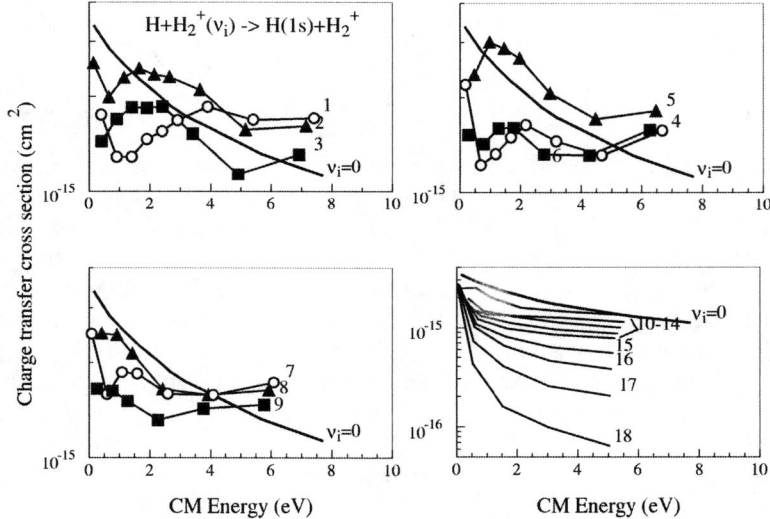

Figure 6. Charge transfer from vibrationally excited states of H_2^+.

states (including the ground one), the characteristic increase of the cross section toward lower energies is expected, and is not strongly dependent on the initial state. The cross sections from high vibrational states are suppressed due to depletion of these states to the vibrational continuum (dissociation).

The CT cross sections from the 13th excited state of H_2^+ to vibrational states of H_2 in collision $H+H_2^+$ are shown in Fig. 7, for various collision energies. As expected, the distribution peaks around quasi-resonance with the 11th excited state of H_2. Figure 8 presents the dissociation cross sections, for all initial vibrational states of H_2. Comparison with the TSH calculations shows a good agreement for higher vibrational states at the higher end of the collision energy range. We note that this process includes both direct dissociation into the dissociative continuum of H_2, as well as charge transfer dissociation into the continuum of H_2^+. These two channels are of the same order of magnitude. Concerning the energy spectra of the dissociating fragments,

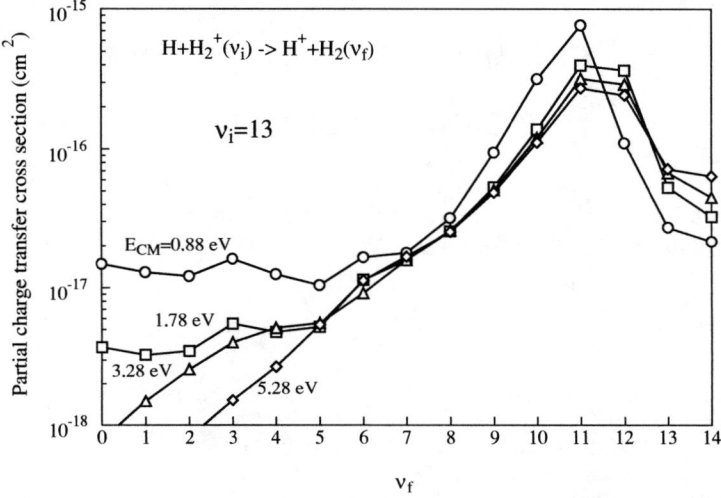

Figure 7. State resolved cross sections for charge transfer to excited states of H_2.

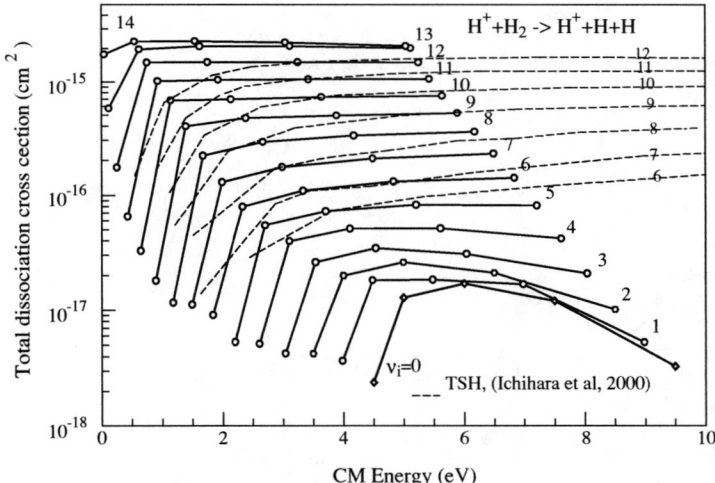

Figure 8. Cross section for dissociation from various excited states of H_2.

these have the characteristic cusp at the continuum edge, which is more pronounced for the lower collision energies.

The database produced can be found at *www-cfadc.phy.ornl.gov* and contains all the processes mentioned above for the $H+H_2^+$ and H^++H_2 collision systems in the form of partial and total, initial- and final-vibrational-state-resolved cross sections. This represents currently the most comprehensive set of quantum-mechanically obtained inelastic data for collisions that involve hydrogen atoms, ions and molecules. Still, in order to study rovibrationally resolved processes at low collision energies with various isotopic combinations of H and, with resolved particle rearrangement channels, the rotational dynamics have to be fully accounted for. This will be the subject of our future work.

ACKNOWLEDGMENT

We acknowledge support from the US Department of Energy, Office of Fusion Energy Sciences, through Oak Ridge National Laboratory, managed by UT-Battelle, LLC under contract DE-AC05-00OR22725.

REFERENCES

1. Krstić, P. S., and Schultz, D. R., *Atomic and Plasma-Material Data for Fusion* **8**, 1 (1998).
2. Krstić, P. S., and Schultz, D. R., *J. Phys. B* **32**, 3485 (1999)
3. Krstić, P. S., and Schultz, D. R., *Phys. Rev. A* **60**, 2118 (1999).
4. Krstić, P. S., and Schultz, D. R., *J. Phys. B* **32**, 2415 (1999).
5. Schultz, D. R., and Krstić, P. S., *Phys. Plasma* **9**, 61 (2002).
6. Krstić, P. S., Schultz, D. R., and Janev, R. K., *Phys. Scripta* **T96**, 61 (2002).
7. Hunter, G., and Kuriyan, M., *Proc. R. Soc. London* **353**, 575 (1977);
8. Hodges, R. R., and Breig, E. L., *J. Geophys. Res.* **96**, 7697 (1991).
9. Hunter, G., and Kuryan, M., *At. Data Nucl. Data Tables* **25**, 287 (1980).
10. Jamieson, M. J., Dalgarno, A., Zygelman, B., Krstić, P. S., and Schultz, D. R., *Phys. Rev. A* **61**, 4701 (2000); *Phys. Rev. A* **62**, 9902 (2000).
11. Pack, T., *J. Chem. Phys.* **60**, 633 (1974).
12. Baer, M., editor, *Theory of Chemical Reaction Dynamics,* Boca Raton, Florida, Vols. I and II (1985).
13. Baer, M., Niedner-Schatteburg, G., and Toennies, J. P., *J. Chem. Phys.* **91**, 4196 (1989).
14. Herman, V., Schmidt, H., and Linder, F., *J. Phys. B* **11**, 493 (1978).
15. Bachman, P., and Reiter, D., *Contr. Plasma Phys.* **35**, 45 (1995).
16. Phelps, A. V., *J. Phys. Chem. Ref. Data* **19**, 653 (1990).
17. Ichihara, A., Iwamoto, O., and Janev, R. K., *J. Phys. B* **33**, 4747 (2000).
18. Krstić, P. S., *Phys. Rev. A*, submitted (2002).
19. Schinke, R., and McGuire, P., *Chem. Phys.* **31**, 291 (1978).
20. Giese, C. F., and Gentry, W. R., *Phys. Rev. A* **10**, 2156 (1974).
21. Schinke, R., *Chem. Phys.* **24**, 379 (1977).
22. Holliday, M. G., Muckerman, J. T., and Friedman, L., *J. Chem. Phys.* **54**, 1058 (1971)
23. Linder, F., Janev, R. K., and Botero, J., in *Atomic and Molecular Processes in Fusion Edge Plasmas,* edited by R. K. Janev, Plenum Press, New York, 1995, p. 397.

3rd International Conference on Atomic and Molecular Data and Their Applications
Gatlinburg, Tennessee, April 24-27, 2002

PARTICIPANTS

Helmar G. Adler
Osram Sylvania

Kanti M. Aggarwal
The Queen's University of Belfast

Alejandro Aguilar
University of Nevada, Reno

Miron Ya Amusia
Racah Institute of Physics

James F. Babb
Harvard-Smithsonian
  Center for Astrophysics

James E. Bailey
Sandia National Laboratories

Connor Ballance
Rollins College

Mark E. Bannister
Oak Ridge National Laboratory

Avraham Barshalom
NRCN & Artep

Klaus Bartschat
Drake University

Donald R. Beck
Michigan Technological University

Kenneth L. Bell
Queen's University of Belfast

Paul M. Bergstrom
National Institute of Standards
  and Technology

Anand Bhatia
NASA/Goddard Space Flight Center

Richard John Blackwell-Whitehead
Imperial College

Matthias Born
Philips Research Laboratories

Linda Brown
Jet Propulsion Laboratory

H. K. Carter
Oak Ridge National Laboratory

Roberto Celiberto
Politecnico di Bari

Robert E.H. Clark
International Atomic Energy Agency

James P. Colgan
Auburn University

Michael Crisp
U.S. Department of Energy

Robert W. Crompton
Australian National University

John J. Curry
National Institute of Standards
  and Technology

Rami Doron
Naval Research Laboratory

Anatoly Faenov
MISDC of VNIIFTRI

Pierre P. Fauchais
University of Limoges

Steven Federman
University of Toledo

Zineb Felfli
Clark Atlanta University

Charlotte F. Fischer
Vanderbilt University

Jeffrey R. Fuhr
National Institute of Standards
  and Technology

Tom Gorczyca
Western Michigan University

Donald C. Griffin
Rollins College

Khaled Hassouni
C.N.R.S. - L.I.M.H.P.

Charles C. Havener
Oak Ridge National Laboratory

Alan Hibbert
Queen's University of Belfast

Rainer Hippler
University of Greifswald

Denis Humbert
LPGP, Universite Paris-Sud

Roger Hutton
Lunds University

Verne L. Jacobs
Naval Research Laboratory

Ratko Janev
Macedonian Academy of
 Sciences and Arts

Sveneric Johansson
Lund Observatory

Takako Kato
National Institute for Fusion Science

Konstantinos Katsonis
Universite Paris XI

Yong-Ki Kim
National Institute of Standards
 and Technology

Kate Kirby
Harvard-Smithsonian Center
 for Astrophysics

Marcel Klapisch
ARTEP, Naval Research Laboratory

Herbert F. Krause
Oak Ridge National Laboratory

Predrag S. Krstic
Oak Ridge National Laboratory

Hirotaka Kubo
Japan Atomic Energy Research
 Institute

Robert L. Kurucz
Harvard-Smithsonian Center
 for Astrophysics

Mark J. Kushner
University of Illinois
 at Urbana Champaign

Graeme Lister
Osram Sylvania Inc.

Stuart David Loch
Auburn University

Robert R. Lucchese
Texas A&M University

Joseph Macek
Oak Ridge National Laboratory

Steven T. Manson
Georgia State University

William C. Martin
National Institute of Standards
 and Technology

Ronald H. McKnight
U.S. Department of Energy

Brendan M. McLaughlin
The Queen's University of Belfast

Claudio Mendoza
NASA Goddard Space Flight Center

Lowell Morgan
Kinema Research & Software

Alfred Z. Msezane
Clark Atlanta University

Alfred Mueller
Universitat Giessen

Izumi Murakami
National Institute for Fusion Science

Hooshang Nikjoo
Medical Research Council

Karen Olsen
National Institute of Standards
 and Technology

James K. Olthoff
National Institute of Standards
 and Technology

Joseph Oreg
NRCN

Patrick Palmeri
NASA Goddard Space Flight Center

Stephane Pasquiers
LPGP, Universite Paris-Sud

Michael S. Pindzola
Auburn University

Milun J. Rakovic
Oak Ridge National Laboratory

Yuri V. Ralchenko
Weizmann Institute of Science

Frank B. Rosmej
GSI-Darmstadt

Zenonas Rudzikas
State Institute of Theoretical Physics
  and Astronomy

John R. Rumble, Jr.
National Institute of Standards
  and Technology

Edward B. Saloman
National Institute of Standards
  and Technology

Seiji Samukawa
Tohoku University

Craig J. Sansonetti
National Institute of Standards
  and Technology

Jean Sansonetti
National Institute of Standards
  and Technology

Akira Sasaki
Advanced Photon Research Center

Daniel Wolf Savin
Columbia Astrophysics Laboratory

David R. Schultz
Oak Ridge National Laboratory

Phillip M. Sheridan
University of Arizona

Apostolos Siskos
University Paris XI

Timothy J. Sommerer
General Electric Research

Phillip C. Stancil
The University of Georgia

Keizou Suzuki
Hitachi Research Laboratory

Hiroshi Tanaka
Sophia University

Swaraj Tayal
Clark Atlanta University

Larry H. Toburen
East Carolina University

Dmitry Uskov
P N Lebedev Physical Institute

Randy Vane
Oak Ridge National Laboratory

Philippe F. Weck
The University of Georgia

W. Phil West
General Atomics

Wolfgang Wiese
National Institute of Standards
  and Technology

Nigel John Wilson
The Queen's University of Belfast

Carl Winstead
California Institute of Technology

Jean-François Wyart
Laboratoire Aimé Cotton, CNRS

Oleg Zatsarinny
Western Michigan University

Honglin Zhang
Los Alamos National Laboratory

Lucy M. Ziurys
University of Arizona

Author Index

A

Ali, R., 144
Apponi, A. J., 154
Aubreton, J., 85

B

Bakshayev, N. N., 253
Bartschat, K., 192
Behar, E., 125
Bergstrom, Jr., P. M., 5
Born, M., 35

C

Capitelli, M., 263
Celiberto, R., 263
Cho, H., 233
Cormier, M., 111

D

Doron, R., 125
Doschek, G. A., 125

E

Elchinger, M. F., 85
Evans, T. E., 171

F

Faenov, A. Y., 253
Fauchais, P., 85
Feldman, U., 125

G

Gagarin, S. V., 253
Gianturco, F. A., 213

Gicquel, A., 61
Goldsmith, B., 171
Griem, H. R., 185

H

Hassouni, K., 61
Ho, S., 75

J

Janev, R. K., 263

K

Kitajima, M., 233
Komosko, V. V., 253
Krstić, P. S., 277
Kubo, H., 161
Kurucz, R. L., 134
Kuznetsov, K. S., 253

L

Laricchiuta, A., 263
Lister, G. G., 48
Loboda, P. A., 253
Lucchese, R. R., 213

M

Magunov, A. I., 253
Markelenkov, S. A., 253
McKoy, V., 241
Motret, O., 111
Müller, A., 202

N

Nikjoo, H., 14

O

Olson, R. E., 171

P

Pasquiers, S., 111
Petunin, S. A., 253
Pikuz, T. A., 253
Popova, V. V., 253

R

Raković, M. J., 144
Rat, V., 85
Rudzikas, Z. R., 221

S

Samukawa, S., 95
Schultz, D. R., 144, 277
Shiiki, M., 75
Shinpaugh, J. L., 23

S (cont.)

Skobelev, I. Y., 253
Stancil, P. C., 144
Suzuki, K., 75

T

Tanaka, H., 233
Toburen, L. H., 23

U

Uehara, S., 14
Uemura, N., 75

W

Wang, J. G., 144
West, W. P., 171
Winstead, C., 241

Z

Ziurys, L. M., 154